Aesthetic Perspectives in Jaw Surgery

Springer
New York
Berlin
Heidelberg
Barcelona
Budapest
Hong Kong
London
Milan
Paris
Singapore
Tokyo

Aesthetic Perspectives in Jaw Surgery

Harvey M. Rosen, M.D., D.M.D.
Clinical Associate Professor of Surgery
University of Pennsylvania School of Medicine
Philadelphia, PA 19104
USA
and
Chief, Division of Plastic Surgery
Pennsylvania Hospital
Philadelphia, PA 19106
USA

with Forewords by
James L. Ackerman, D.D.S.
Linton A. Whitaker, M.D.

With 573 illustrations

Springer

Harvey M. Rosen, M.D, D.M.D.
Clinical Associate Professor of Surgery
University of Pennsylvania School of Medicine
Philadelphia, PA 19104
USA
and
Chief, Division of Plastic Surgery
Pennsylvania Hospital
Philadelphia, PA 19106
USA

Library of Congress Cataloging-in-Publication Data
Rosen, Harvey M.
Aesthetic perspectives in jaw surgery / Harvey M. Rosen.
p. cm.
Includes bibliographical references and index.
ISBN 0-387-98304-X (hardcover : alk. paper)
1. Jaws—Surgery. 2. Chin—Surgery. 3. Surgery, Plastic.
I. Title
[DNLM: 1. Jaw—surgery. 2. Chin—surgery. 3. Surgery, Plastic—methods. WU 101 R813a1997]
RD526.R67 1997
817.5′22—dc21 97-26331

Printed on acid-free paper.

Production supervised by Lesley Poliner and managed by MATRIX Publishing Services; manufacturing supervised by Rhea Talbert.
Typeset by MATRIX Publishing Services, York, PA.
Printed and bound by Friesens, Altona, Manitoba, Canada
Printed in Canada.

9 8 7 6 5 4 3 2 1

ISBN 0-387-98304-X Springer-Verlag New York Berlin Heidelberg SPIN 10635158

To my darling Tish,

whose natural and unaffected beauty has inspired

the aesthetic concepts herein put forth.

You are a joy to behold.

Foreword: An Orthodontic Viewpoint

by James L. Ackerman, D.D.S.

Having followed Dr. Harvey Rosen's career for 25 years, I am delighted to comment from an orthodontist's perspective about this landmark text. My close collaboration with Dr. Rosen began shortly after he had completed his fellowship in craniofacial surgery at the Hospital of the University of Pennsylvania. Since that time, I have worked extensively with him in the treatment of patients who required surgery for the correction of their orthodontic problems and of patients who required orthodontics as an adjunct to the surgical correction of their facial deformities. For both of these patient populations, Dr. Rosen was the first to propose that orthognathic surgery is primarily aesthetic surgery. This concept challenged the traditional view in orthodontics that dental occlusion was the principal consideration in jaw surgery. Although Dr. Rosen considered the establishment of good dental occlusion as one of the important goals of surgical orthodontics, he emphasized that overall facial appearance was the highest priority in orthognathic surgery. He published this concept and lectured about it for many years, but only recently has this idea become widely accepted.

At the Symposium on Reconstruction of Jaw Deformity, the first symposium sponsored by the Plastic Surgery Educational Foundation and held at the Children's Hospital of Philadelphia in October, 1976, I presented the then-prevailing orthodontic view that "Jaw deformity will inevitably affect the integumental profile of the individual and will lead to a relative imbalance of the nose, lips and chin. Dental occlusion can be used as a reference for defining these distortions." It was only a few years later that Dr. Rosen pointed out to me that this orthodontic viewpoint was conceptually flawed. Although there is an intimate interrelationship between the skeletal foundation of the face and overlying soft tissue, it is the soft tissue that must influence and in certain circumstances dictate the appropriate skeletal displacements to achieve the most pleasing facial appearance.

This idea was in sharp contrast to the accepted cephalometric treatment planning methodology in which the dental occlusion and skeletal cephalometric values based on normative data dictated how the various skeletal units should be altered. Dr. Rosen was not calling for the abandonment of radiographic cephalometrics as a clinical tool, but instead, he emphasized that the physical examination of the patient's face was the key to diagnosis and that the surgeon should frequently increase the skeletal support as a method of favorably influencing the overlying soft tissue. The "cookbook approach" of utilizing cephalometric analyses to establish treatment goals in orthognathic surgery was actually the antithesis of the correct treatment planning method.

Dr. Rosen also exploded perhaps the most cherished orthodontic myth dating back to the turn of the century (Edward H. Angle), namely, that facial beauty results from facial proportionality. Later, the cephalometricians attempted to determine ideal facial proportion based on normative standards as had artists, such as Albrecht Dürer, centuries ago. Dr. Rosen observed that facial disproportion, at least compared with normative standards, sometimes creates the most beautiful faces. There are many illustrations in this text that support this point.

It was soon realized that, to achieve these principles, skeletal elements had to be surgically displaced in ways that were traditionally considered to be unphysiologic and, as a result, potentially unstable. Dr. Rosen began to perform operations in which he either altered the cant of the occlusion or created vertical facial elongation, both of which were considered to be heretical. His use of porous block hydroxyapatite as a bone graft substitute along with rigid fixation created stable results in circumstances that previously would have led to dramatic relapse. Each time he proposed a procedure for one of our mutual patients that was clearly in violation of the conventional wisdom in orthodontics, I would express great skepticism. In time, I learned the valuable lesson that conventional wisdom often lags significantly behind newer concepts and techniques.

Dr. Rosen has contributed an enormous amount to our understanding of the relationship between the facial skeletal foundation and the soft tissue facial mask and how the facial skeleton can be altered surgically to achieve optimal facial aesthetics, excellent dental occlusion, and stability. Each year, in lectures and publications, there seems to be greater acceptance of his ideas. I have no doubt that, in the future, some of his concepts will be universally accepted. I hope that, by the publishing of this text, these concepts will be more widely disseminated and that he will receive the credit that he so richly deserves. My hope for readers is that they learn as much from this text as I have from Dr. Rosen.

Foreword: A Surgical Viewpoint

by Linton A. Whitaker, M.D.

With the plethora of information available on facial surgery, it is extraordinary for an experienced surgeon to read a book in today's world and feel that it is not only exceptionally well written, but that it also brings clarity and new perspective to a subject that has been dissected from nearly every angle. Harvey Rosen's book does all of this.

This is a philosophical treatise, of a near-biographical nature, representing the apogee in conceptualization and execution of jaw surgery. It is based on broad experience, along with close observations, leading to conclusions by a masterful observer of details and synthesizer of information. These abilities, coupled with his extremely careful and exceptional surgical ability, have made Harvey Rosen widely recognized as the standard-bearer in the world of jaw surgery today.

Following the precepts of developing a philosophy first, then making certain it has truth to back it up, a methodology results. In this instance, the philosophy of soft tissue dominating over occlusion in jaw surgery has evolved over the last two decades, but it is brought into its clearest relief yet in these pages. In accord with this philosophy, soft tissue goals are set first, then the plan is developed for the bone to achieve these goals. This has, in the author's presentation, sometimes resulted in unusual concepts, most dramatically exemplified by the idea of skeletal expansion as the procedure of choice whenever possible, often when others would be doing skeletal contraction. The results are often dramatic. The philosophy has been followed by and integrated with careful planning and subsequent execution of the surgical procedure, keeping in mind at all times that jaw surgery is fundamentally aesthetic surgery.

The book clearly presents the thoughts behind the methodology as illustrated in the initial descriptive chapters. Subsequent detailed patient examples demonstrate the carefully thought out concepts. The principles of aesthetics so well described are brought to us through a combination of classic concepts of beauty dating at least to the time of Nefertiti in ancient Egypt, progressing through classic descriptions of plastic surgical goals developed by Gonzales-Ulloa and others, and by subsequent physical anthropologic studies and surveys in the last decade. Coupled with these classic and modern concepts of attractiveness, now nearly universal, has been a background of development of surgical technique and technology that have allowed the joining of conceptualization and execution, to its present advanced state.

From Robert Ivy, first Professor of Plastic Surgery at the University of Pennsylvania and a physician and dentist; through George M. Dorrance, whose focus on jaws culminated in the book *The History of Treatment of Fractured Jaws* in 1941; and followed by the influence of Paul Tessier, who transmitted so many of his ideas to me that were modified and passed on to Harvey Rosen, who has in turn refined them, ideas for the book evolved. At the same time that recent ideas about appearance developed, the techniques and technology of surgery were evolving. Beginning with Gillies in the late 1940s with his classic midface disjunction procedure, followed by Obwegeser's description of advancement genioplasty with Trauner in 1957 and his subsequent sagittal split osteotomy procedure, the basis of jaw surgery was established. In 1957, Tessier did his first Le Fort III type osteotomy, followed by an intracranial approach to more extensive craniofacial surgery in 1961. Subsequently, rigid fixation, especially stimulated by Hans Luhr, established greater predictability with jaw surgery, and the advances of artificial bone substitutes provided greater flexibility.

Against this background, and with a deep understanding of it, Harvey Rosen formulated his ideas and methods, borrowing from and adding to developments in aesthetic surgery of the facial skeleton developed by this author and others. Harvey Rosen has established a new standard in jaw surgery and conveys it in the clearest possible terms in this book. It will be welcomed by all who work in the field and take the time to read it and consider it carefully.

Preface

Orthognathic surgery represents the ultimate aesthetic operation. It provides one with the opportunity to dramatically and permanently enhance facial balance and proportion. It is perhaps the only instance in all of plastic surgery in which a severely deformed individual can be transformed into an extraordinarily attractive one. Most commonly performed in the adolescent or young adult, surgical results can have profoundly beneficial effects over the lifetime of the patient. It is a surgical opportunity that must not be wasted; it must be planned to achieve its full aesthetic potential.

Biotechnical advances, developed in the 1970s and clinically employed in the 1980s, have changed traditional concepts of orthognathic surgery. The routine use of rigid fixation and nonresorbable bone substitutes have transformed skeletal displacements once thought to be highly unstable into highly predictable ones. This, in turn, has allowed the treatment planning pendulum to swing in the direction of aesthetic enhancement and away from stability concerns. The preeminent role of the occlusal result can no longer be justified.

This book represents a personal aesthetic philosophy of jaw surgery that has evolved over a 15-year time frame. It is a philosophy that is based on three underlying aesthetic principles that are repeatedly emphasized throughout this work. The first is that the skeletal support of the facial soft tissues is a critical factor in determining the aesthetic success or failure of orthognathic surgery. Accordingly, soft tissue goals must influence the direction and extent of skeletal displacements. Second, to achieve these soft tissue goals, it is generally more predictable to expand (enlarge) skeletal volume than it is to contract (diminish) skeletal volume. Third, efforts to expand the facial skeleton frequently result in facial skeletal disproportion, which, under the right aesthetic circumstances, can be extremely attractive.

By its very nature, aesthetic judgments are highly subjective. It is for this reason that the reader may not totally agree with the author's approach. One can hardly take exception with such disagreement. What is beautiful or attractive to one individual may not be so regarded by another. However, it is easy to reach consensus on what beauty is not as opposed to what it may be. Accordingly, it is easier to define that which is not physically attractive. Likewise, it is easier to provide concepts as to how not to define beauty. For instance, it cannot be defined in quantitative terms or in the context of what a "normal face" should ideally be. Beauty is simply not synonymous with normalcy.

This work should be considered more as a narrative than as a text. It is certainly not meant to be employed as a reference or an atlas. Its format is arranged to achieve a storylike effect, that is, what I consider to be a philosophical means to achieve an aesthetic end. The underpinnings of every chapter serve this end. For instance, technical aspects are included only if they can make an aesthetic difference. It is hoped this will demystify orthognathic surgery and reduce it to its simplest and purest element, that is, aesthetic surgery. Its goals will have been met if it can influence younger surgeons not in how to do it, but in what to do and why to do it at all.

Unless otherwise stated, all patients in this book were surgically treated by myself.

Harvey M. Rosen
Philadelphia, Pennsylvania

Acknowledgments

Numerous individuals deserve recognition for their roles played in the conceptual development and actual publication of this book.

The academic background from which these aesthetic concepts were nurtured was influenced by many throughout my career, first as a resident in general surgery and then in plastic surgery at the Massachusetts General Hospital and then as a craniofacial fellow at the University of Pennsylvania.

Although the "old guard" at the MGH in the 1970s, consisting of Drs. Bradford Cannon, John Constable, John Remensnyder, and Michael Lewis, was by no means considered a bastion of aesthetic surgery, the reconstructive lessons so well taught by them and the clinical judgment that characterized their surgical care served as the foundation. As a new attending surgeon at the MGH in the late 1970s, James May began to play a major role in resident training. He taught me to think logically about a problem, critically question the possible solutions, have the courage to execute a surgical plan once having made a decision, and do it with technical excellence.

As a fellow in craniofacial surgery at the University of Pennsylvania in 1980, I interacted with two individuals whose guidance greatly impacted my thinking about jaw surgery and subsequently my career. Linton Whitaker influenced me to think of orthognathic surgery as aesthetic surgery and to consider the appearance of the soft tissue as paramount. By example, he taught that aesthetic surgery was, to a large extent, both intuitive and subjective. He continually challenged me to be innovative and imaginative. The concepts expressed in this book are, in part, a reflection of his thinking. James Ackerman, then Chief of the Dental Service at Children's Hospital of Philadelphia, taught me the basics of surgical orthodontics. More importantly, he had the confidence and courage to support the concept of a plastic surgeon performing "bread and butter" orthognathic surgery, an idea for which he was criticized within the community. He continually allowed me the freedom to treat our patients based on aesthetic concerns at the risk of potentially creating an unstable result. This "go for broke" attitude was only possible because of his self-confidence and the great respect and admiration in which he was held among his peers.

To all of these mentors, I am deeply indebted.

I would similarly like to acknowledge all of the orthodontists, particularly those whose patients appear in this book, with whom I have worked over the years. They too have afforded me the latitude to treat their patients as I deemed aesthetically appropriate.

Finally, the entire production staff at Springer-Verlag deserves recognition. However, I would like to individually thank Esther Gumpert, Senior Medical Editor at Springer-Verlag, who had confidence in this project from its inception.

Harvey M. Rosen
Philadelphia, Pennsylvania

Glossary of Symbols and Abbreviations

The following abbreviations and symbols are utilized in the cephalometric tracings depicted throughout this book.

AHOM	Anterior horizontal osteotomy of mandible, that is, osseous genioplasty
ANS	Anterior nasal spine
BSSRO	Bilateral sagittal split ramus osteotomy
G	Glabella
Me	Menton
Pg	Pogonion
Go	Gonion
Sn	Subnasale
SNA	Sella to Nasion to Pt. A ∢
SNB	Sella to Nasion to Pt. B ∢
$\underline{1\|1}$	Maxillary central incisors
$\underline{6\|6}$	Maxillary first molars
$\overline{1\|1}$	Mandibular central incisors
$\overline{6\|6}$	Mandibular first molars
→	Reposition anteriorly
←	Reposition posteriorly
↑	Reposition superiorly
↓	Reposition inferiorly

Contents

Principles

Part 1

The first section of this book presents aesthetic principles on which treatment planning is based. Because only one occlusal result is desirable, the great variability in treatment planning derives from attempts to achieve the best aesthetic result. Herein lies the challenge and the art of orthognathic surgery: making decisions of what to do. It is a challenge because it must be based on the aesthetic sense of the surgeon and not on rigid cephalometric guidelines. If the proper aesthetic judgments are made, a treatment plan is derived from which no deviation need occur. An aesthetic road map can be brought into the operating room, and the operation becomes a mere technical exercise. The focus is, therefore, properly shifted away from how to do it, and emphasis is placed on what to do and why to do it.

Treatment Planning: Aesthetic Goals

CHAPTER 1

The traditional and time-honored goals of any orthognathic surgical correction have attempted to address both stability and aesthetic concerns. Stability issues relate to establishing a healthy, functional occlusion, whereas aesthetic considerations have focused on normalizing facial balance and proportions as much as possible.

Although these traditional goals were laudable ones, this treatment planning approach had always been fundamentally flawed for two basic reasons. The first was that the most aesthetic correction was not necessarily the most stable correction. As a result, these mutual goals of stability and aesthetics became competing goals, and more often than not, facial aesthetics were sacrificed in the name of skeletal stability. Throughout the years, the classic example of this had been in the Class III skeletal patient who was treated with mandibular setback surgery regardless of its aesthetic consequences because for many years moving the mandible posteriorly was deemed to be more stable than moving the maxilla anteriorly.

The second problem with these traditional treatment planning goals was that aesthetic judgments were largely based on normative data derived from both skeletal cephalometric analyses (Fig. 1-1) and soft tissue cephalometric analyses (Fig. 1-2). The error in this approach is that normalcy is simply not synonymous with beauty. For instance, there are many individuals who have so-called "normal" facial proportions as in Fig. 1-2, and yet they are not considered beautiful. On the other hand, we are all aware of strikingly attractive individuals who have facial measurements and proportions that fall outside these normal ranges. One can begin to understand why this traditional treatment planning philosophy involving a balance between aesthetics and stability and basing aesthetic decisions on normative cephalometric data was significantly flawed.

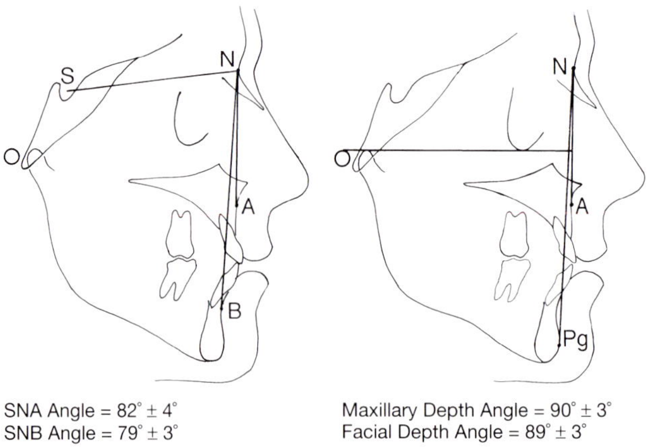

FIG. 1-1 Standard skeletal cephalometric measurements describing normal sagittal positions of the jaws. Left, SNA and SNB angular measurements relate the maxilla and mandible to the cranial base. Normal values are 82° ± 4° and 79° ± 3°, respectively. Right, maxillary depth and facial depth angles relate the maxilla and mandible to the Frankfort horizontal. Normal values are 90° ± 3° and 89° ± 3°, respectively.

In the early 1980s, several biotechnical advances became available to surgeons that should have significantly changed our treatment planning perceptions. The most important of these was the use of rigid internal fixation to stabilize osteotomy sites. Without question, rigid fixation affords real short-term advantages including the elimination of the need for intermaxillary fixation. This, in turn, provides for decreased temporomandibular joint stiffness, improved oral hygiene, and an improved sense of well-being among

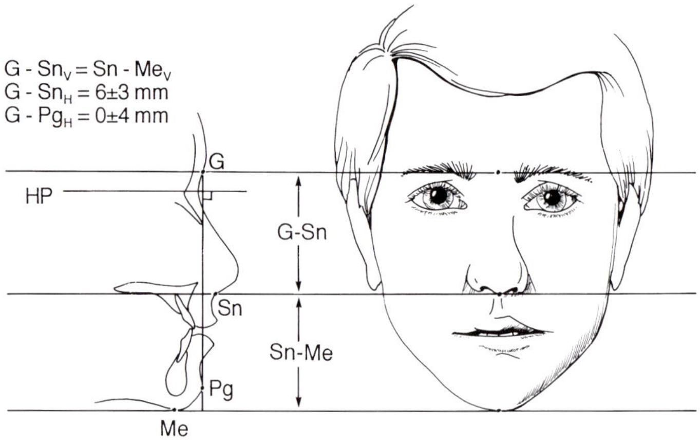

FIG. 1-2 Standard soft tissue cephalometric measurements. Left, a vertical perpendicular dropped from glabella, the horizontal distance from which is measured at subnasale, Sn, and pogonion, Pg. Normal values are 6 ± 3 mm and 0 ± 4 mm, respectively. Right, normal vertical facial proportions: glabella to subnasale is equal to subnasale to menton, Me.

surgical patients. Indeed, many individuals are more accepting of undergoing a surgical correction utilizing this technique.

There is, however, continuing debate as to whether or not rigid fixation has reduced problems related to skeletal relapse. The initial hope that rigid fixation would entirely eliminate the problems associated with skeletal relapse, that is, movement at the osteotomy site, has not been realized. However, it is true that rigid fixation has significantly reduced mean relapse rates for many types of skeletal displacements, but more importantly, it has made these same types of skeletal displacements much more predictable; that is, the expected range of any degree of skeletal relapse has been dramatically narrowed. For instance, mandibular advancement surgery is not very predictable if the relapse at 1 year in a given series of patients ranges from 0 to 6 mm despite a mean relapse of only 1.5 mm. On the other hand, the procedure is significantly more predictable with the same mean relapse of 1.5 mm but with a narrowed range of relapse of 0 to 2 mm.

Enhanced predictability is what rigid fixation has clearly achieved. This has been demonstrated both for maxillary impaction and advancement surgery employing the LeFort I osteotomy and for mandibular advancement surgery employing the bilateral sagittal split ramus osteotomy. It has most recently been demonstrated for mandibular advancement surgery accompanied by significant counterclockwise rotational movement of the mandibular body with the point of rotation being at the mandibular osteotomy. For years, this had been considered heresy in orthognathic surgery, but it is now a very predictable operation when rigid fixation techniques are employed.

Perhaps the only operation in which gross skeletal instability remained despite rigid fixation was when one attempted to inferiorly reposition the maxilla as in the treatment of the short-face patient. It remained for the second major biotechnical advance, nonresorbable bone substitutes, to solve this problem. In particular, porous block hydroxyapatite has many biomechanical properties that make its use in this context ideal. Unlike autogenous bone grafts, it does not lose volume over time, and unlike other alloplastic materials, it does not lead to resorbtion of adjacent host bone. Its compressive strength becomes more than adequate to withstand masticatory loads once its interstices have become ingrown with autogenous tissue, and when its use is combined with rigid fixation techniques. It, therefore, has become the ideal material to use in interpositional fashion to place into the resulting osteotomy gap once the maxilla is inferiorly repositioned. The long-term predictability of this operation employing porous block hydroxyapatite has not been duplicated when autogenous bone grafts and rigid fixation have been utilized.

Recently, Major et al have demonstrated good stability of inferiorly repositioned maxillae employing rigid fixation and autogenous cortical bone grafts harvested from the iliac crest. Follow-up was a mean 16 months. Despite this new data, the author still prefers porous block hydroxyapatite because long-term follow-up (up to 9 years; see Chapter 10) is now available in patients treated with hydroxyapatite. In addition, the morbidity associated with harvesting cortical bone from the iliac crest is extremely high.

With the advent of rigid fixation techniques and bone graft substitutes, our entire perception of orthognathic surgery and its goals should have dramatically changed. Clearly, it was no longer acceptable to sacrifice facial aesthetics in an effort to restore a normal occlusion. More importantly, orthognathic surgery could now be truly envisioned as aesthetic surgery. One could now present to patients a surgical treatment plan that was thought to be in their aesthetic best interest without having undue concerns that what was being presented to them would be inherently unstable.

The patient demonstrated in Fig. 1-3 is a good example of this philosophy. She is 18 years old and presents for surgical correction of a Class III malocclusion. One can readily observe the rather flat profile. Examination of her occlusion and cephalometric radiograph demonstrates that the maxilla can be advanced only 4 mm to achieve a Class I relationship. Although this will improve her appearance, the extent of the improvement will be modest at best. It is in this situation that one should consider double-jaw surgery because a two-jaw procedure eliminates the constraints that the occlusion places on the aesthetic result. Accordingly, a treatment plan that will include a 10 mm advancement of the maxilla is suggested to the patient because it is in her aesthetic best interest and because it can be accomplished with a reasonable degree of predictability as it relates to skeletal stability. This will obviously necessitate a 6 mm advancement of the mandible to achieve a Class I occlusion. In addition, a 5 mm vertical elongation of the

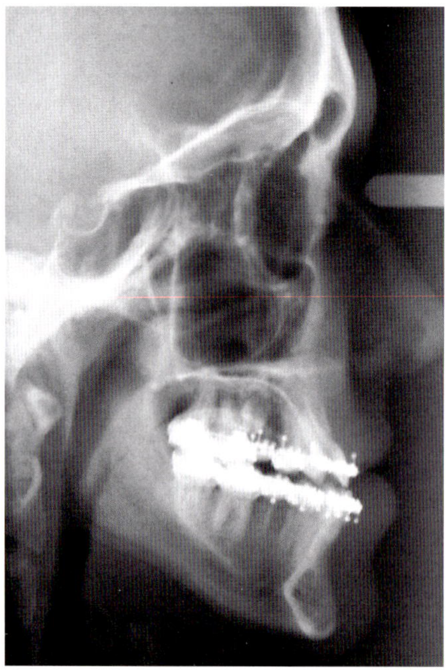

A

FIG. 1-3 Lateral cephalometric radiograph revealing a modest 4-mm discrepancy of the maxilla and mandible. The skeletal diagnosis is maxillary deficiency.

PREOPERATIVE **POSTOPERATIVE**

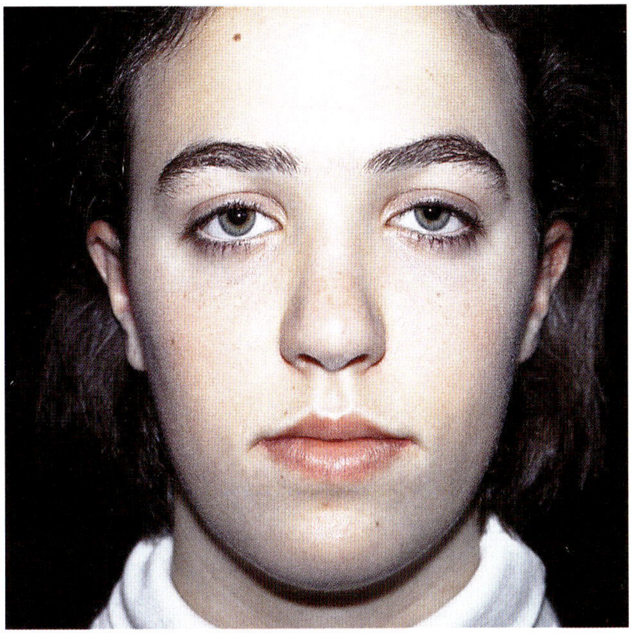

B

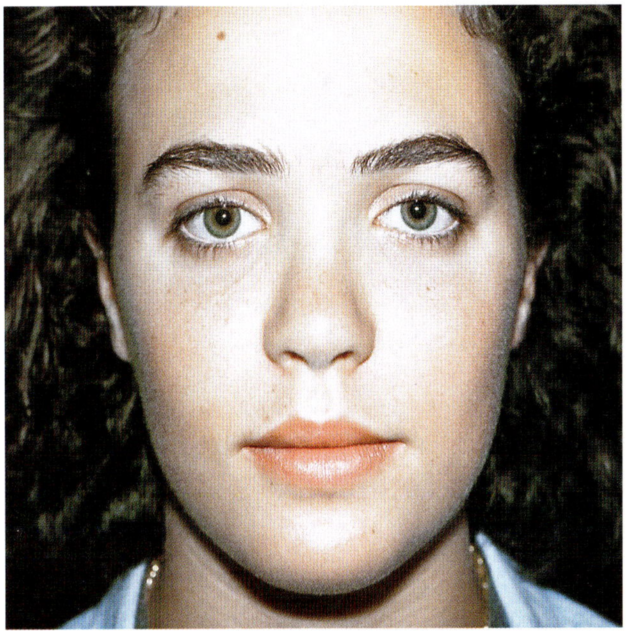

D

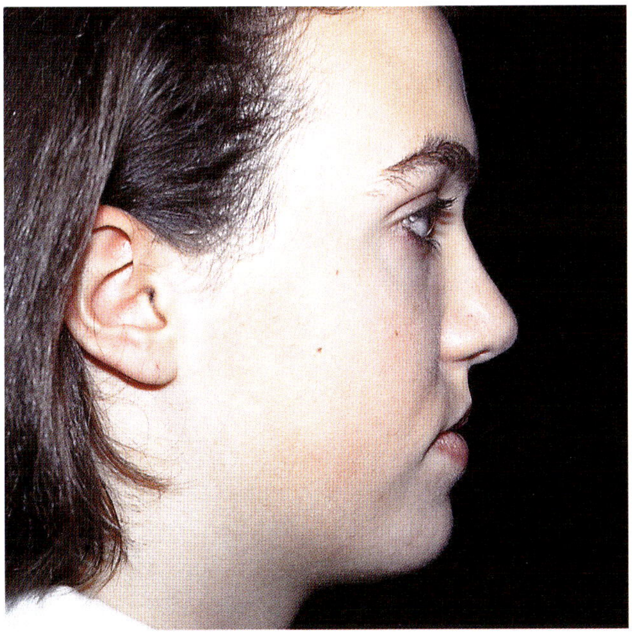

C

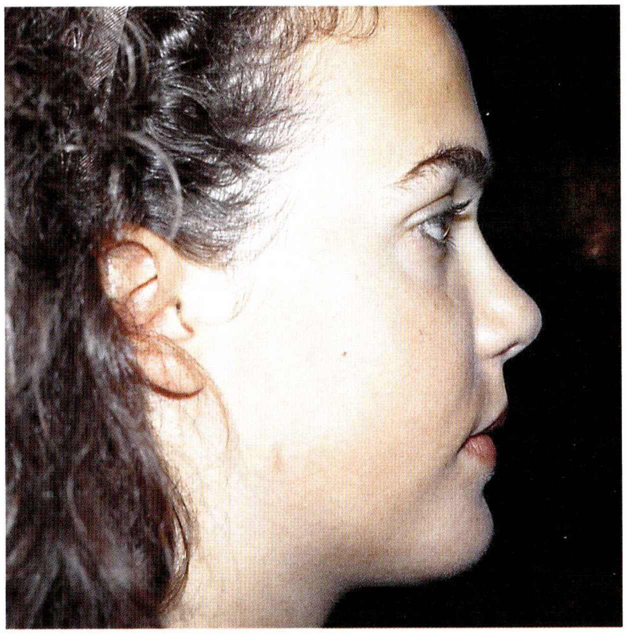

E

FIG. 1-3 (*Continued*) This patient was treated with maxillary advancement and mandibular advancement and vertical lengthening genioplasty. Maxillary advancement only, with or without genioplasty, would have provided an unacceptable result.

PREOPERATIVE | POSTOPERATIVE

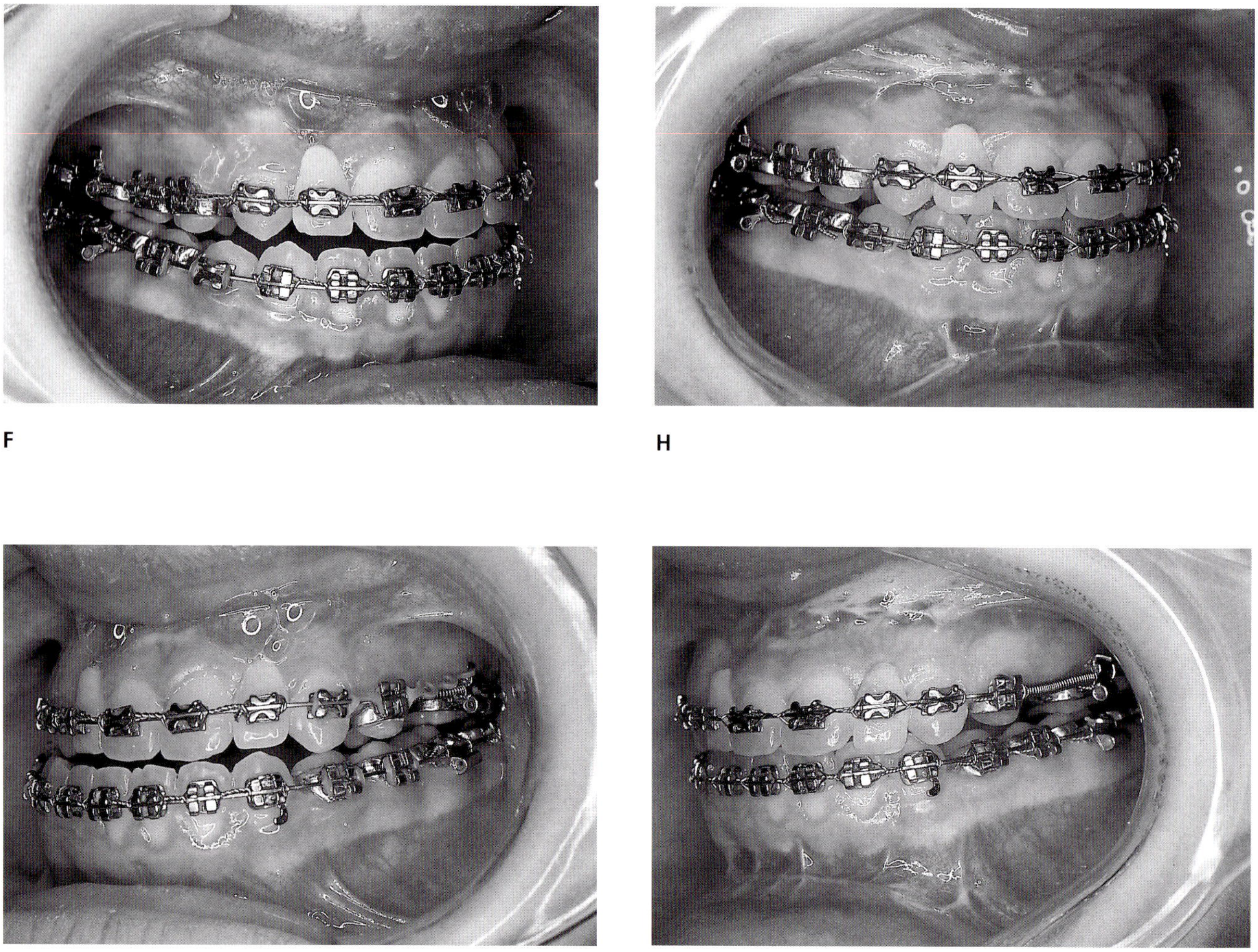

F H G I

Fig. 1-3 *(Continued)* Occlusal results. Orthodontist: Dr. Ralph Apuzzio

symphysis utilizing an osseous genioplasty with interpositional placement of hydroxyapatite is also planned. This approach provides for an aesthetic result in both the frontal and profile views that could not have been achieved with single-jaw surgery. Importantly, the occlusal result is the same whether single- or double-jaw surgery is undertaken. For purposes of this discussion, the occlusal result is considered to have served as a means to an aesthetic end rather than as an end in and of itself.

If this is the treatment planning philosophy that one is willing to pursue, it then makes little sense to rely on skeletal diagnoses to treatment plan patients. It must be remembered that skeletal diagnoses are derived from skeletal cephalometric data that is compared to normative data derived from "normal" individuals, not necessarily individuals who are considered beautiful or attractive. It bears repeating that normalcy is not necessarily synonymous with beauty. In addition, skeletal cephalometric data are limited to the midsagittal plane and give us no information regarding the soft tissues.

A good example of the deficiencies of treatment planning based on skeletal cephalometric data is seen in Fig. 1-4. This patient was referred with a clinical diagnosis of mandibular prognathism having been made. Examination of her cephalometric radiograph confirms this diagnosis. Despite this data, this patient would be poorly served aesthetically if one were to reposition the mandible posteriorly. Instead, the maxilla was advanced and vertically elongated to give her greater exposure of her anterior maxillary teeth. The large mandible and excessively steepend mandibular plane was camouflaged by a vertical reduction genioplasty. The occlusal result is the same regardless of what treatment planning approach would have been followed.

Although soft tissue cephalometric data (Fig. 1-2) is more clinically relevant than skeletal data in arriving at an aesthetic treatment plan, it, too, lacks clinical significance. Soft tissue data suffers from the fact that they are limited to the sagittal midline and, most importantly, are derived from "normal" individuals. Here again lies the fallacy in employing these data to determine surgical treatment planning.

Alternatively, the best possible approach to aesthetic treatment planning relies on a thorough description of the facial morphology of the patient through the most basic of all clinical modalities, the physical examination. This is meant to be a "qualitative" assessment of how one can best alter the facial morphology to the maximum aesthetic benefit of the patient. The word qualitative is emphasized to highlight the fact that the visual impression of certain facial characteristics may be disparate from quantitative

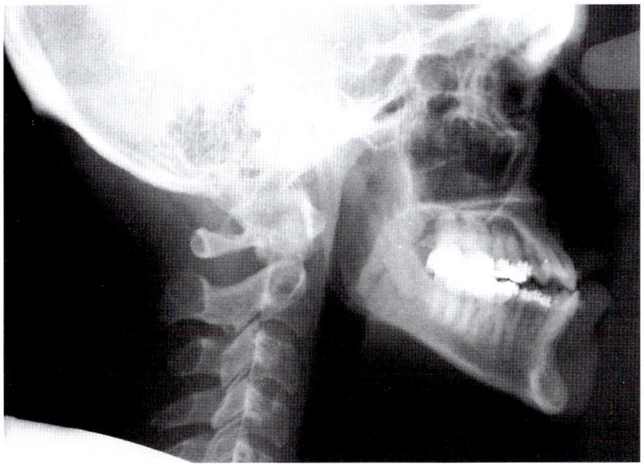

A

FIG. 1-4 Lateral cephalometric radiograph of patient with mandibular prognathism and normally postioned maxillae.

PREOPERATIVE | POSTOPERATIVE

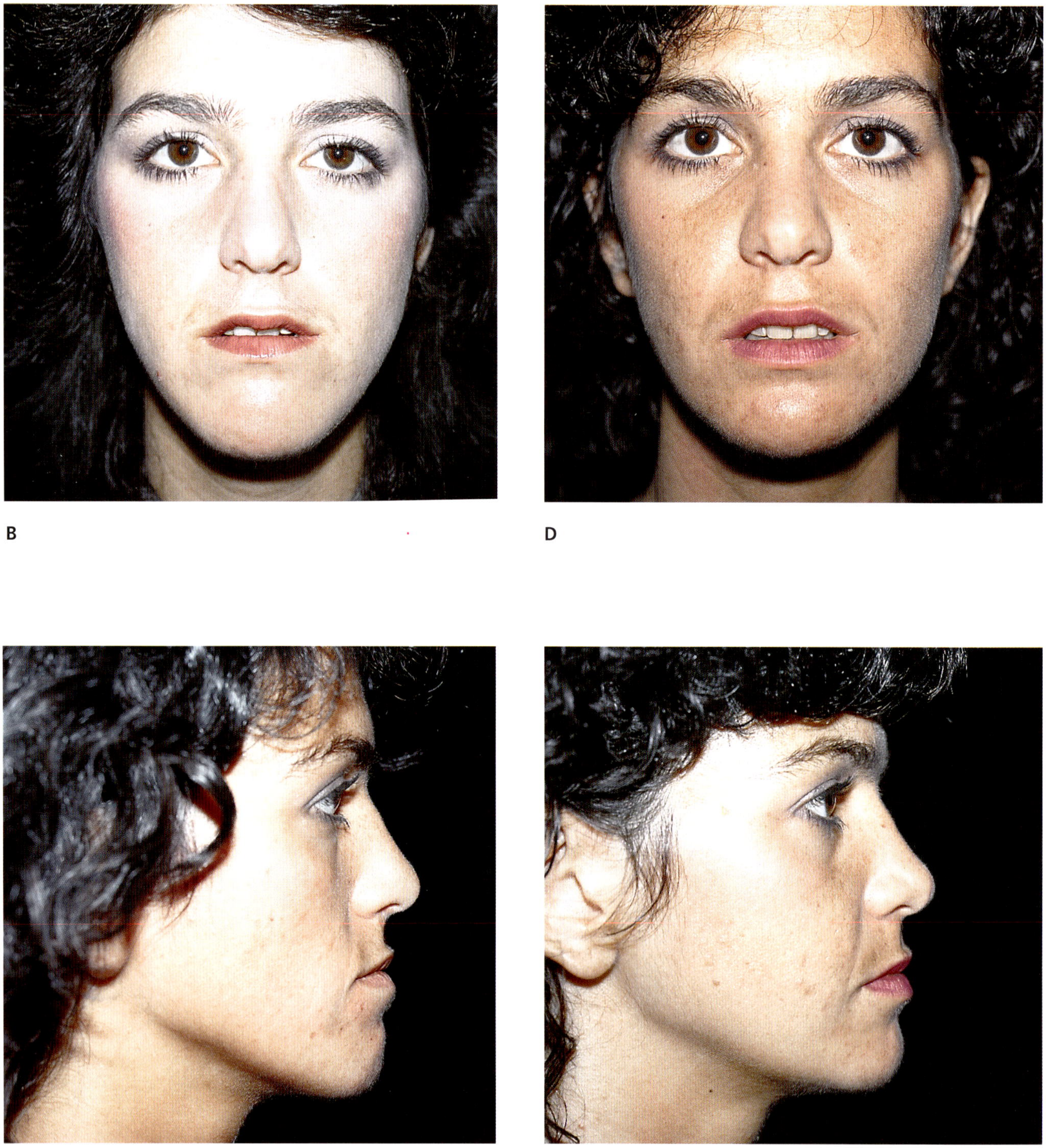

B | D

C | E

FIG. 1-4 (*Continued*) The patient is treated with maxillary advancement and lengthening, and vertical reduction genioplasty. D: Notice the increased incisal exposure. From: Rosen, H.: Maxillary advancement for mandibular prognathism: Indications and rationale. *Plast Reconstr Surg* 87:823, 1991. Reprinted with permission.

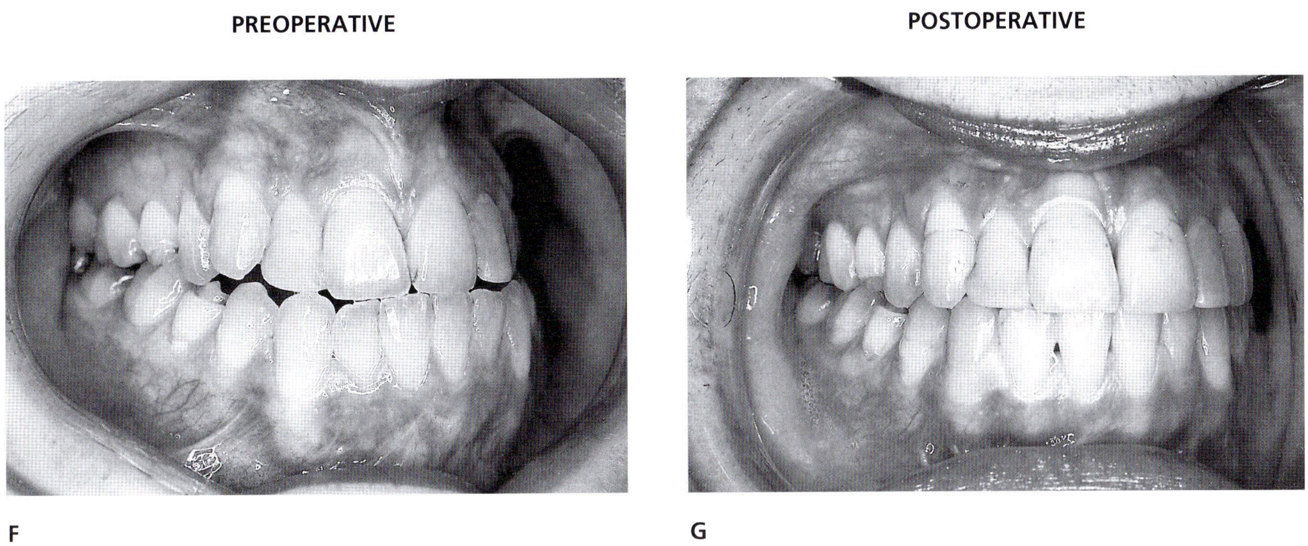

FIG. 1-4 (*Continued*) Occlusal results. Orthodontist: Dr. James Ackerman

measurements. For example, a quantitatively long lower face relative to the midface may not visually appear to be so. Most importantly, this qualitative assessment must take into account the very dynamic interrelationship between the underlying skeletal foundation and overlying facial mask, and how one can affect the appearance of the other.

A description of facial morphology should include the following. In the frontal view, the two most critical considerations are overall symmetry of the face and vertical proportionality. It is, of course, important to be cognizant of any facial asymmetries in the upper face or midface, but it is those of the lower face that are most clinically relevant in jaw surgery. The most common of these are soft tissue chin point deviations from the midline, discrepant vertical positioning of the left versus the right inferior border of the mandible, and occlusal cants or tilts that may secondarily cause oral commissure tilting or only may be obvious with full smiling. A qualitative assessment of the height of the lower face, as visualized from menton to subnasale, relative to the height of the midface, as visualized from subnasale to glabella, is then undertaken. Although these distances can be measured with a caliper, it is the qualitative visual assessment of whether the lower face appears to be too long or too short that is most relevant. Other qualitative assessments in the frontal view related to vertical proportionality include the presence or absence of circumoral lip strain, the depth of the labiomental fold (it may be excessively deepened or completely effaced), and the relationship of the incisal edge of the maxillary teeth to the relaxed vermilion border of the upper lip. This last determination is much more important than the extent of gingival show with full smiling because the latter also involves the dynamics of the individual smile in addition to structural considerations. For instance, one would never want to shorten the anterior maxilla because of excessive gingival show when there was only 4 to 5 mm of tooth show with the lips relaxed. The last assessment in the frontal view is the visual impression of nasal base width, taking into account how it may be affected by maxillary surgery.

In the lateral view, the two basic assessments include overall facial divergence and facial convexity or concavity (Fig. 1-5). Facial divergence is determined by the sagittal position of soft tissue pogonion relative to a vertical line dropped from glabella. If pogonion visibly appears to be posterior to this line, the face diverges posteriorly. If it appears to be anterior to this line, the face diverges anteriorly. Overall facial convexity or concavity is a description of the profile by a line connecting glabella to subnasale to pogonion. These points may subtend a line that is straight or one that is convex or concave. An additional, and perhaps more important, consideration is the description of the parasagittal profile; that is, what are the sagittal relationships of the soft tissues from the infraorbital area to the paranasal area to the parasymphyseal area, and does this provide one with a general impression of a convex or concave appearance? A profile that is straight with neutral divergence is described as an orthognathic profile. The most aesthetic profiles are convex both in the midsagittal and parasagittal planes.

Descriptions of normal facial divergence derived from direct anthropometric measurements are available (Fig. 1-6). These data are taken from normal (Class I occlusions without prior orthodontic treatment) Caucasian, North American males and females. Mean values of normal facial divergence in both males and females are negative with the face being posteriorly divergent; that is, soft tissue pogonion is posterior to a vertical perpendicular dropped from glabella. Anteriorly divergent profiles are typically not seen in either sex until one approaches two standard deviations above the mean.

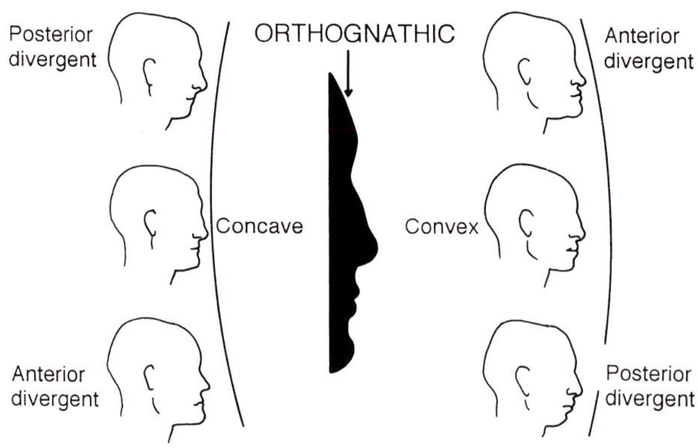

FIG. 1-5 Morphologic description of the profile. The profile can diverge anteriorly, diverge posteriorly, or be described as orthognathic if pogonion lies on a vertical perpendicular dropped from glabella. The profile can also be described as convex, concave, or straight if one relates glabella to subnasale to pogonion. From: Graber, T. and Swain, B.: Current Orthodontic Concepts and Techniques. 2nd Edition. W.B. Saunders Co., Philadelphia. 1975.

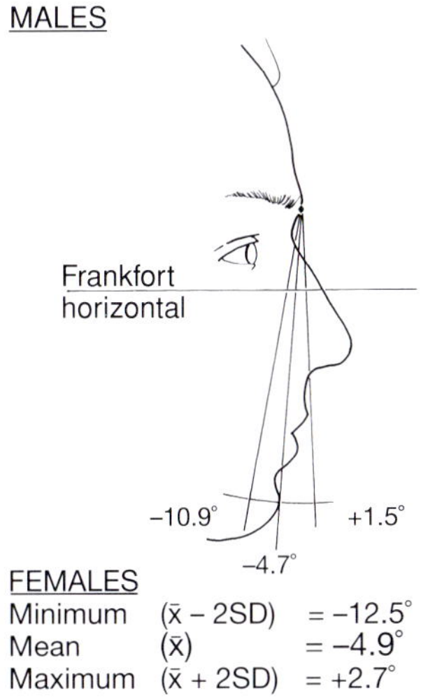

FIG. 1-6 Direct anthropometric measurements depicting normal degrees of facial divergence in Caucasian males and females. Note that "normal" faces in both sexes diverge posteriorly. From: Farkas, L.: Inclination of the facial profile: Art versus reality. *Plast Reconstr Surg* 75:509, 1985. Reprinted with permission.

The last and most important factor in the description of the facial morphology is a qualitative assessment of the attributes and deficiencies of the overlying soft tissues, that is, the facial mask. In particular, are the soft tissues well supported by the underlying skeletal foundation? Are the soft tissues redundant or deficient in both absolute and relative terms? Are the soft tissues thin and delicate and refined, or are they thick and fleshy? This last assessment must consider both cutaneous and subcutaneous layers. All of these soft tissue factors have a profound impact on the overall visual impression of the face. There is a sense of highlighting, definition, angularity, and refinement when the facial mask is well supported by the skeletal foundation. With a lack of skeletal support, the face is devoid of these features and, in the extreme, can be described as amorphous.

This basic concept is important in the treatment planning process because it is the support of the facial soft tissues that is, perhaps, the most critical factor that determines the aesthetic success or failure of many orthognathic surgical procedures. It is for this reason that normative cephalometric data and anthropometric data as described above should be ignored in surgical planning. On the contrary, the best way to provide for a properly supported facial mask is to achieve adequate (not necessarily normative) sagittal projection of the mid- and lower face and adequate height of the lower face. In this context, *adequate* is defined as displacing the jaws as much as neccessary to achieve the desired soft tissue response. Ironically, therefore, soft tissue considerations dictate skeletal proportions. This is in marked contrast to the traditional concept of orthognathic surgery in which one attempts to achieve normal facial measurements and proportions and, in so doing, hopes that the soft tissues will respond in a favorable fashion. This is a naive approach, because it is now understood that the soft tissue response to many types of skeletal displacements made in an effort to achieve facial normalcy are, at best, unpredictable and unreliable.

Accordingly, it is more advisable to set qualitative goals for the soft tissue and, having set those goals, to allow them to dictate facial skeletal displacements and, therefore, facial proportions. This approach is well illustrated in Fig. 1-7. This patient is referred for surgical correction of a Class III malocclusion. The frontal appearance can be described as symmetric with normal vertical proportionality; that is, the height of the lower face approximates that of the midface. Nasal base width is excessive. The profile is characterized by neutral divergence (the soft tissue chin point falls on a vertical perpendicular dropped from glabella) with overall slight facial convexity. However, the soft tissues are inadequately supported by the underlying skeletal foundation despite these so called "normal" facial proportions in both the vertical and antero-posterior planes. There is a lack of zygomatic highlighting with no submalar hollowing, there are jowls and submental fullness, and there is poor definition along the inferior border of the mandible. There is a total absence of facial angularity and refinement. Following presurgical orthodontics, a lateral cephalometric radiograph demonstrates an occlusal relationship that will allow for a 4 mm advancement of the maxilla. This is entirely inadequate to achieve a well-supported facial mask.

Despite the chin being in normal sagittal position, a surgical treatment plan involving advancement of both maxilla and mandible as well as the chin is proposed. In addition, despite preexisting normal vertical propor-

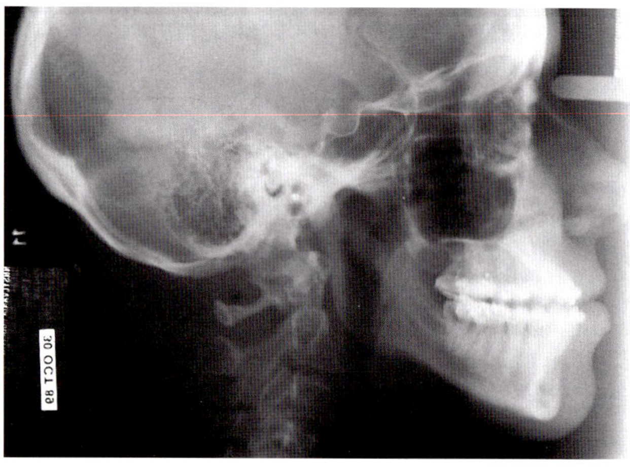

A

FIG. 1-7 Lateral cephalometric radiograph revealing a modest 4-mm sagittal discrepancy of maxilla and mandible.

PREOPERATIVE | **POSTOPERATIVE**

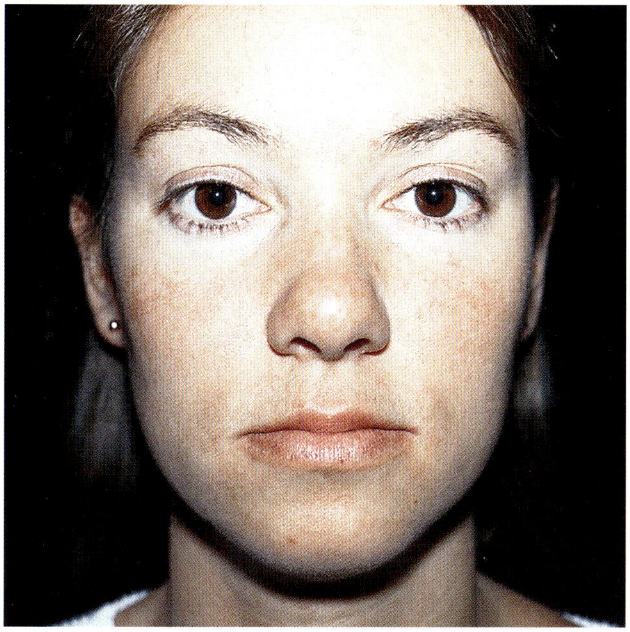

B

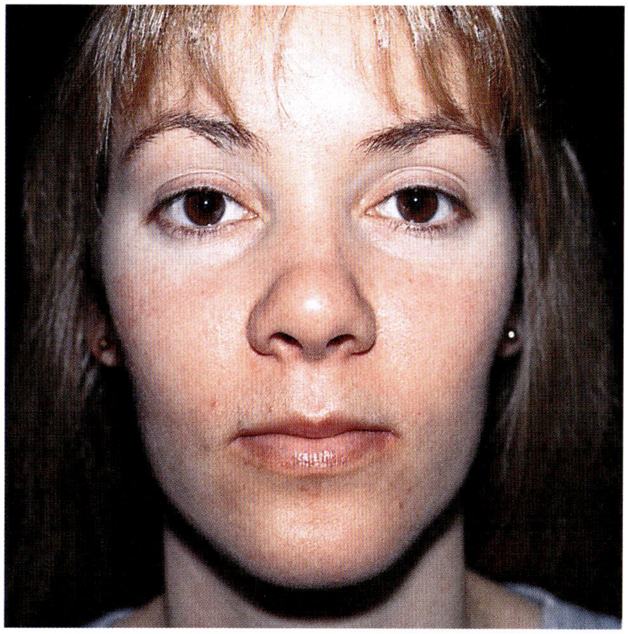

D

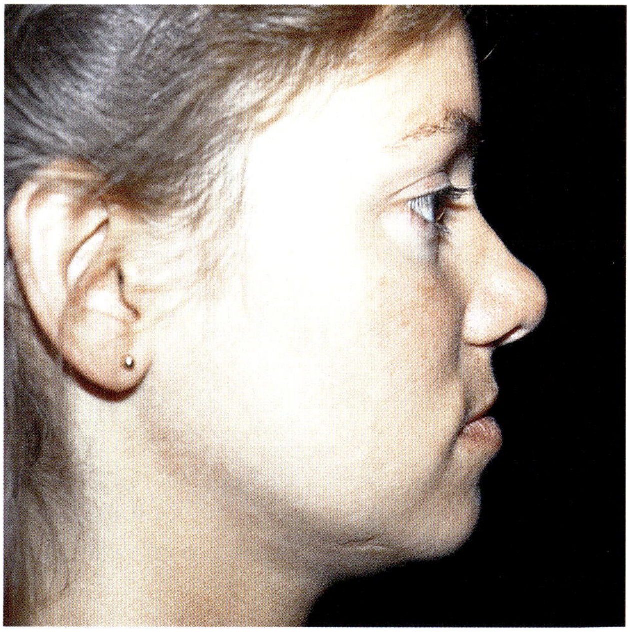

C

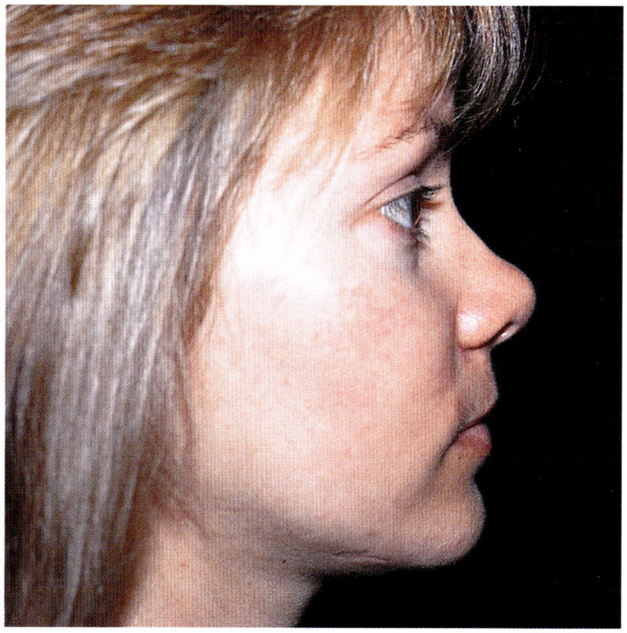

E

FIG. 1-7 *(Continued)* The patient is treated with double-jaw surgery to advance the maxilla, mandible, and chin, and vertically elongate the maxilla and chin to create a long face and one with excessive anterior divergence. This was done to achieve enhanced skeletal support of the soft tissues. From: Rosen, H.: Facial skeletal expansion: Treatment strategies and rationale. *Plast Reconstr Surg* 89:798, 1992. Reprinted with permission.

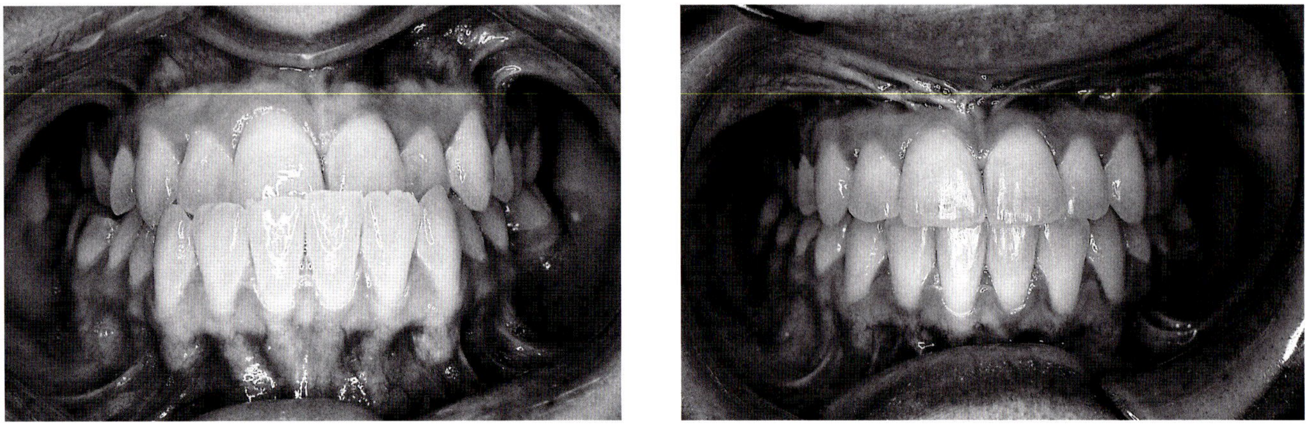

FIG. 1-7 (*Continued*) Occlusal results. Orthodontist: Dr. Julian Freeman

tions, inferior repositioning of the maxilla as it is advanced and vertical elongation of the chin as it is advanced are also planned. This treatment plan is designed to achieve maximum soft tissue support.

Postoperatively, in the frontal view there is a marked enhancement of soft tissue definition with a well-defined zygomatic complex, a submalar hollow, and the absence of jowls. This has been achieved at the expense of creating vertical facial disproportion; that is, the lower face is now quantitatively too long relative to the midface. In the profile view, there is enhanced definition of the inferior mandibular border and improved angularity in the cervical mandibular region. However, the face now projects to an excessive degree, at least compared to normative standards, at the level of both subnasale and pogonion. Sagittal disproportion has been created. This case illustrates how soft tissue considerations have dictated skeletal proportions.

This result can be contrasted to the one seen in Fig. 1-8. This patient had undergone correction of a Class III malocclusion with mandibular setback surgery performed by another surgeon. The occlusal result is a stable, functional one with normal Class I relations. Unfortunately, occlusal considerations have dominated the treatment planning process, and her facial aesthetics have been completely sacrificed. As a result of a reduction in overall skeletal facial volume, there is now inadequate soft tissue support with relative soft tissue excess and a complete absence of facial angularity, highlighting, and definition. There is a gross disequilibrium between soft tissue mass and skeletal mass. Although this is an unusual and extreme illustration of this principle, other more common and subtle examples of this phenomenon are frequently observed in orthognathic surgery.

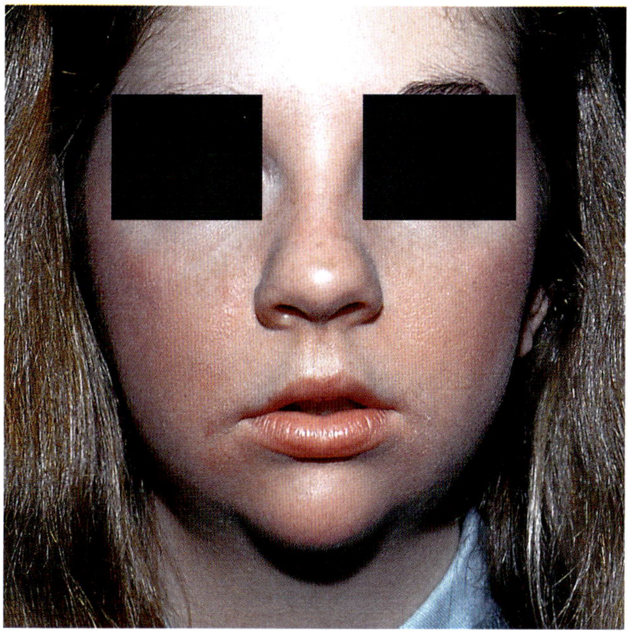

A

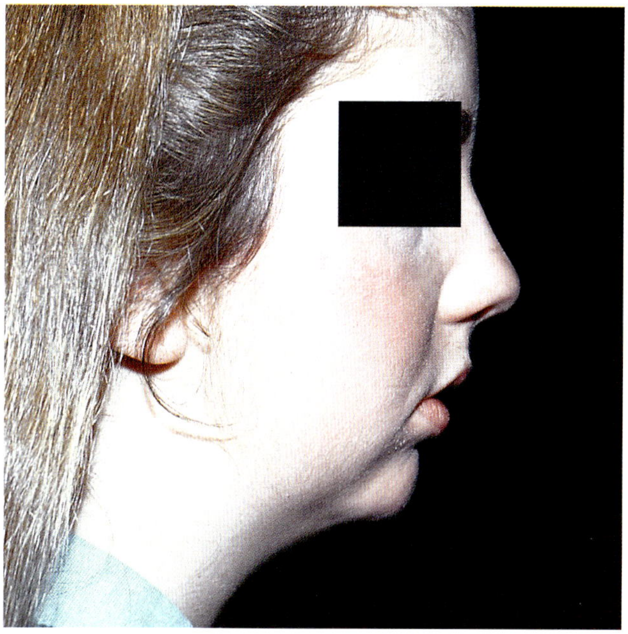

B

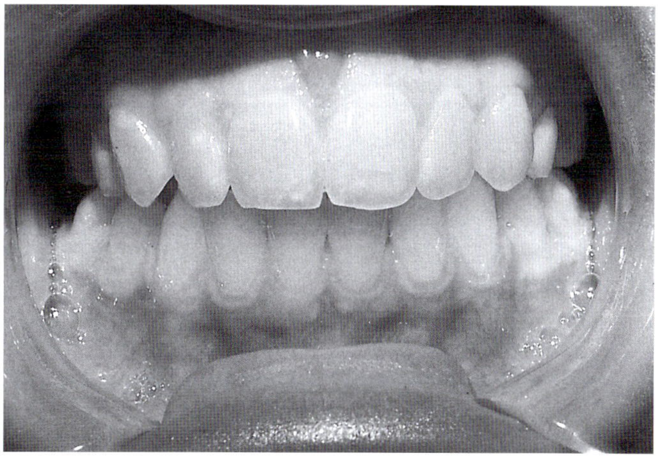

C

FIG. 1-8 A patient who underwent mandibular setback surgery to correct a Class III malocclusion. There is a severe lack of skeletal support and soft tissue excess secondary to a reduction in facial skeletal volume. C: The occlusal result.

The patient in Fig. 1-9 was treated by myself. She was referred for surgical correction of a Class I, anterior open-bite deformity. The patient has a long lower face with a moderate degree of lip strain. The surgical correction involved posterior maxillary impaction, to allow the mandible to autorotate at the condyle to close the open bite, and vertical reduction genioplasty. A total decrease of lower face height of 8 mm was achieved. Postoperatively, there is a stable Class I occlusal result with closure of the

PREOPERATIVE POSTOPERATIVE

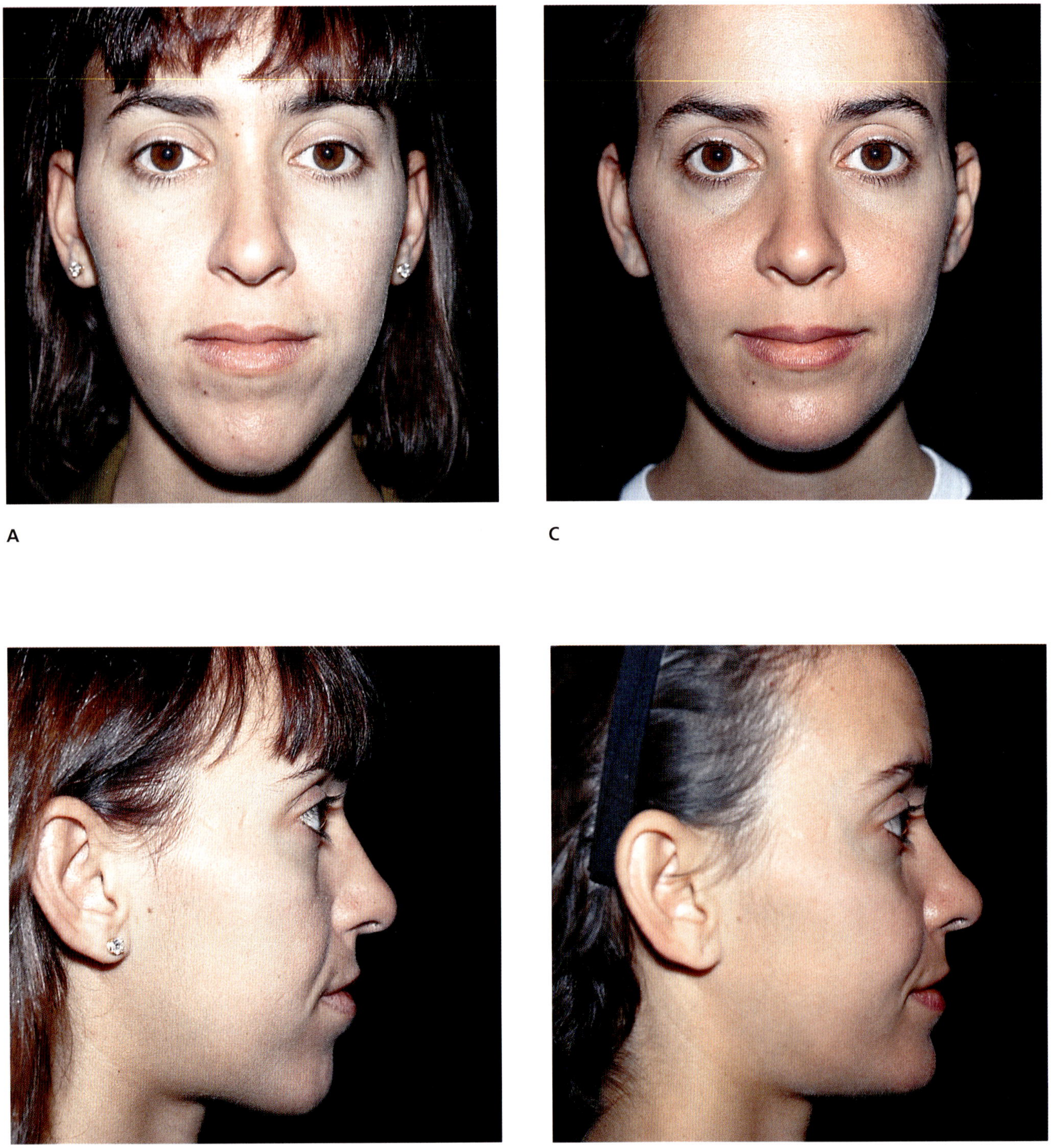

A C

B D

FIG. 1-9 A patient who underwent superior repositioning of the maxilla and osseous genioplasty to treat an anterior open bite. There is loss of soft tissue support in the midface secondary to vertical facial reduction, creating an exaggeration of the nasolabial fold.

PREOPERATIVE

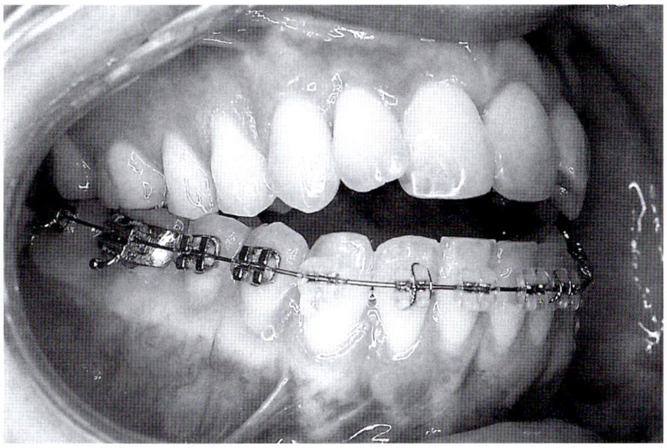

E

POSTOPERATIVE

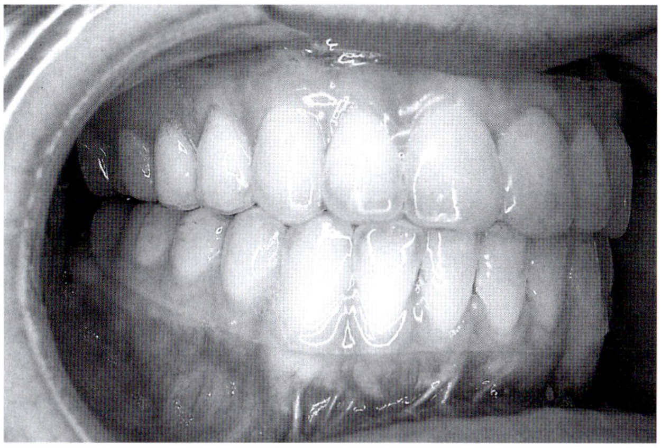

F

FIG. 1-9 (*Continued*) Occlusal results. Orthodontist: Dr. Richard Drummond

anterior bite. In the frontal view, there is an acceptable aesthetic result with an obvious reduction in the height of the lower face and the elimination of the lip strain. However, close inspection of the lateral postoperative view demonstrates an accumulation of soft tissue in the midcheek region just lateral to the nasolabial fold. There is now relative soft tissue excess in this region secondary to vertical facial reduction. Skeletal reduction has exceeded the ability of the soft tissues to contract, and as a result, there is now a deepening of the nasolabial fold. This soft tissue response to vertical facial reduction is not infrequently observed following surgical correction of the long-face patient. You only need to look for it! Fig. 1-10 demonstrates another long-face patient before and after surgical correction of an anterior open bite. As a result of reduction of skeletal volume in the vertical dimension, the patient has developed jowls with a squaring off of the lower face and a much older appearance.

The underlying principle herein described is that contraction (decrease in volume) of the facial skeleton results in a relative soft tissue excess because of the finite ability of the facial soft tissues to contract. In fact, the soft tissues have lost foundational support, which, in turn, results in a loss of facial highlighting, definition, and angularity. Stated differently, one loses the perception of skeletal definition as viewed through the facial mask. This adverse response of the soft tissues to skeletal contraction is not without precedent in the surgical literature. It is frequently seen in the nasal tip and supratip area following reduction rhinoplasty, as critically reviewed by Sheen. Van der Deusen referred to this phenomenon in the face and upper lip following maxillary setback surgery. Both Kawamoto and Rosen have described the detrimental soft tissue response as limiting factors in reduction genioplasty and mandibular setback surgery. It has been eloquently

PREOPERATIVE POSTOPERATIVE

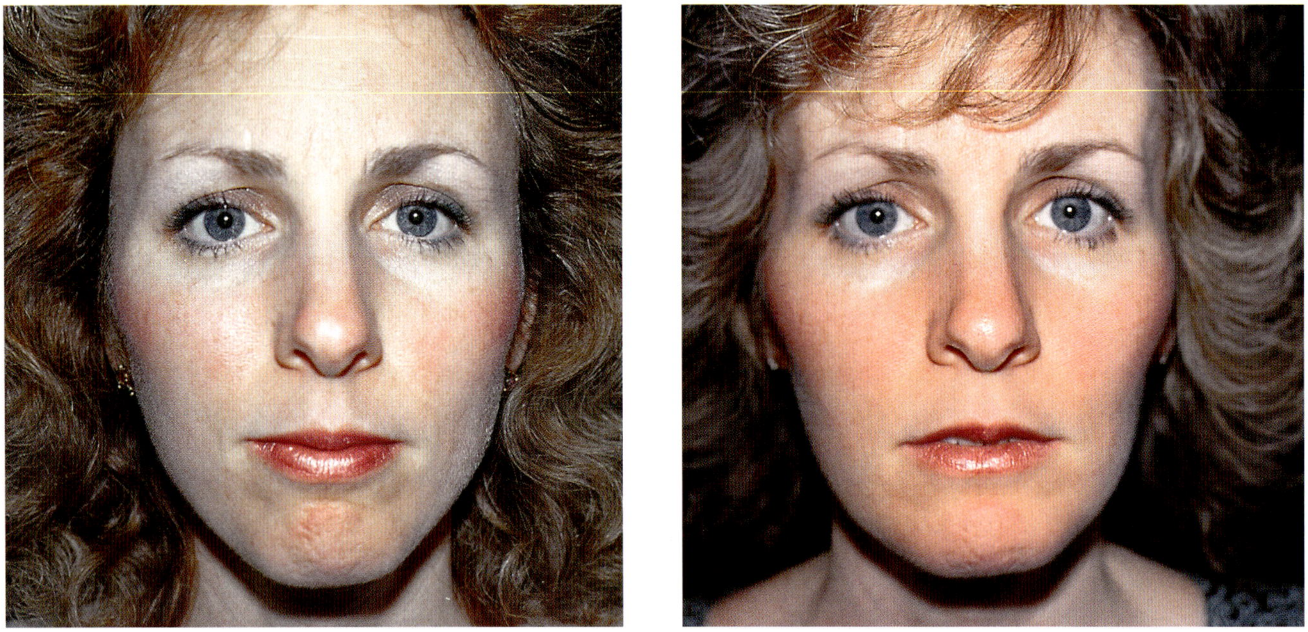

A B

FIG. 1-10 A patient who underwent superior repositioning of the maxilla to treat an anterior open bite. There is loss of soft tissue support in the lower face secondary to vertical facial reduction, creating a jowly appearance.

stated by Freihofer to be the reason for a premature aged appearance after posterior repositioning of the anterior maxilla.

On the other hand, the soft tissue response is typically a favorable one following surgical expansion (enlargement) of the facial skeleton. This is well illustrated in the patient seen in Fig. 1-11 following a bidimensional expansion of the facial skeleton in the sagittal and vertical dimensions to correct a short face and small mandible. As one expands the facial skeleton, the facial mask is expanded, and any soft tissue excess is minimized or eliminated. As a result, it enhances overall facial angularity and highlighting.

Accordingly, when presented with the aesthetic option to either expand or contract the facial skeleton, I now always choose to increase skeletal volume. This clear-cut choice is predicated on two very compelling reasons: (1) It can now be done with a reasonable degree of clinical predictability in terms of skeletal relapse, and (2) the soft tissue response to skeletal expansion is a predictable and favorable one. This, in turn, fulfills our aesthetic goals of jaw surgery, that is, to provide for a well-supported facial mask by achieving adequate projection of the mid- and lower face and adequate height of the lower face.

In summary, the pendulum for treatment planning in orthognathic surgery has now shifted to achieving the best possible aesthetic outcome. Orthognathic surgery is aesthetic surgery. The occlusal result, in this au-

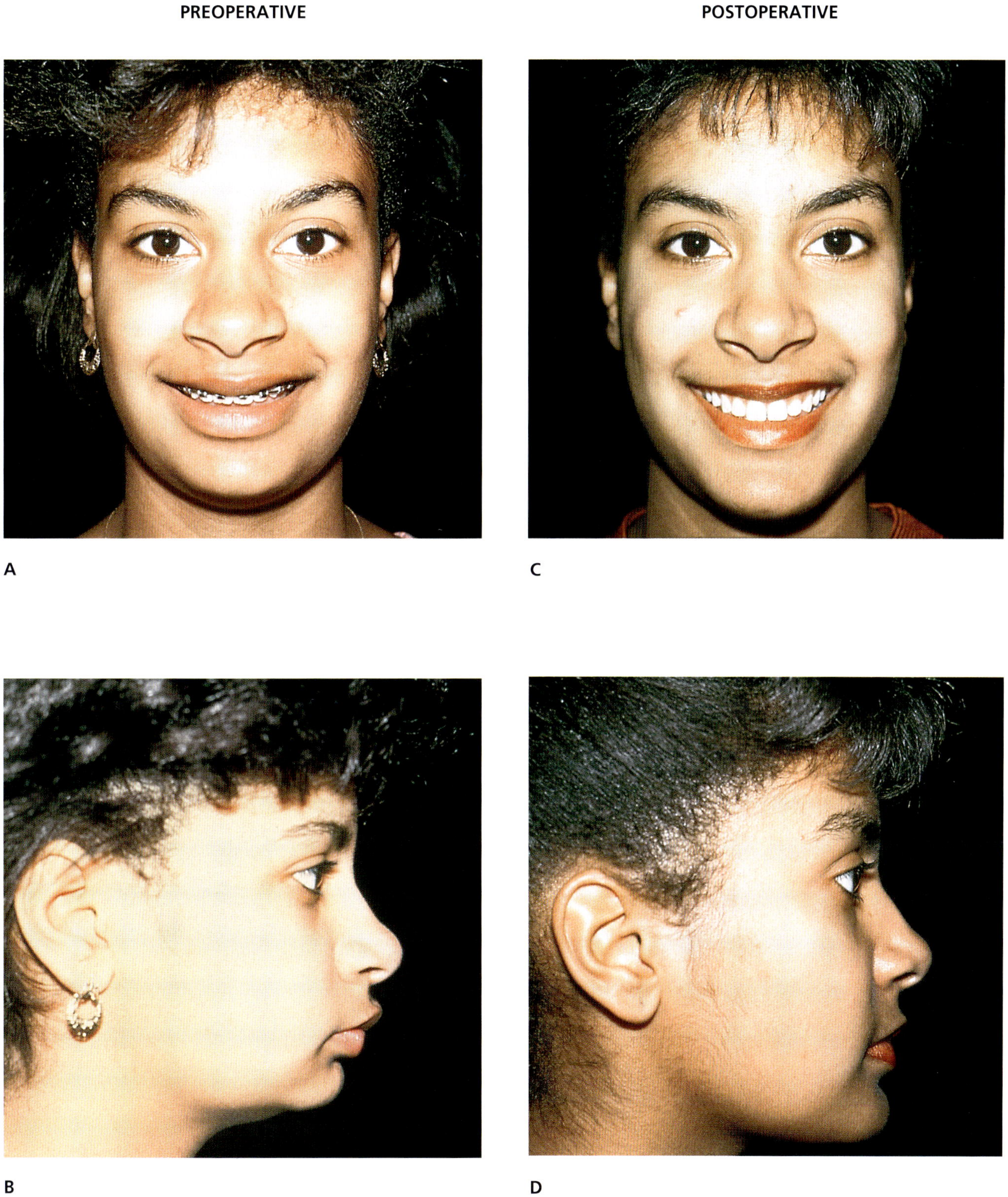

FIG. 1-11 A short-face patient treated with double-jaw surgery to expand the facial skeleton both sagittally and vertically. Note exquisite enhancement of the soft tissue support creating an angular, defined appearance. From: Rosen, H.: Definitive surgical correction of vertical maxillary deficiency. *Plast Reconstr Surg* 85:215, 1990. Reprinted with permission.

thor's opinion, should now be conceptualized as a means to an aesthetic end. It is a step, albeit a necessary one, along the path to achieving our aesthetic goals. The occlusal result must never be considered as an end in and of itself.

Authors Note: The aforementioned statements regarding the goals of orthognathic surgery must be put in proper perspective. Strictly speaking, orthognathic surgery must still be considered functional in nature; that is, it corrects an abnormal occlusion in the overwhelming majority of patients, thereby improving masticatory function and oral health. However, there is only one ideal occlusal result that should be achieved, and that is a normal one. Regardless of which jaw or jaws are moved, the occlusal result should be the same. However, the aesthetic outcome may vary tremendously from unacceptable to acceptable to outstanding depending on which jaw or jaws are moved and to what extent they are moved. It is for this reason that treatment planning efforts must focus on the aesthetic outcome. Accordingly and pragmatically, orthognathic surgery is best considered aesthetic surgery.

Statements in this chapter regarding predictability and skeletal stability must also be put in perspective. Prior to the advent of rigid fixation, skeletal instability primarily related to movement at the osteotomy site and/or reseating of the condyle following condylar displacement. As these events happened relatively early in the postoperative period, observations made regarding skeletal stability at 1 year following surgery were clinically relevant. However, rigid fixation has minimized movement at the osteotomy site as a cause for skeletal relapse. Factors now involved are condylar displacement, observed in the relatively early postoperative period, as well as condylar resorption and skeletal remodeling. These latter two factors may occur for years after surgery. It is, therefore, perhaps premature to make dogmatic statements regarding long-term skeletal stability in orthognathic surgery.

Suggested Reading

1. Carpenter, C.W., Nanda, R.S., and Currier, G.F.: The skeletal stability of LeFort I downfracture osteotomies with rigid fixation. *J Oral Maxillofac Surg* 47:922, 1989.
2. Farkas, L.G., Sohm, P., Kolar, J.C., Katic, M.J., and Munro, I.R.: Inclinations of the facial profile: Art versus reality. *Plast Reconstr Surg* 75:509, 1985.
3. Friehofer, H.P.: Reversing segmental osteotomies of the upper jaw. *Plast Reconstr Surg* 96:88, 1995.
4. Kawamoto, H.K., Jr.: Discussion of reduction mentoplasty. *Plast Reconstr Surg* 70:151, 1982.
5. Kierl, M.J., Nanda, R.S., and Currier, G.F.: A 3-year evaluation of skeletal stability of mandibular advancement with rigid fixation. *J Oral Maxillofac Surg* 48:587, 1990.
6. Legan, H.L. and Burstone, C.J.: Soft tissue cephalometric analysis for orthognathic surgery. *J Oral Surg* 38:744, 1980.

7. Major, P.W., Philippson, G.E., Glover, K.E., and Grace, M.: Stability of maxilla downgrafting after rigid or wire fixation. *J Oral Maxillofac Surg* 54:1287, 1996.
8. Ricketts, R.M.: Cephalometric analysis and synthesis. *Angle Orthod* 31:141, 1961.
9. Rosen, H.: Porous block hydroxyapatite as an interpositional bone-graft substitute in orthognathic surgery. *Plast Reconstr Surg* 83:985, 1989.
10. Rosen, H.: Definitive surgical correction of vertical maxillary deficiency. *Plast Reconstr Surg* 85:215, 1990.
11. Rosen, H.: Maxillary advancement for mandibular prognathism: Indications and rationale. *Plast Reconstr Surg* 87:823, 1991.
12. Rosen, H.: Occlusal plane rotation: Aesthetic enhancement in mandibular micrognathia. *Plast Reconstr Surg* 91:1231, 1993.
13. Rosen, H.: Facial skeletal expansion: Treatment strategies and rationale. *Plast Reconstr Surg* 89:798, 1992.
14. Rosen, H.: Aesthetics in facial skeletal surgery. *Perspectives in Plastic Surgery* 6:1, 1993.
15. Sheen, J.H.: Problem Noses. In *Aesthetic Rhinoplasty*. St. Louis: Mosby, 1978, p. 370.
16. van der Deusen, F.N. and Egyedi, P.: Premature aging of the face after orthognathic surgery. *J Craniofac Surg* 18:335, 1990.
17. Wardrop, R.W. and Wolford, L.M.: Maxillary stability following downgraft and/or advancement procedures with stabilization using rigid fixation and porous block hydroxyapatite implants. *J Oral Maxillofac Surg* 47:336, 1989.
18. Wolford, L.M., Chervello, P.D., and Hilliard, F.W.: Occlusal plane alterations in orthognathic surgery. *J Oral Maxillofac Surg* 51:730, 1993.

Treatment Planning: The Role of Facial Disproportion

CHAPTER

2

Occasionally, in our zeal to expand the facial skeleton to provide adequate soft tissue support, faces may be made abnormally large, at least compared to normative standards, in both the sagittal and vertical dimensions. Accordingly, the following question must be addressed: Can facial disproportion in the sagittal and/or vertical dimensions be attractive and, if so, under what circumstances? Let us first examine the sagittal dimension.

As mentioned in Chapter 1, normative soft tissue cephalometric data would suggest that the ideal sagittal position of soft tissue pogonion falls on a vertical perpendicular from glabella ± 4 mm. Normative anthropometric data, as described by Farkas, demonstrated the normal Caucasian face to diverge posteriorly, and it was not until one deviated from the norm by more than two standard deviations that pogonion fell anterior to glabella, that is, the face became anteriorly divergent. However, if one looks at the profile photograph of a model taken from a fashion magazine (Fig. 2-1), pogonion is well in advance of these normal parameters; that is, there is an excessive degree of anterior facial divergence. This anteriorly divergent profile falls outside the range of normalcy, but the profile is still remarkably attractive. Why is this? Excessive degrees of anterior facial divergence can be not only acceptable but beautiful provided that the anterior divergence is accompanied by the following morphologic features:

1. There must be an overall convex profile in the midsagittal plane from glabella to subnasale to pogonion.
2. There must be a convex soft tissue relationship in the parasagittal plane from the infraorbital region to the paranasal region to the parasymphyseal region.
3. There must be a relatively normal maxillomandibular relationship as evidenced by the sagittal positions of the upper and lower lip relative to one another. Ideally, one would like

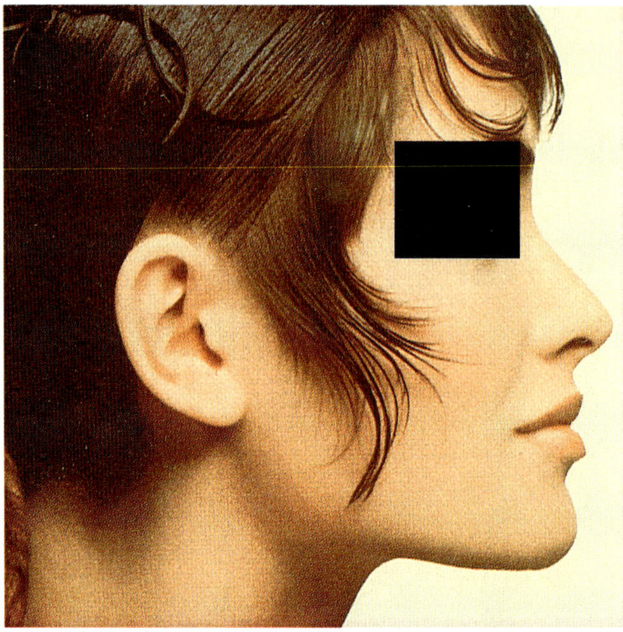

FIG. 2-1 A fashion model demonstrating excessive degrees of anterior facial divergence. However, there is overall facial convexity, good labial relations, and an adequately projecting nasal dorsum and tip.

to see the upper vermillion slightly in advance of the lower vermilion or, at the very least, with an equal degree of prominence.

4. There must be adequate projection of the nasal dorsum and nasal tip. This is important because the apparent size of the nose diminishes as one moves the mid- and lower face forward. If the size of the nose cannot accommodate increased projection of the upper and/or lower jaw, an unaesthetic appearance will result.

Although excessive anterior facial divergence can be an attractive feature of a model's face, can it be so in more average looking individuals whom we treat on a routine basis? Providing the aforementioned criteria are met, I believe it can be. The patient profiles in Fig. 2-2 are before and after photographs of a patient presenting for surgical correction of a Class III malocclusion. He was treated with advancement of both the maxilla and the mandible as well as advancement genioplasty. Postoperatively, there is an excessive degree of anterior facial divergence relative to normative standards, but it is seen in association with overall midsagittal and parasagittal convexity, good relationships of the upper and lower lips, and an adequately projecting nose. The patient seen in Fig. 2-3 presented with a Class III anterior open bite and was treated with advancement of the maxilla and the mandible and advancement genioplasty. Postoperatively, there is a marked degree of excessive anterior facial divergence in the presence of overall facial convexity, good labial position, and an adequately projecting nose. Excessive anterior facial divergence *can be* attractive.

PREOPERATIVE POSTOPERATIVE

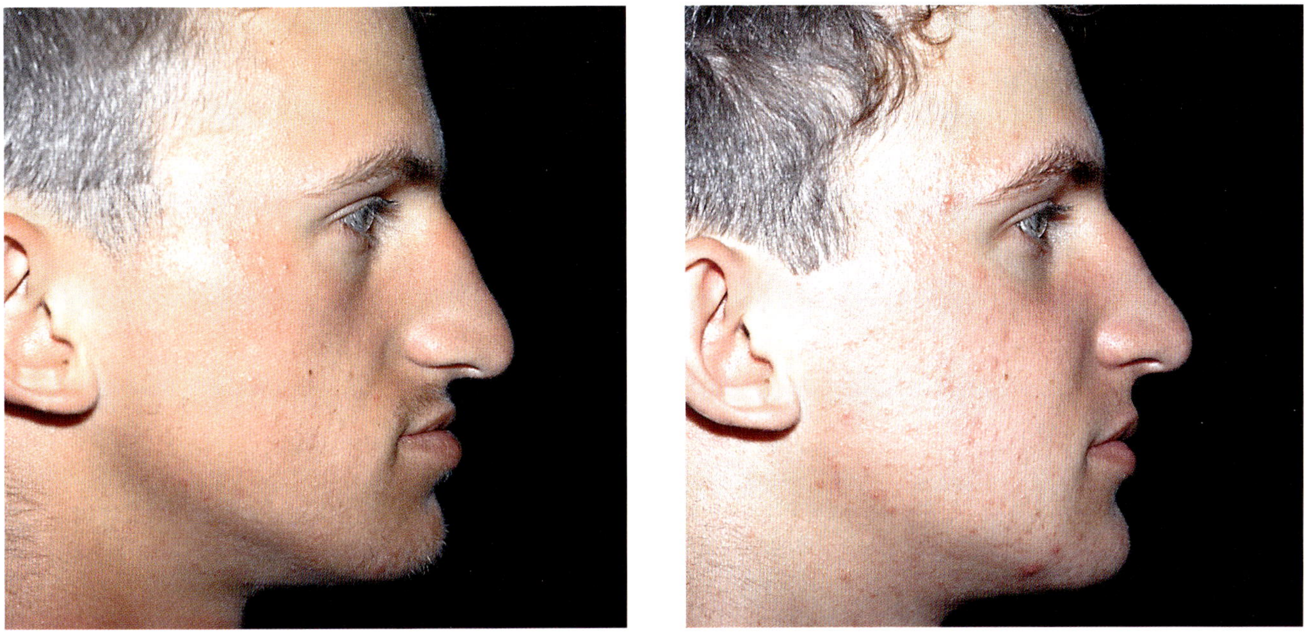

A B

FIG. 2-2 A Class III patient treated with maxillomandibular advancement. There is now excessive anterior divergence. Note the overall convex facial profile.

PREOPERATIVE POSTOPERATIVE

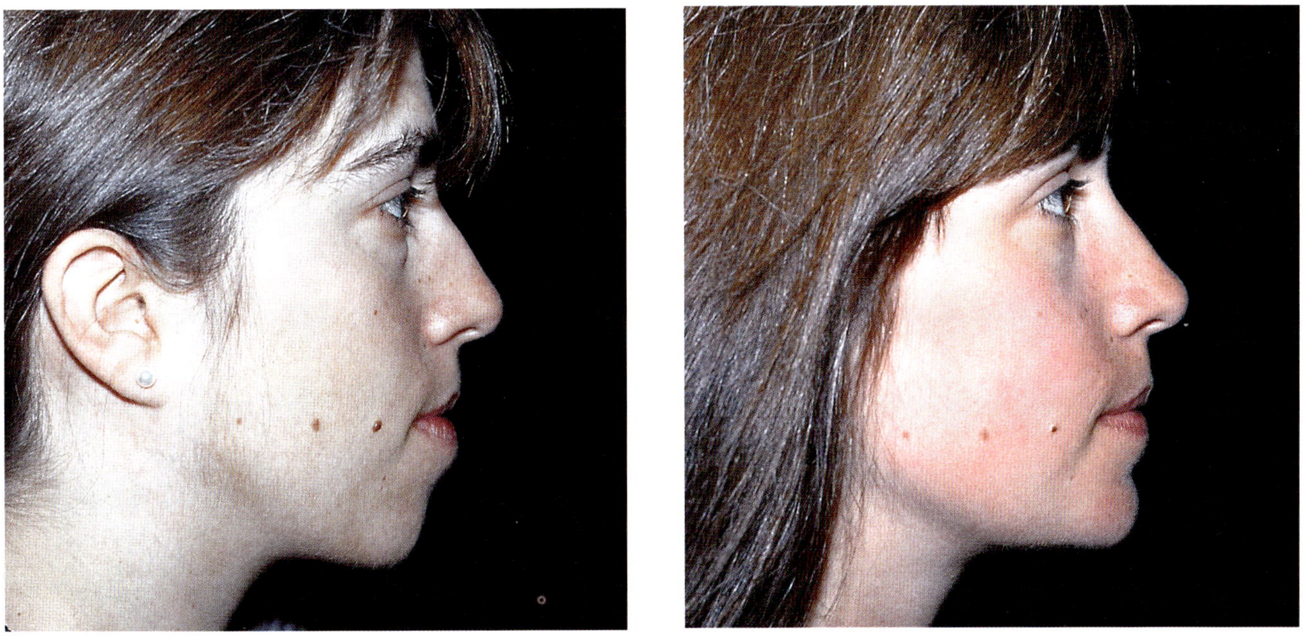

A B

FIG. 2-3 Another Class III patient treated with maxillomandibular advancement. There is now excessive anterior facial divergence and overall facial convexity.

What about the vertical dimension? Can excessive height of the lower face (subnasale to menton) relative to the midface (glabella to subnasale) be aesthetically pleasing? Provided that it is unaccompanied by the aesthetic stigmata or visual tip-offs of a long face, quantitative vertical facial excess can be extraordinarily attractive (Fig. 2-4). This photograph of a model demonstrates excessive height of the lower face relative to the midface, but because her quantitatively long lower face is unaccompanied by the visual stigmata of a long face, her appearance is attractive. These stigmata include:

1. Excessive lip strain in an attempt to achieve lip seal.
2. An excessive degree of gingival show with full smiling.
3. An absent or exceedingly shallow labiomental fold.

Excessive maxillary incisal show with the lips at rest (traditionally thought to be greater than .3 mm) should definitely not be included among the stigmata of the quantitatively long face. When unaccompanied by lip strain and excessive degrees of gingival show, incisal exposure of up to 6 or even 7 mm in a female patient can be very appealing. Excessive maxillary incisal show, in and of itself, should never be an indication for intruding the maxilla.

Can quantitative, vertical facial excess be as attractive in our patients as in the model in Fig. 2-4? The patient in Fig. 2-5 was referred for surgical correction of a Class II malocclusion. She obviously has a short lower face as well. Surgical correction included vertical maxillary lengthening,

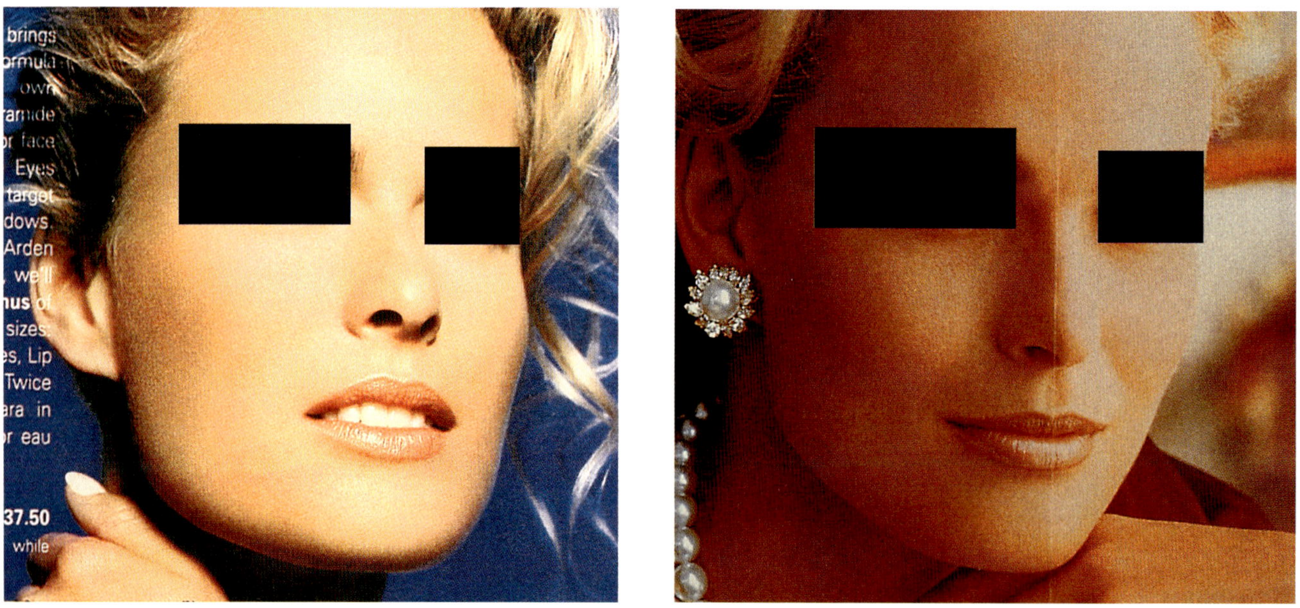

A

B

FIG. 2-4 A fashion model with a quantitatively long lower face. This is not visually objectionable because there is no lip strain and the labiomental fold is well defined. Note approximately 7 mm of incisal exposure.

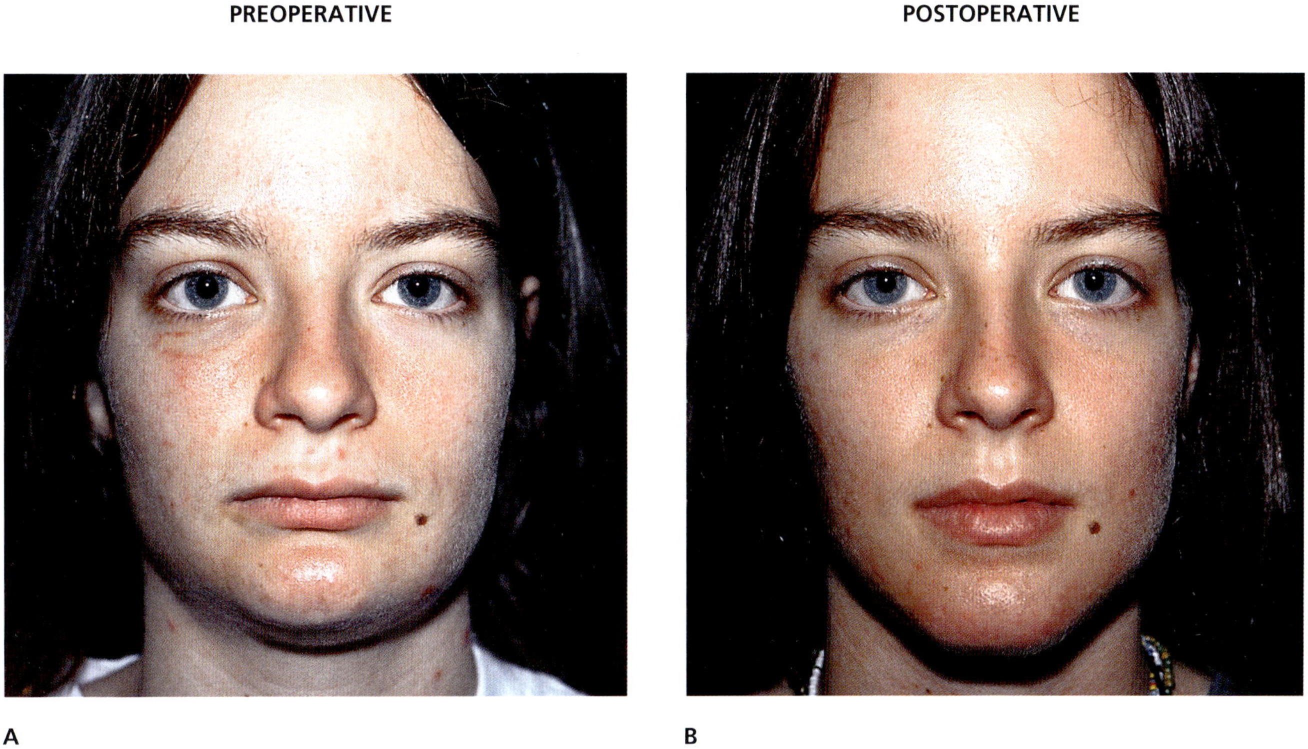

FIG. 2-5 A short-face patient intentionally overcorrected to achieve a long lower face. However, there is no lip strain, there is not excessive gingival show if the patient were to smile, and there is a well-defined labiomental fold. Although the lower face is now quantitatively long, it does not visibly appear to be so.

mandibular advancement and lengthening, and osseous genioplasty to vertically lengthen the chin. The treatment plan was intentionally designed to overcorrect her vertical deficiency, creating vertical facial excess. Despite this, her appearance is a very attractive one because her now long face is unaccompanied by any of the visual tipoffs that our eye is used to associating with the long-face patient: there is no lip strain with attempted lip closure; there is not excessive gingival show if the patient were to smile; and the labiomental fold, although less deep postoperatively, has been preserved.

The depth and definition of the labiomental fold is critically important to the visual perception of the height of the lower facial third. The fold runs perpendicular to the long axis of the lower face and, therefore, breaks up the vertical axis of the lower face and deemphasizes the visual impression of vertical facial excess. A shallow or absent fold can make a quantitatively normal lower facial third visibly appear to be too long. Surgical lengthening of the lower face will almost always cause the labiomental fold to become less deep. It is an important limiting factor in the extent to which one can lengthen the face.

The patient in Fig. 2-6 presents another example of the surgical creation of a quantitatively long lower face based on the effort to enhance soft tissue support. This patient presented with a Class III malocclusion and a

PREOPERATIVE POSTOPERATIVE

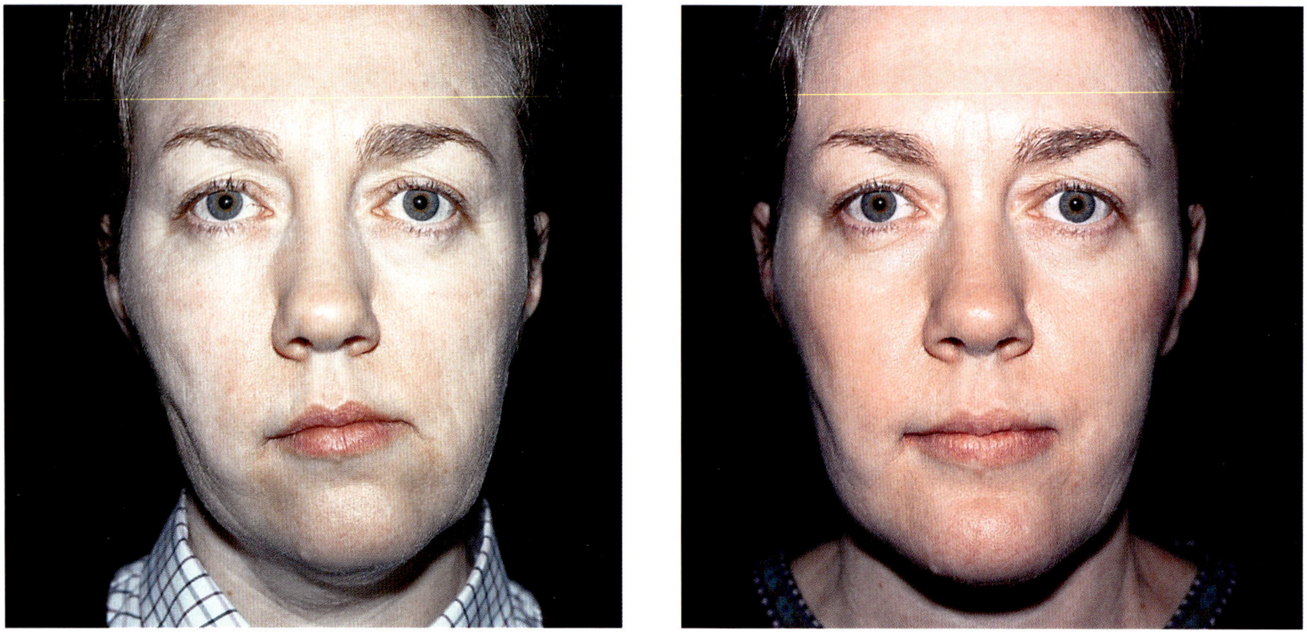

A B

FIG. 2-6 A facial asymmetry patient intentionally overelongated in the lower face to restore facial symmetry. This was done to achieve enhanced support of the soft tissues.

significant facial asymmetry. Her surgical correction involved lateral shifting of the maxilla and the mandible to the right as well as lengthening the normal left side of the face rather than shortening the, already quantitatively long, right side of the face. Despite quantitative vertical facial excess, one does not have the visual impression of a face that is long because of the absence of lip strain and the presence of a labiomental fold.

We can now return to the question posed at the beginning of this chapter. Can facial disproportion in the sagittal and/or vertical dimension be attractive? The answer is clearly a definitive yes. Despite this, however, most of the unaesthetic results that we continue to see in orthognathic surgery today are due to the fact that this very basic aesthetic principle goes unheeded. Perhaps it is because surgeons are uncomfortable with the notion of making faces disproportionately large. Most surgeons continue to feel more secure relying on the more traditional parameters of facial normalcy and, therefore, continue to place undo emphasis on normative anthropometric and cephalometric data. As a result, the tendency is either to overcontract or underexpand the facial skeleton. Consequently, the soft tissue response to surgery is less than desirable, and a suboptimal aesthetic outcome results.

One merely needs to broadly categorize the different dentofacial deformities that are surgically treated to illustrate this trend. The first morphologic type is the Class II individual who typically presents with excessive facial convexity and a posteriorly divergent profile. Typically, these patients are underexpanded, and adequate projection of the lower face relative to the soft tissue requirements is never achieved. The second category would be the Class III individual with a flat or concave profile and an anteriorly divergent face, depending on the skeletal etiology. Historically, these individuals have been overcontracted because an excessive reliance has been placed on mandibular setback surgery by itself or in combination with maxillary advancement surgery to achieve a normal occlusion. This, too, results in inadequate projection of the mandible and lower face with subsequent lack of skeletal support. The third category would be the short-face patient. It is common for these patients to be surgically underexpanded, and therefore, they remain undercorrected with inadequate height of the lower face, resulting in deficient soft tissue support. The last morphologic category is the long-face patient. It is in this type of patient where the tendency to overcontract the facial skeleton is most commonly seen, creating aesthetic disasters with the patients appearing older and possibly edentulous. There is little question that the aesthetic, prudent course of action in this patient category is to leave the patient undercorrected, minimizing contraction of the facial skeleton in the vertical dimension. The common denominator among all these patient categories is a disequilibrium between skeletal and soft tissue volumes.

In summary, the three underlying principles to aesthetic surgical treatment planning are the following:

1. The support of the facial soft tissues is a critical factor in determining the aesthetic success or failure of orthognathic surgery. Accordingly, soft tissue goals, to a large extent, must influence the direction and extent of skeletal displacements.
2. To achieve appropriate soft tissue support, it is more predictable to expand skeletal volume than it is to contract skeletal volume.
3. Efforts to expand the facial skeleton frequently result in facial skeletal disproportion, which, under the right aesthetic circumstances, can be extremely attractive.

It behooves the surgeon to remember that a "normally" balanced and proportioned face is only as attractive as the extent to which its soft tissues are supported. If we must choose between normal facial proportions and soft tissue support, it is the latter that must take precedence. This assumes that the aesthetic guidelines described have been followed. However, despite adhering to these guidelines, the extent to which we can enlarge the maxillomandibular complex is not limitless. Unfortunately, these limitations are nonquantitative and subjective in nature. Where is the boundary between attractive qualitative facial disproportion and a lower face that visibly appears to be too large and possibly even cumbersome? Herein lies the art, judgment, and challenge of orthognathic surgery.

Suggested Reading

1. Farkas, L.G., Sohm, P., Kolar, J.C., Katic, M.J., and Munro, I.R.: Inclinations of the facial profile: Art versus reality. *Plast Reconstr Surg* 75:509, 1985.
2. Rosen, H.: Maxillary advancement for mandibular prognathism: Indications and rationale. *Plast Reconstr Surg* 87:823, 1991.
3. Rosen, H.: Facial skeletal expansion: Treatment strategies and rationale. *Plast Reconstr Surg* 89:798, 1992.
4. Rosen, H.: Bimaxillary Surgery. In M. Habal (Ed.), *Advances in Plastic and Reconstructive Surgery*. Chicago: Mosby Year Book, 10:93, 1993.
5. Rosen, H.: Aesthetics in facial skeletal surgery. *Perspectives in Plastic Surgery* 6:1, 1993.
6. Rosen, H.: Aesthetic guidelines in genioplasty: The role of facial disproportion. *Plast Reconstr Surg* 95:463, 1995.
7. Salyer, K.E., Vasconez, H.C., Genecov, E., and Kolar, J.: Analysis of facial proportions in twenty beautiful faces. *Plast Surg Forum* 10:129, 1987.
8. Whitaker, L.A.: Aesthetic contouring of the facial support system. *Clin Plast Surg* 16:815, 1989.

Treatment Planning: Prediction Tracings and Model Surgery

CHAPTER 3

Prediction Tracings

Once a treatment plan has been reached based on the physical examination of the patient, prediction tracings are helpful in determining the final position of soft tissue pogonion in both the sagittal and vertical dimensions. Accordingly, they help dictate the aesthetic desirability of simultaneous jaw repositioning and osseous genioplasty. Situations in which this is most clinically applicable are those involving mandibular rotational changes in the sagittal plane with the point of rotation being the mandibular condyle and/or the ramus osteotomy. Regardless of the point of rotation, the aesthetic consequences of mandibular rotation are profound. With counterclockwise mandibular rotation, as is most frequently seen in anterior open-bite and/or long-face correction, soft tissue pogonion is displaced superiorly and anteriorly. Another way of conceptualizing this is that one loses vertical dimension in the lower face but gains sagittal projection at soft tissue pogonion. In addition, the mandibular plane is flattened, and as a result, the gonial angle visually appears to be less obtuse, although it is really unchanged. Lastly, the mandibular incisors are uprighted, although this is a very subtle effect. The aesthetic changes accompanying mandibular clockwise rotation, regardless of the point of rotation, are just the opposite. The most dramatic of these involves inferior and posterior displacement of soft tissue pogonion; that is, one loses sagittal projection of the lower face, but the vertical dimension of the lower face is augmented. In addition, the mandibular plane is steepened, and as a result, the gonial angle appears to have opened. This again is only an illusion. Lastly, and least apparent, there is flaring of the mandibular incisor teeth.

The only significant aesthetic impact of maxillary rotation in the sagittal dimension is the change in the axial inclination of the maxillary incisor teeth. With counterclockwise rotation, the effect is incisal flaring, whereas clockwise rotation causes incisal uprighting. Of course, rotational changes of the maxilla are accompanied by rotational mandibular changes, so the overall aesthetic effects of maxillary rotation can be dramatic. The only exception to this latter statement is surgical closure of an anterior open bite secondary to maxillary clockwise rotation to effect inferior repositioning of the maxillary anterior teeth.

Isolated Mandibular Surgery

Mandibular rotational changes with their accompanying aesthetic alterations occur in several clinical situations. The first of these is single-jaw surgery involving the mandible in which the mandibular body is rotated about the ramus osteotomy. This most frequently occurs in mandibular advancement surgery to correct a Class II malocclusion in which excessive overbite remains following the presurgical phase of orthodontics. In this instance, clockwise rotation of the mandibular body must occur to keep the posterior bite from opening because of excessive incisal guidance (Fig. 3-1). Because positional forward posturing of the jaw in the clinical setting cannot be accompanied by this rotational movement, a prediction tracing is necessary to determine its extent and, therefore, the position of soft tissue pogonion. In this situation, pogonion comes forward much less than one would anticipate. In addition, it helps predict the aesthetic advisability of simultaneous advancement genioplasty (Fig. 3-2).

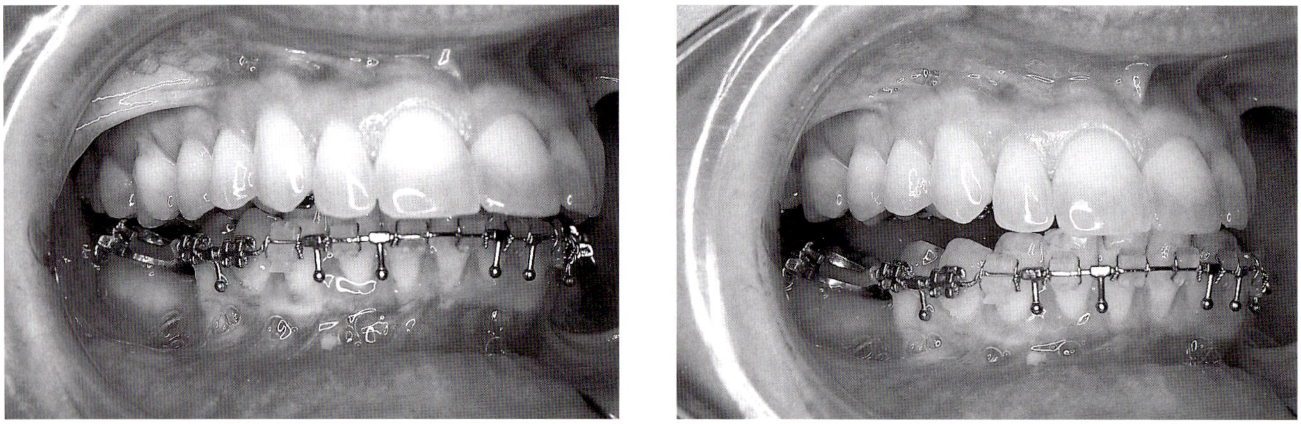

A B

FIG. 3-1 A: A Class II, Division 1 anterior deep-bite malocclusion. B: As the mandible is postured forward, incisal guidance causes the posterior bite to open. To prevent this from happening during surgery, the distal mandibular segment must be rotated clockwise as it is advanced.

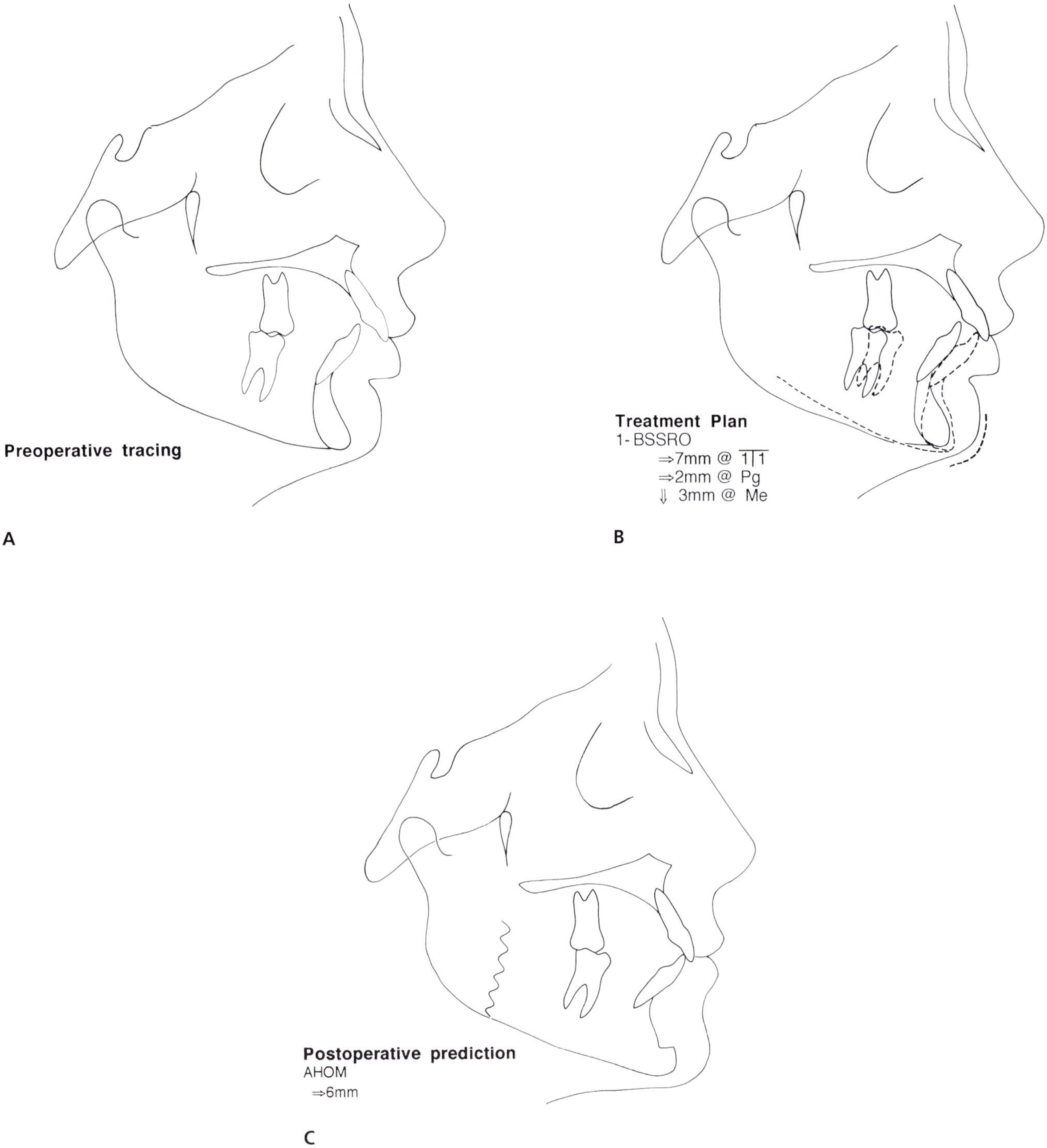

FIG. 3-2 A: A cephalometric tracing of a deep-bite, Class II patient. B: As the mandible is advanced, it is rotated clockwise to keep the posterior bite from opening. As a result, the chin only comes forward 2 mm despite the incisor teeth advancing 7 mm. The lower face is also lengthened 3 mm. Note also the flaring of the mandibular incisors and steepening of the mandibular plane. C: An osseous genioplasty to advance the chin 6 mm will be necessary to achieve an acceptable result.

Occasionally, mandibular advancement surgery may be accompanied by counterclockwise rotation of the mandibular body to close an anterior open bite. In this instance, pogonion will advance a greater amount than anticipated but at the expense of reducing height of the lower facial third (Fig. 3-3). In such an instance, simultaneous osseous genioplasty to increase the vertical dimension of the chin may be indicated. Again, a prediction tracing is helpful in predicting this need. Counterclockwise rotation of the mandibular body may also accompany mandibular setback surgery to correct a Class III malocclusion. Similar aesthetic considerations must apply.

For all of the aforementioned mandibular skeletal displacements, an accurate prediction tracing is predicated on simulating an overbite/overjet relationship that can be clinically achieved at the time of surgery. Therefore, tracings should be correlated with hand holding of the final surgical models in the desired final occlusion.

Isolated Maxillary Surgery

Single-jaw surgery involving Le Fort I osteotomy to superiorly reposition the maxilla remains the cornerstone of treatment of the long-face patient with or without an anterior open bite. Such maxillary displacement is always accompanied by counterclockwise mandibular autorotation with the point of rotation in the vicinity of the mandibular condyle. Such rotation is invariably accompanied by superior and anterior displacement of soft tissue pogonion. The extent of this displacement as well as the aesthetic advisability of simultaneous advancement genioplasty is best determined with a prediction tracing (Fig. 3-4).

Double-Jaw Surgery

There are certain circumstances in double-jaw surgery in which one intentionally alters the occlusal plane. As a result, changes occur in soft tissue pogonion position, mandibular plane angle, gonial angle, and axial inclination of incisors. The most common circumstance in which this occurs is the Class II, vertical maxillary excess patient. In this instance, the prediction tracing is desirable to determine the extent of mandibular advancement and, possibly, advancement genioplasty that will be necessary following autorotation of the mandible.

Perhaps the most useful indication to alter the occlusal plane in an effort to improve aesthetic results occurs in the severely micrognathic individual in whom both the mandibular ramus and body of the mandible is deficient. As a result, there is a lack of posterior face height and excessive anterior face height due to an excessive vector of vertical growth in the mandible. This is always accompanied by an abnormally steepened mandibular occlusal plane. Surgical correction must involve counterclockwise rotation of the occlusal plane to achieve adequate projection of the lower face

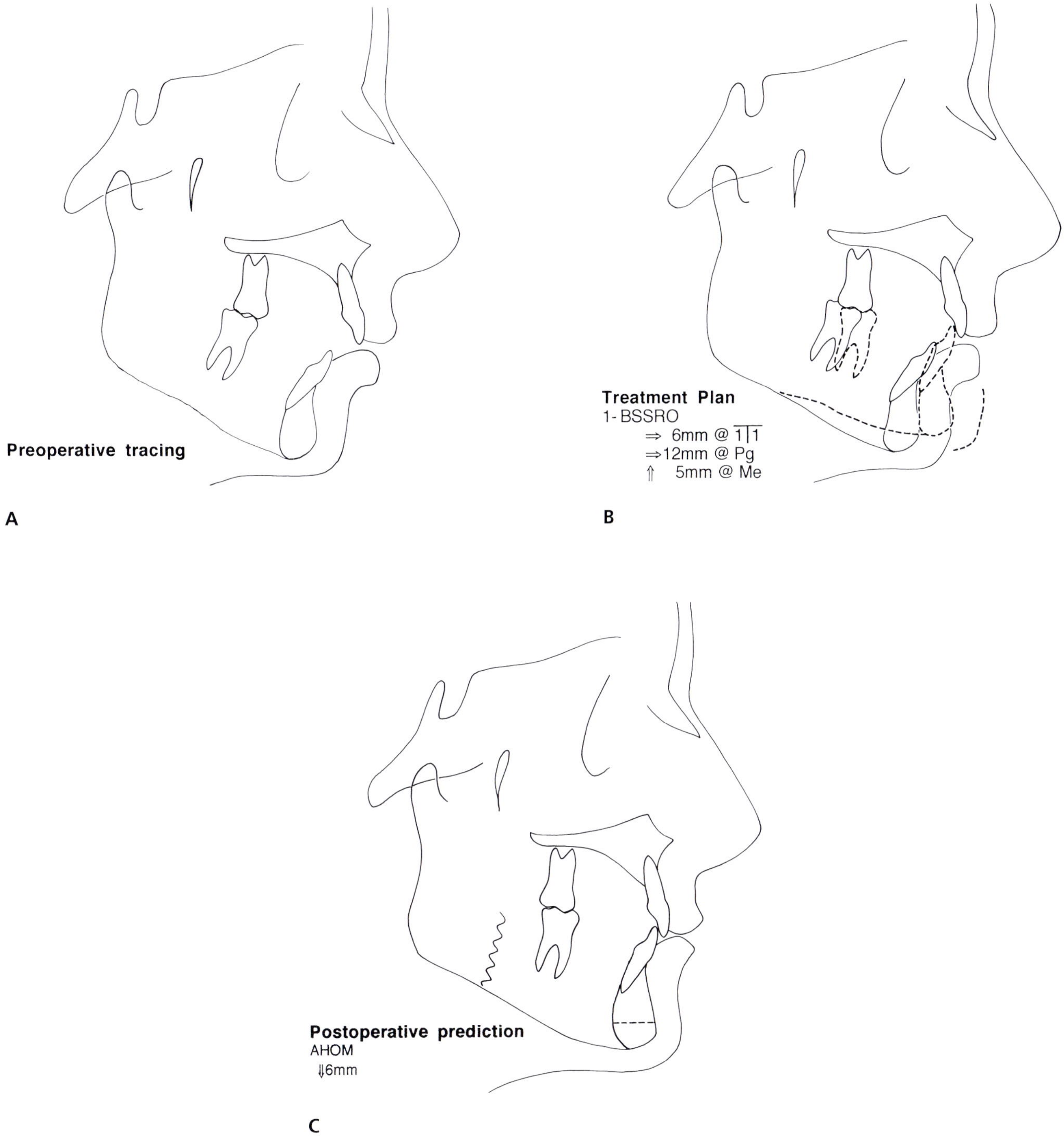

FIG. 3-3 A: A cephalometric tracing of a Class II, open-bite patient to be treated with mandibular advancement. B: As the mandible is advanced, the distal segment must be rotated counterclockwise to close the bite. Note the chin comes forward 12 mm, compared with a 6 mm advancement of the incisors, and there is a loss of 5 mm of lower face height. In addition, the mandibular incisors have been uprighted, and the mandibular plane has been flattened. C: The final prediction tracing demonstrates the need for osseous genioplasty to vertically lengthen the chin to compensate for the loss of lower face height.

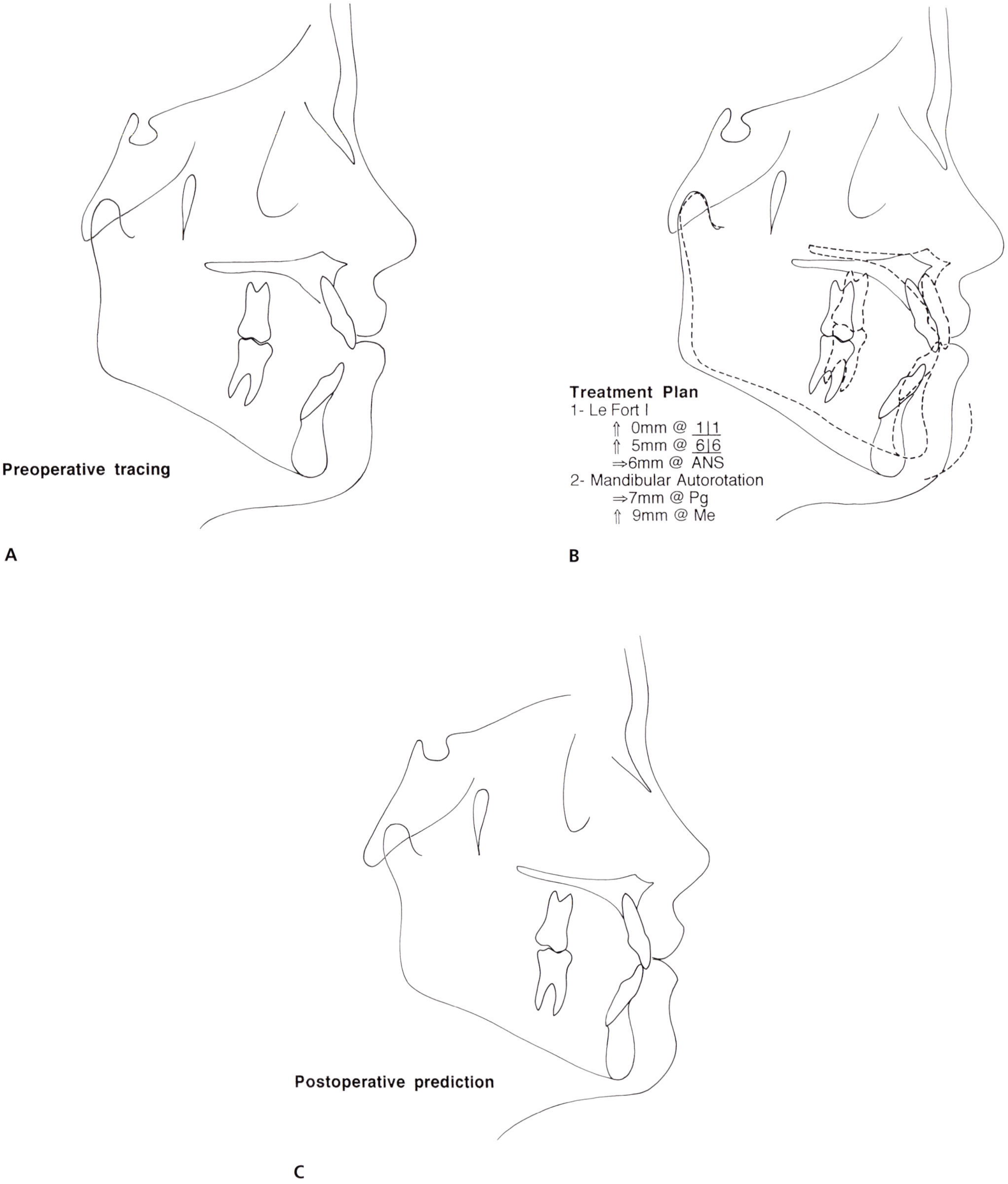

FIG. 3-4 A: A cephalometric tracing of a Class I, open-bite, long-face patient. B: Posterior maxillary impaction surgery to close the open bite results in mandibular autorotation with the chin coming forward 7 mm and superiorly 9 mm. C: A final prediction tracing demonstrates good chin position and no need for osseous genioplasty.

at the level of soft tissue pogonion. This is accomplished by performing double-jaw surgery to counterclockwise rotate both the maxilla and the mandible (Fig. 3-5). A more detailed discussion of the aesthetic implications of this surgical correction can be found in Chapter 9. A similar type of surgical correction could be considered for the Class II individual who has undergone prior orthodontic attempts to correct a malocclusion by extraction of maxillary premolars and retraction of the anterior maxillary teeth. These patients are typically left with inadequate projection of the lower face and excessively uprighted maxillary incisors. Counterclockwise rotation of both the maxilla and the mandible will correct these aesthetic deficiencies. Prediction tracings help ascertain the extent of counterclockwise rotation that will be necessary to achieve adequate projection of the lower face at the level of pogonion.

The final situation in which occlusal plane rotation is desirable to achieve aesthetic goals is in the Class II individual who has a relatively prominent chin. This is frequently accompanied by vertical maxillary deficiency. In these individuals, clockwise rotation of the maxilla and the mandible is frequently desirable to avoid the need for reduction genioplasty (an operation fraught with undesirable results) and still achieve adequate projection of the mandibular dentoalveolus (Fig. 3-6).

Parenthetically, these examples of double-jaw surgery to achieve occlusal plane alteration illustrate the great aesthetic advantage of performing simultaneous maxillary and mandibular repositioning. It completely eliminates the constraints that the occlusion places on the aesthetic results as in single-jaw surgery. It, therefore, allows one to achieve ideal degrees of skeletal volume in both sagittal and vertical dimensions in an effort to produce predictable and favorable soft tissue outcomes. By taking advantage of the ability to alter the occlusal plane, one can further control skeletal volumes.

One last word regarding prediction tracings seems appropriate. An accurate prediction tracing is only as good as the lateral cephalometric radiograph from which it is derived. The patient's mandible must be postured in its most retruded position for meaningful information to be obtained. This is particularly true in the Class II skeletal patient who may be chronically posturing the mandible forward.

Model Surgery

There is one compelling aesthetic reason to perform model surgery. It allows for the accurate positioning of the maxilla in the vertical, transverse, and sagittal dimensions and the accurate orientation of the maxilla as it relates to its tilt in the transverse dimension and sagittal planes. This can be achieved only by employing an intermediate splint fabricated from the mounted surgical models after having performed model surgery on the maxillary cast. It eliminates the need for utilizing direct intraoperative measurements (which are notoriously unreliable) in an effort to accurately po-

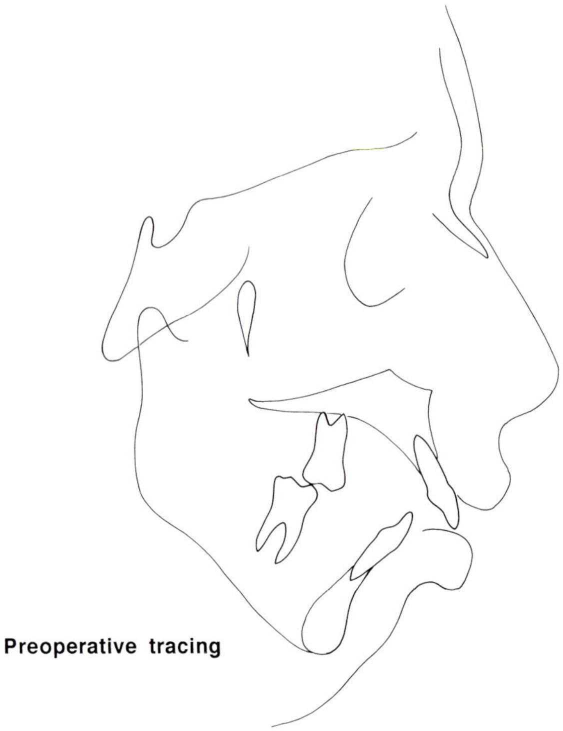

A

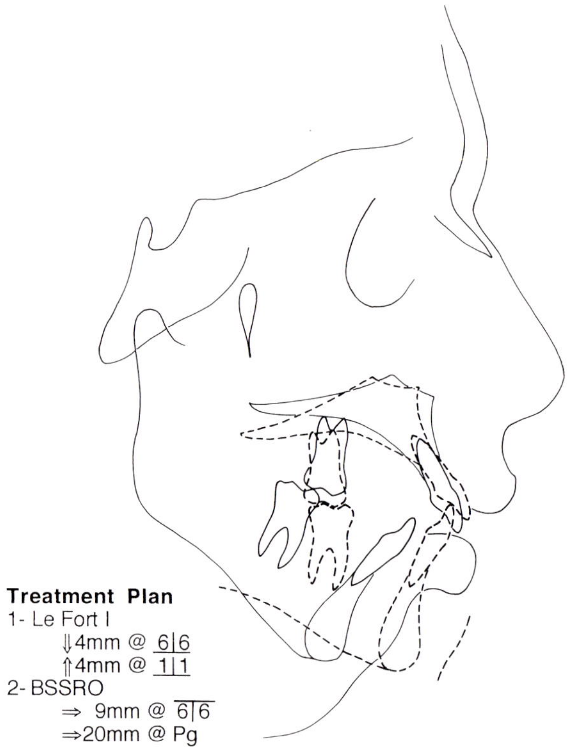

B

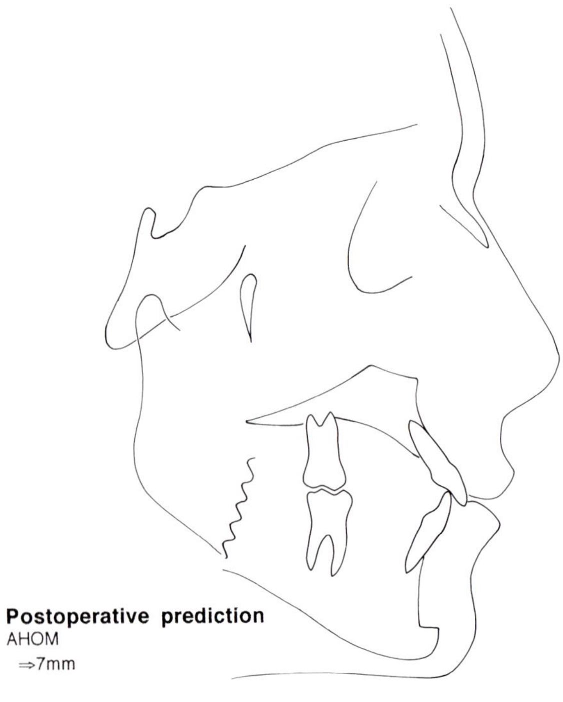

C

FIG. 3-5 A: A cephalometric tracing from a patient with severe mandibular deficiency and steep occlusal plane inclination. Adequate projection of the lower face cannot be achieved without counterclockwise rotation of both maxilla and mandible. B: An operation depicting counterclockwise rotation of both jaws and mandibular advancement. The maxilla is lengthened at the molars and impacted at the incisors. This allows for a differential magnitude of mandibular advancement, as it must be counterclockwise rotated to close the surgically created open bite. Note that the lower molars come forward 9 mm and the chin comes forward 20 mm. C: The final prediction tracing demonstrates the need for additional advancement genioplasty of 7 mm.

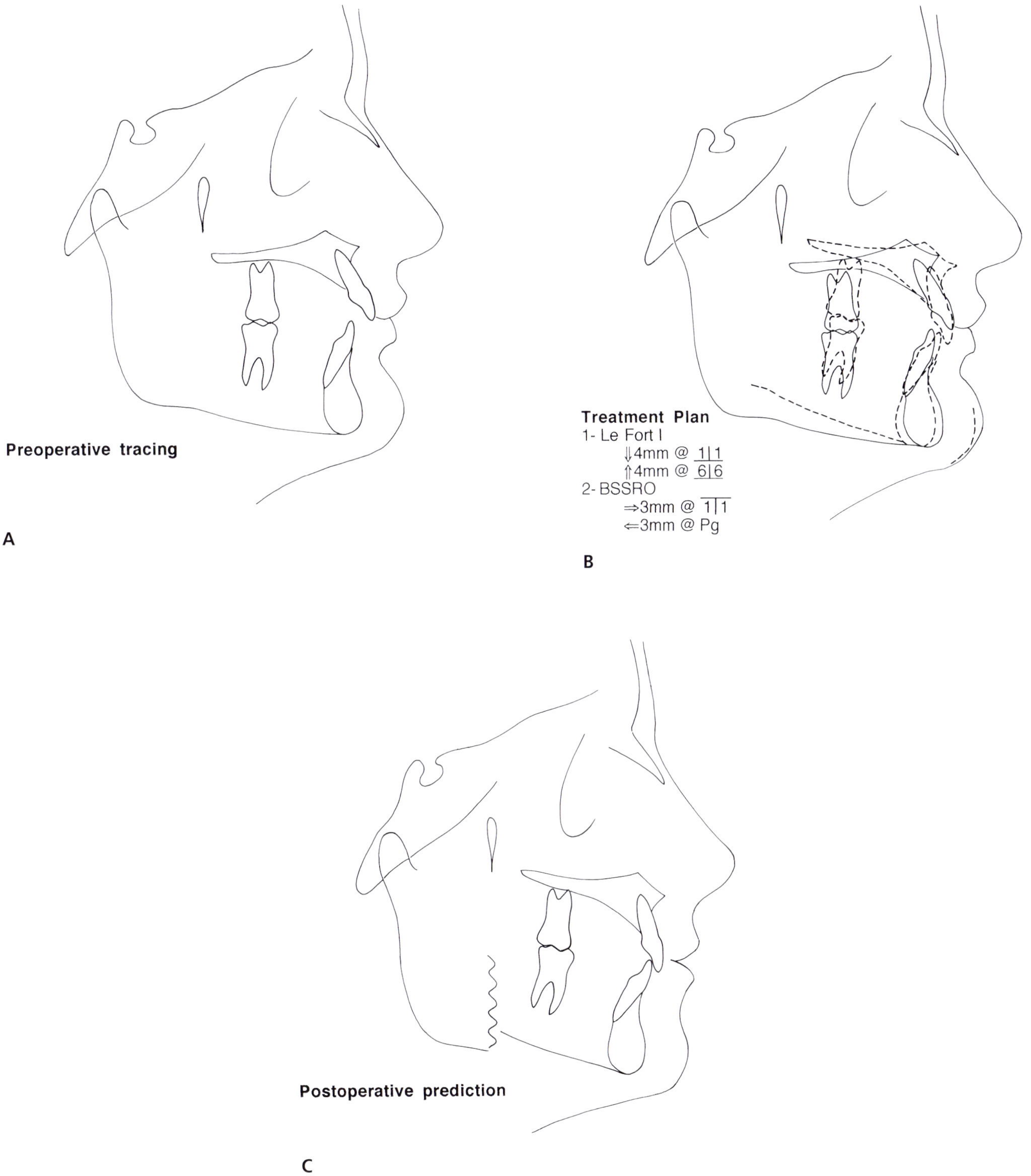

FIG. 3-6 A: A cephalometric tracing from patient with a Class II malocclusion and prominent chin. There is also anterior vertical maxillary deficiency. B: To advance the mandible but keep the chin from becoming more prominent, the maxillary occlusal plane will also be rotated clockwise by lengthening anteriorly and impacting posteriorly. This deepens the bite and necessitates the distal mandibular segment to rotate clockwise as it advances. The lower incisors come forward 3 mm and pogonion goes backward 3 mm. C: A postoperative prediction tracing demonstrates an aesthetic position of the chin without the need for reduction osseous genioplasty.

sition the maxilla. The only intraoperative measurement that must be determined is the vertical position of the anterior maxilla. All other quantitative determinations are built into the intermediate splint. Accurate positioning of the maxilla will then determine placement of the mandible based on the desired occlusal relations.

Model surgery is aesthetically indicated only in double-jaw surgery. Single-jaw surgery does not require this because accurate positioning of the jaw is dictated by the occlusal relationship in which the unoperated jaw serves as a template.

For model surgery to serve its aesthetic function, an accurate face bow transfer must be utilized to accurately position the maxilla in three planes of space. In addition, the mandibular cast must be mounted in relation to the maxillary cast in its most retruded position. Any discrepancy between centric relation and centric occlusion will be built into the intermediate splint and transferred to the patient, resulting in less than desirable maxillary position.

Suggested Reading

1. Bell, W.H., Proffit, W.P., and White, R.P. (Eds.): *Surgical Correction of Dentofacial Deformities*. Philadelphia: Saunders, 1992.
2. Epker, B.N. and Fish, L.C.: *Dentofacial Deformities: Integrated Orthodontic and Surgical Correction*. St. Louis: Mosby, 1986.
3. Rosen, H.: Occlusal plane rotation: Aesthetic enhancement in mandibular micrognathia. *Plast Reconstr Surg* 91:1231, 1993.
4. Wolford, L.M., Chervello, P.D., and Hilliards, F.W.: Occlusal plane alterations in orthognathic surgery. *J Oral Maxillofac Surg* 51:730, 1993.

Surgeon–Patient Communications

CHAPTER 4

Orthognathic surgical patients can potentially undergo major and dramatic changes in their appearances. Although these changes are intended to be positive, the patient's perception of these changes must coincide with those of the surgeon. This is true in both a qualitative and quantitative sense. For purposes of this discussion, qualitative perceptions refer to directions of skeletal displacements, whereas quantitative perceptions refer to the magnitude of such displacements. The ability of the surgeon and the patient to communicate with one another must be present to establish the similarity and/or differences of these aesthetic perceptions. This, in turn, has obvious psychologic implications as it relates to the postoperative emotional adjustments of the patient.

Aesthetic goals are clearly delineated during the initial visit of the patient. At this time, the patient's perception of any aesthetic deficiencies and/or goals is ascertained. It is also at this time that the author chooses to commit himself to an aesthetic treatment plan based solely on the morphologic appearance of the patient and how he can best enhance it. This global treatment plan includes (1) which jaw or jaws are to be repositioned and (2) the vector or direction of skeletal displacements. The decision to perform single- versus double-jaw surgery is almost always determined at this initial visit. The decision to perform adjunctive osseous genioplasty is most frequently determined at this time as well. This basic surgical blueprint of qualitative skeletal displacements is rarely different from the surgical procedure actually performed, which may take place from one to one and one-half years later.

Once the goals of the patient are ascertained and a surgical treatment plan is defined, it becomes clear whether the aesthetic goals of the patient will be achieved with the recommended course of action, that is, whether the aesthetic perceptions of patient and surgeon coincide.

Contrasting Aesthetic Perceptions

The most common qualitative perceptual aesthetic difference is observed in the patient referred for surgical correction of a Class III skeletal deformity. It is not uncommon for the patient's perception of the problem to be a disproportionately large mandible. Yet the surgical recommendation may be to advance the maxilla. Imagine the added consternation of the patient when a double-jaw procedure is recommended to advance the maxilla and set back the mandible or, even worse, to advance both the maxilla and mandible. In these instances, the surgeon must be able to convey to the patient the aesthetic advantages of the recommended method of establishing the appropriate degree of facial convexity, adequate facial projection, and skeletal support of the facial soft tissues.

Another encountered qualitative perceptual difference is in the female, Class II, short-face patient. Although the patient can readily appreciate the aesthetic need for mandibular advancement, she is occasionally surprised by the recommendation to vertically elongate the maxilla "simply" to expose the incisor teeth to a greater degree. Her reluctance to accept this may be due to the perceived added complexity of a double-jaw procedure as compared to a mandibular advancement procedure. Regardless of the reason, the aesthetic implications of inferior maxillary displacement must be put in its proper perspective: it not only exposes the maxillary teeth but it vertically expands the entire lower facial skeleton. It must be explained to the patient that this, in turn, will have an enormous impact on the support of the perioral soft tissues and help rejuvenate the entire lower face.

Failure of the surgeon to effectively explain the aesthetic desirability of the recommended surgical plan and to resolve these qualitative perceptual differences represents a significant obstacle in achieving the mutual goals of an optimal aesthetic result and a satisfied patient. It is under this circumstance that the author may consider to decline to render surgical care to the patient.

Quantitative perceptual differences usually involve the magnitude of projection of the mid- and/or lower face. This frequently becomes an issue in the Class II, long-face patient. These patients should almost always be treated with double-jaw surgery (see Chapter 9) so that occlusal constraints are eliminated in determining the extent to which the maxilla and mandible are to be advanced. The degree of advancement then becomes a pure aesthetic judgment. It once again behooves the surgeon to adequately explain the aesthetic rationale of adequate projection of the face as it relates to overall facial convexity and soft tissue support. Quantitative perceptual differences, unlike qualitative ones, rarely cannot be resolved.

Assessment of Risk/Benefit Ratio

If one is to be able to justify basing treatment planning decisions on aesthetic goals, the associated risks must be proportionately minimal. The presentation of any surgical treatment plan must be accompanied by a forthright discussion of the associated potential complications. It is only in this way that a reasonable assessment of the risk/benefit ratio can be made. The patient can then decide whether the aesthetic benefits are worth taking any inherent surgical risks. This discussion also takes place at the initial visit.

Patients are informed that the most commonly encountered complication in orthognathic surgery is injury to the inferior alveolar nerve during the sagittal split ramus osteotomy procedure. They are told that approximately 20% of patients have a permanent sensory deficit of the lower lip. Additionally, they are informed that this also affects the sensibility of the chin. This potential risk is then placed in its proper perspective as related by previous patients who have sustained this complication; that is, this is considered a relatively minor problem that, in retrospect, with rare exception, would not have deterred individuals from having the procedure performed over again. In other words, the benefits they derived from the surgery are far in excess of this sensory deficit. Nevertheless, this potential risk is repeatedly mentioned to patients because of its high incidence. It is this singular risk that they must be willing to accept if they are to undergo mandibular surgery.

Although there are certainly other potential complications of maxillary and/or mandibular surgery that are reviewed with patients, they are all infrequently encountered and, accordingly, are not emphasized. Maxillary surgery is associated with such minimal risk that it is enthusiastically recommended to patients to achieve what many would consider to be rather modest skeletal displacements. The most common complication this author has encountered performing maxillary surgery is an anterior tooth that requires endodontic treatment because of the proximity of a fixation screw. This has occurred in approximately 1% to 2% of patients.

Communication Aids

Several modalities are available to assist the surgeon in conveying aesthetic objectives to the potential patient. Because dramatic changes in the physical appearance of the patient may occur, it is important that the patient comprehend the magnitude of such change.

Prediction Tracings

Prediction tracings have maintained a time-honored role in communicating aesthetic goals to patients. They have the advantage of being quickly and easily performed with relatively little expense, and they can, with a reasonable degree of accuracy, predict the qualitative result of the proposed treatment plan. It has been my experience that patients find this modality to be productive and even reassuring in their perception of treatment planning goals. It also allows the patient to visualize predicted results of different surgical approaches. In my practice, this pragmatically relates to the option of performing simultaneous adjunctive genioplasty in addition to maxillary and/or mandibular surgery. When reviewing prediction tracings with patients prior to the presurgical phase of orthodontics, they must be advised that the predicted results can be substantially different from those that may be actually achieved. This obviously depends on the extent to which orthodontic decompensation of the anterior dentition occurs.

Prediction tracings suffer from the obvious deficiency of being a two-dimensional drawing that can only demonstrate intended profiles in the mid-sagittal plane. It gives the patient no information about the parasagittal profile or the frontal result. In addition, it is essentially devoid of any display of the soft tissue skeletal support with the exception of the lips and the submental morphology. Nevertheless, this simple tool has proven to be an effective communication aid narrowing the gap between surgical aesthetic goals and patient perceptions and expectations.

Patient Photographs

The review of before and after photographs of patients who have undergone previous orthognathic surgery has been an invaluable aid to demonstrate to patients the enormous potential for significant alteration in one's appearance. Though this is not done for all patients, I have insisted in reviewing such photographs with those patients in whom dramatic changes in facial morphology are expected to occur. This modality has proven to be most useful in patients who will undergo large increases in skeletal volume of the lower face. This includes the severe Class II, long-face patient and the Class II, short-face patient. It is only with review of such photographs that these patients can be expected to comprehend the magnitude and totality of such change and, therefore, be emotionally prepared. Additionally, it offers the patients the opportunity to visualize such changes in not only the profile but also the frontal view. Lastly, it is perhaps the best technology that is available for demonstrating accurate changes in soft tissue characteristics secondary to skeletal support enhancement. Review of such photographs is also done at the initial consultation. This timing is preferred so that the patient has a clear understanding of the aesthetic objectives from the inception of the patient–surgeon relationship.

Criticism of this approach has included creation of unreasonable patient expectations that cannot possibly be achieved and, therefore, risk patient disappointment. This argument is countered with the realization that facial skeletal surgery, particularly when it increases skeletal volume, is quite predictable and reproducible as it relates to the aesthetic outcome. Herein lies the legitimacy of this communication modality with skeletal surgery as compared to its apparent risks when reviewing results that have been achieved with soft tissue aesthetic surgery. Its critics must bear this distinction in mind.

Computer Imaging

Computer imaging is, without question, a high-tech patient communication aid with which many patients identify and which they, in fact, request. Although this may represent a potentially powerful marketing tool, I have yet to be convinced that it, in and of itself, significantly adds to one's surgical volume. It also suffers from many of the previously mentioned disadvantages of other modalities. Like prediction tracings, it plays its greatest role in demonstrating a profile result. Its clinical applicability in the frontal view remains severely limited. The accuracy of its prediction is only as good as its software program and, therefore, is limited by the vagaries of the soft tissue response to certain skeletal displacements. Despite signed disclaimers to the contrary, the before and after computer image of the patient implies a specific result in the mind of the patient.

For all these reasons, I have yet to incorporate this technology in the communication process with the patient and continue to rely on verbal images, prediction tracings, and a review of patient photographic material in an effort to generate reasonable and accurate patient aesthetic perceptions.

Patient Response to Surgery

The emotional response of the vast majority of patients undergoing orthognathic surgery is positive. Given the potential magnitude of facial morphologic change, I remain pleasantly surprised at the very low incidence of adverse psychologic adjustments among patients. One can possibly explain this on the basis of the overwhelmingly positive or favorable perception of physical change among patients and the adequate preoperative preparation of the patient as it relates to both the quality and magnitude of such change. An additional surprising observation has been the lack of correlation between patient age and emotional response. One would expect older individuals to have greater difficulties in adjusting to structural changes in their facial appearance. This has not proven to be the case among my patients.

I have, however, observed some weak correlation (and this is purely anecdotal) between the type of skeletal correction and adverse emotional adjustments following surgery. Without question, short-face patients undergoing large increases in the vertical dimension of the lower face have most frequently had difficulty in adjusting to their new appearance, regardless of how positive they perceive the change. This is perhaps due to the associated change in the emotional countenance of the patient that is perceived by others. The short-face appearance is frequently thought to be associated with sadness and/or anger. Following surgical correction, this emotional expression is changed to one of greater contentment. It is not difficult to understand how the different reaction of others to the patients' appearance can potentially cause difficulties in their adjustment to their new appearance. Because of these observed adjustment problems, accurate patient understanding of the expected aesthetic change is repeatedly emphasized in the preoperative period in this patient population. Fortunately, we should not feel obligated to attempt to prepare the patient for the reaction of others.

Occasionally, there have been difficulties in the postoperative period generated by the reaction of family members. These adverse reactions have primarily been observed in male spouses of patients and in siblings of the same sex and of similar age to the patient. An enormous enhancement of one's appearance has the obvious potential of changing the emotional interactions between spouses and siblings alike. Once again, it is beyond the responsibility of the surgeon to alter his preoperative communications with the patient or the proposed surgical plan based on the possibility of such adverse family reactions to the altered appearance and/or personality of the patient.

Suggested Reading

1. Ackerman, J.L. and Proffit, W.R.: Communication in orthodontic treatment planning: Bioethical and informed consent issues. *The Angle Orthodontist* 65:253, 1995.
2. Kijak, H.A. and Bell, R.: Psychosocial complications in surgery and orthodontics. In W.R. Proffit and R.P. White (Eds.): *Surgical Orthodontic Treatment.* St. Louis: Mosby, 1991, pp. 71–92.
3. Kijak, H.A., Hobal, T., and West, R.A.: Psychologic changes in orthognathic surgery patients: A 24-month follow-up. *J Oral Surg* 42:506, 1984.
4. Lanigan, D.T., Hey, J.H., and West, R.A.: Major vascular complications following orthognthic surgery: Hemorrhage associated with Le Fort I osteotomies. *J Oral Maxillofac Surg* 48:561, 1990.
5. Lanigan, D.T., Hey, J.H., and West, R.A.: Aseptic necrosis following maxillary osteotomies: Report of 36 cases. *J Oral Maxillofac Surg* 48:142, 1990.
6. Peterson, L.J. and Topazian, R.: Psychological considerations in corrective maxillary and mandibular surgery. *J Oral Maxillofac Surg* 34:157, 1976.
7. Phillips, C., Hill, B., and Connor, C.: The influence of video imaging on patient's perceptions and expectations. *The Angle Orthodontist* 65:263, 1995.
8. Sarver, D.M.: Videoimaging: The pros and cons. *The Angle Orthodontist* 63:167, 1993.
9. Turvey, T.A.: Intraoperative complications of sagittal osteotomy of the mandibular ramus: Incidence and management. *J Oral Maxillofac Surg* 43:504, 1985.

Aesthetic Orthodontic Considerations in the Surgical Patient

CHAPTER 5

The surgeon performing jaw surgery must have some fundamental understanding of the aesthetic results that can be obtained with orthodontic camouflage of underlying skeletal deformities. It is only in this way that he or she can adequately counsel prospective patients as to the aesthetic advantages and disadvantages of a surgical correction over an orthodontic correction. Secondly, if a surgical approach is chosen, it behooves the surgeon to understand the aesthetic implications of the orthodontic preparation of the patient. Accordingly, the surgeon and orthodontist can more fully cooperate in achieving the best possible aesthetic correction.

Orthodontic Camouflage of Skeletally Based Malocclusions

Certain malocclusions based on skeletal disproportion can be adequately treated by the orthodontist without significantly compromising the aesthetic outcome. Although a surgical correction may provide the best possible aesthetic result, the orthodontic result can also be acceptable in these situations. These patients may also benefit from other modest surgical procedures (such as genioplasty or rhinoplasty) to further enhance their orthodontic results.

In attempting to determine which skeletally based malocclusions can be adequately corrected orthodontically from an aesthetic perspective, the following questions must be addressed:

1. Are the morphologic facial characteristics associated with the given malocclusion aesthetically acceptable?
2. If they are acceptable, will orthodontic tooth movement influence them adversely?

3. If they are unacceptable, either before or after orthodontic correction, can they be enhanced with relatively minor surgical procedures such as genioplasty, rhinoplasty, and/or midface onlay augmentation?

Let us examine the classical morphologic characteristics associated with Class II and Class III malocclusions as well as the long face to address these questions.

Class II Malocclusion

The common morphologic denominator associated with most Class II malocclusions is posterior facial divergence and excessive facial convexity. In those Class II patients with significant skeletal discrepancies, there also may be abnormal sagittal relationships between the upper and lower lip with a rather procumbent lower lip. Lastly, Class II malocclusions can be associated with abnormalities of lower face height that can be either excessive or deficient.

In those Class II patients with modest sagittal discrepancies between the maxilla and the mandible, with reasonably normal lip relations and with no abnormalities of lower face height, orthodontic correction is frequently a viable option (Fig. 5-1). With modest deformities, any residual weakness of the chin can be corrected at a later date with advancement genioplasty

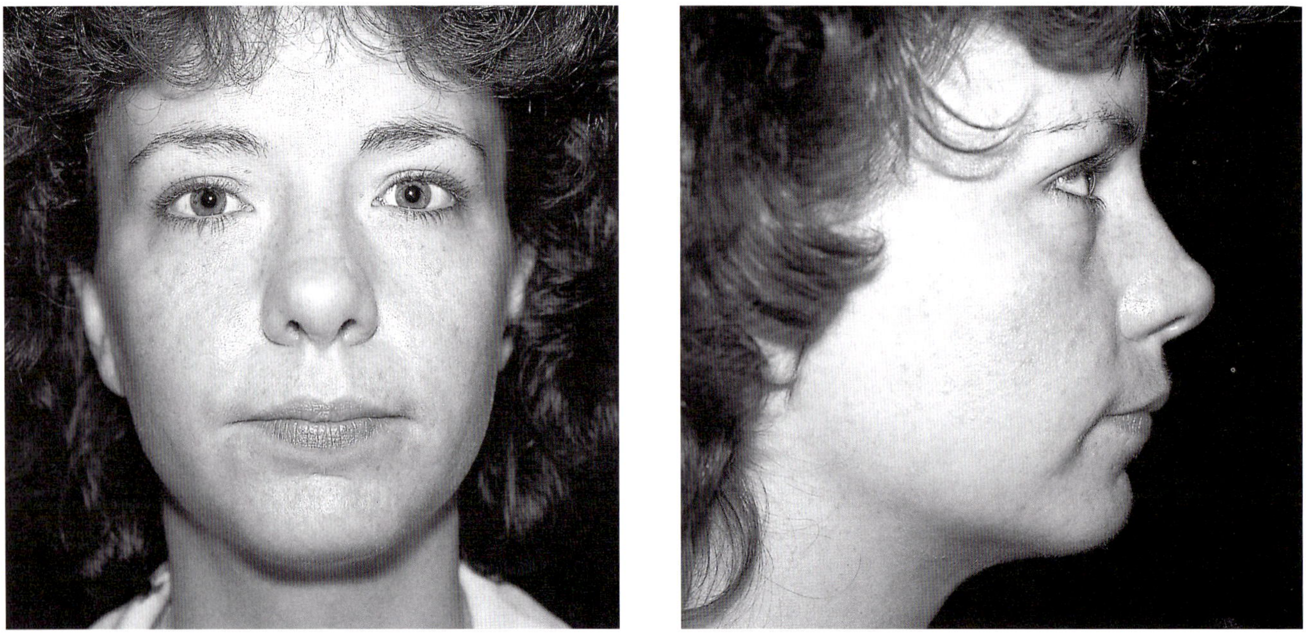

A B

FIG. 5-1 A Class II patient who is a reasonable candidate for orthodontic correction as far as her facial aesthetics are concerned.

provided that the depth of the labiomental fold is not excessive (Fig. 5-2). As the severity of the Class II discrepancy increases, orthodontic correction risks excessive opening of the nasolabial angle and loss of upper lip support as the maxillary incisors are retracted. In addition, the excessive facial convexity and posterior divergence remain (Fig. 5-3). In such patients, advancement genioplasty will provide only minimal improvement.

In those Class II patents associated with abnormalities of lower face height, orthodontic correction becomes more tenuous because there is little control over height of the lower face. With increased lower face height primarily due to vertical maxillary excess, there are many morphologic factors that become indications for jaw surgery. These include excessive tooth and gingival show and circumoral muscular strain in an effort to achieve lip seal. None of these factors can be corrected with orthodontic tooth movement. With lesser degrees of tooth and gingival exposure, an osseous genioplasty may provide for an acceptable aesthetic result. In this situation, the chin must be moved upward as well as forward to reduce overall height of the face (Fig. 5-4). However, if the depth of the labiomental fold is excessive, chin surgery should not be considered an option (see Chapter 11). An additional aesthetic orthodontic consideration in the Class II long-face patient is that any orthodontic retraction of the maxillary teeth may increase the extent of their exposure beneath the relaxed upper lip.

The Class II patient with deficient height of the lower face represents an entirely different set of problems in terms of facial aesthetics. Because skeletal volume is deficient in two dimensions, these faces are always characterized by an inadequate degree of skeletal support to the overlying facial mask. This is clinically manifested in lower lip eversion, deepened labiomental folds, jowls, and submental fullness. Overall, these patients' faces are short and square, a morphologic pattern that is characteristic of older individuals. Certainly, to achieve the best possible aesthetic result in these patients, jaw surgery is an ideal treatment plan. However, for those patients who decline major surgery, orthodontic correction combined with osseous genioplasty can provide for moderate improvement. In these situations, chin surgery must not only advance the chin (to a more modest extent than with the Class II long-face patient) but must also vertically elongate the chin to effect an increase in overall height of the lower face. In so doing, the labiomental fold becomes less deep, and there is a modest improvement in the soft tissue support of the lower face and neck secondary to bidimensional expansion of the lower face (Fig. 5-5). It should be apparent that as the extent of the skeletal discrepancy in any of the aforementioned Class II patients becomes greater, the chances for an acceptable orthodontic correction without major jaw surgery diminishes.

Class III Malocclusion

Unlike the Class II patient, the chances for a successful orthodontic camouflage in the Class III patient, with or without minor surgery, are much

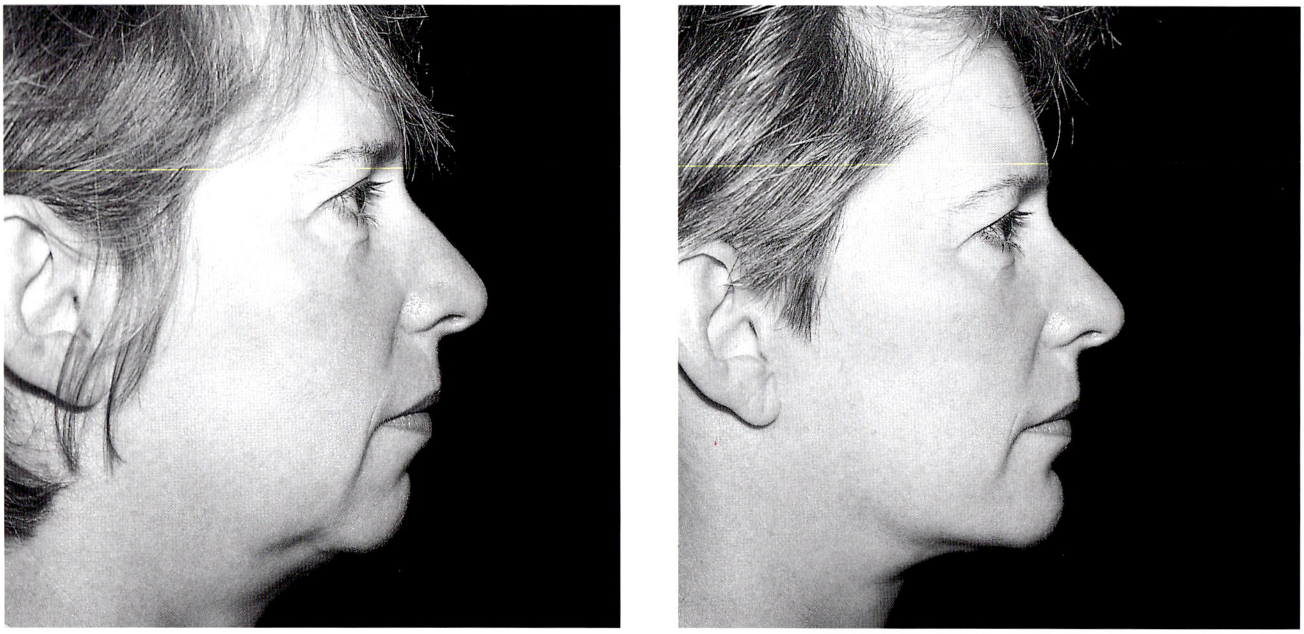

A B

FIG. 5-2 A: A Class II patient treated with orthodontic correction and B: following advancement genioplasty to improve the projection of the chin. Note that the labiomental fold deepens as the chin is advanced. A simultaneous face-lift was performed.

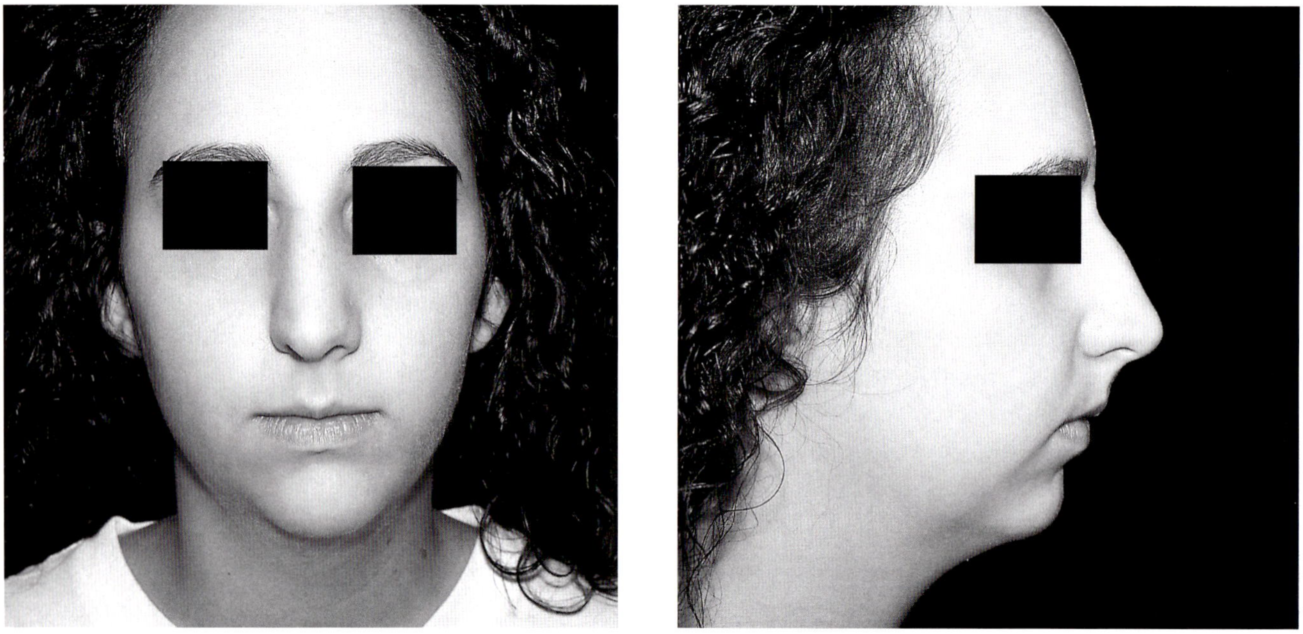

A B

FIG. 5-3 A Class II patient following orthodontic therapy. She refused a surgical correction. There is excessive opening of the nasolabial angle, loss of upper lip support, and unacceptable degrees of facial convexity and posterior divergence.

PREOPERATIVE POSTOPERATIVE

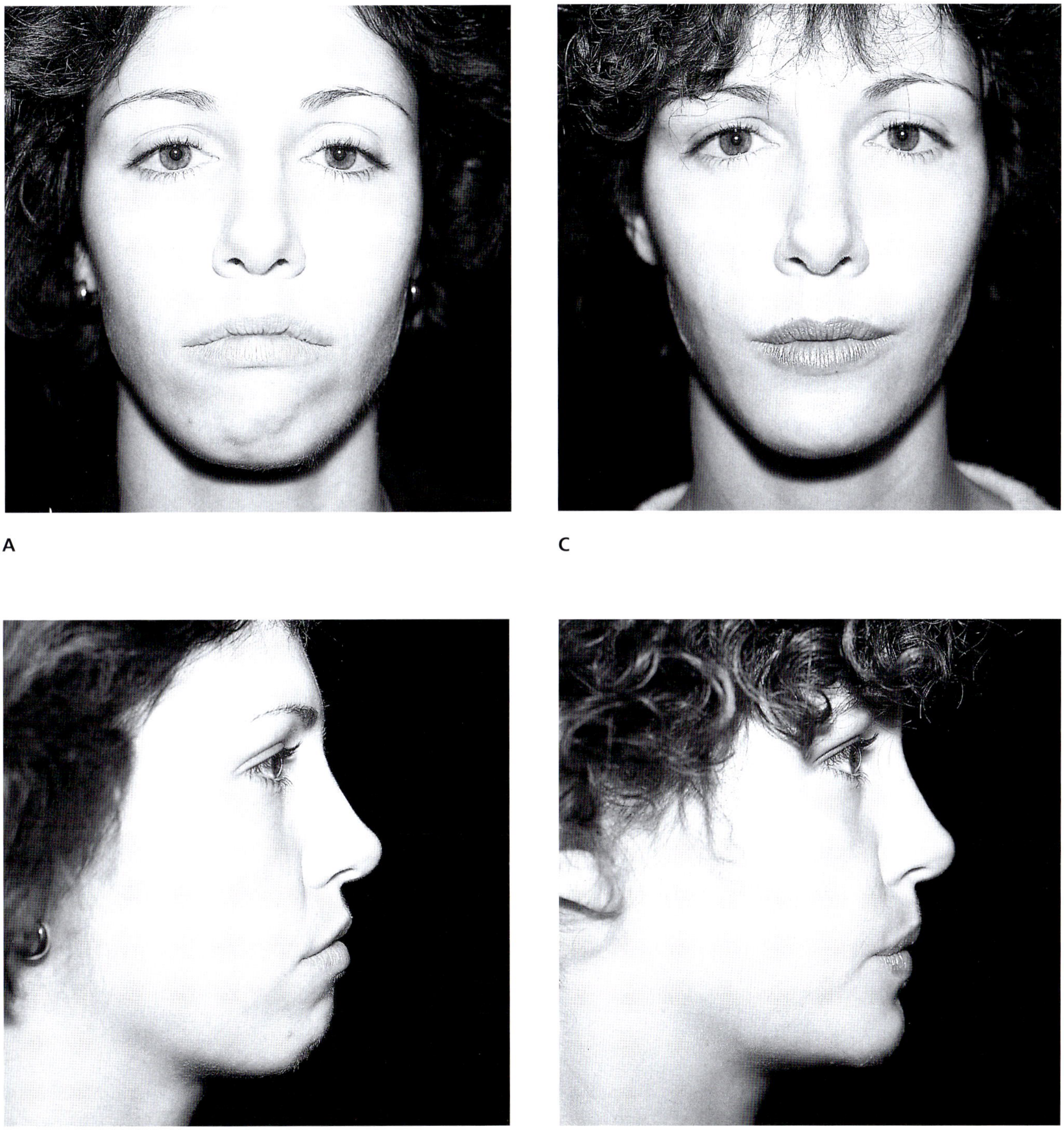

A C

B D

FIG. 5-4 A & B: A Class II, long-face patient following orthodontic correction. Note lip strain, the superiorly displaced soft tissue chin pad, and sagittal weakness of the lower face. C & D: The same patient following osseous genioplasty to vertically reduce and advance the chin. Although there has been a reduction in the vertical dimension of the lower face, the lip strain has not been entirely eliminated.

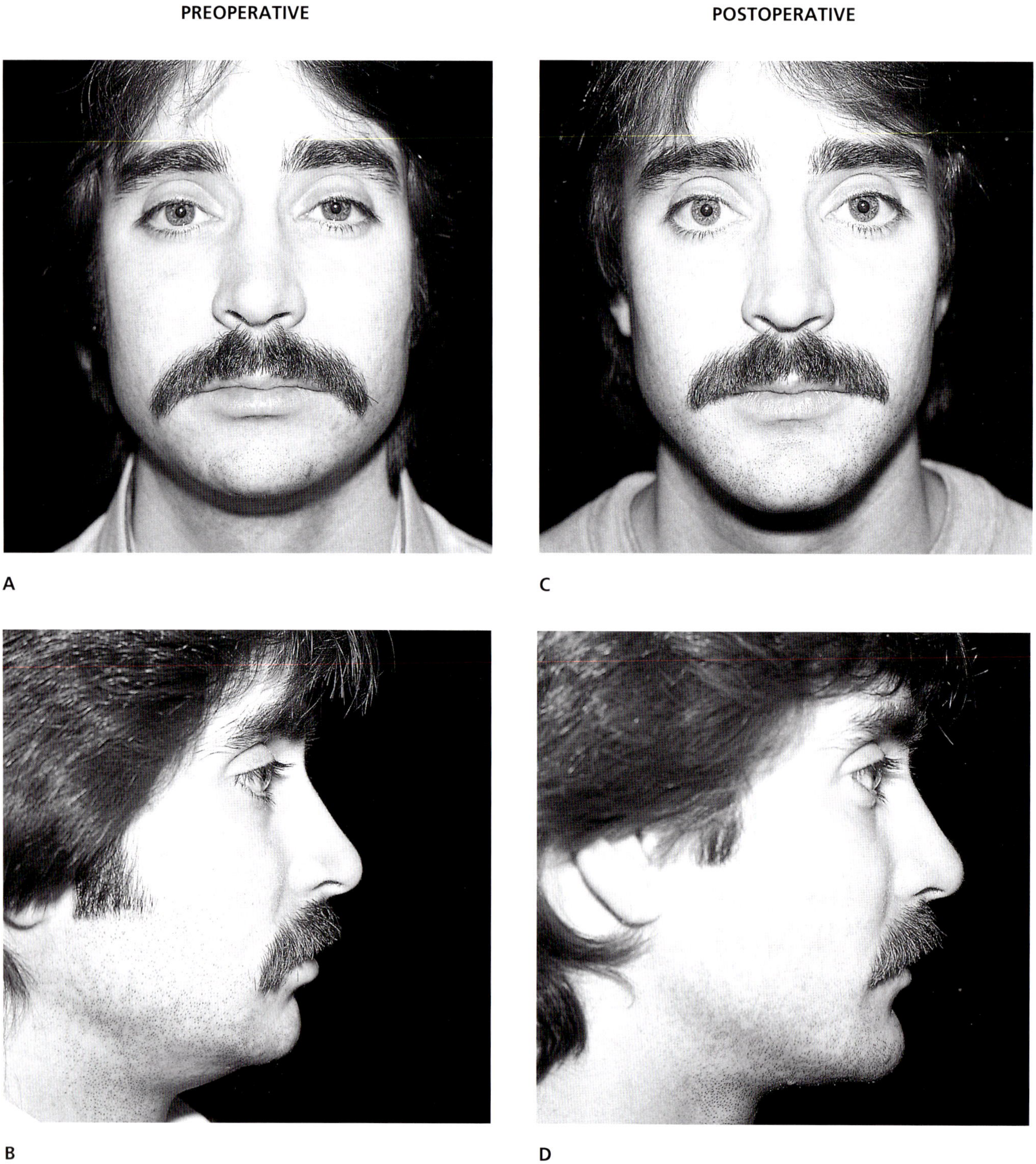

FIG. 5-5 A & B: A Class II, short-face patient following orthodontic correction. Note square, short face and deep labiomental fold. C & D: The same patient following osseous genioplasty to vertically elongate and advance the chin. Note the longer, more oval appearance and a shallower labiomental fold.

less common. This is primarily due to three factors. The first concerns the classical morphologic characteristics seen in most Class III individuals. These include a relatively flat or even concave profile, which may be in association with an anteriorly divergent face. Among North American Caucasians, these facial characteristics are seen much less frequently than overall facial convexity with posterior divergence and, therefore, are much less aesthetically acceptable. Contrast this to many Asian cultures in which these Class III facial characteristics (concavity with anterior divergence) are much more common and more readily accepted. The second reason is that orthodontic attempts to treat the Class III malocclusion more frequently affect facial appearance adversely. For instance, orthodontic retraction of lower incisor teeth appear to make the chin more prominent and create a greater degree of facial flattening in the perioral region (Fig. 5-6). It only accentuates what is already aesthetically unacceptable. This also decreases the depth of the labiomental fold, which visibly exaggerates any preexisting increase in lower face height that may be present. The last reason is that minor surgical procedures such as genioplasty are not very effective in camouflaging this skeletal deformity. Attempts to reduce the sagittal projection of the chin with osseous genioplasty result in poor aesthetic outcomes because of the creation of relative soft tissue excess, the poor correlation of movement of the soft tissue pogonion relative to the hard tissue pogonion, and the problem of palpable, and frequently visible, osseous overlap at the most proximal extent of the osteotomy cut.

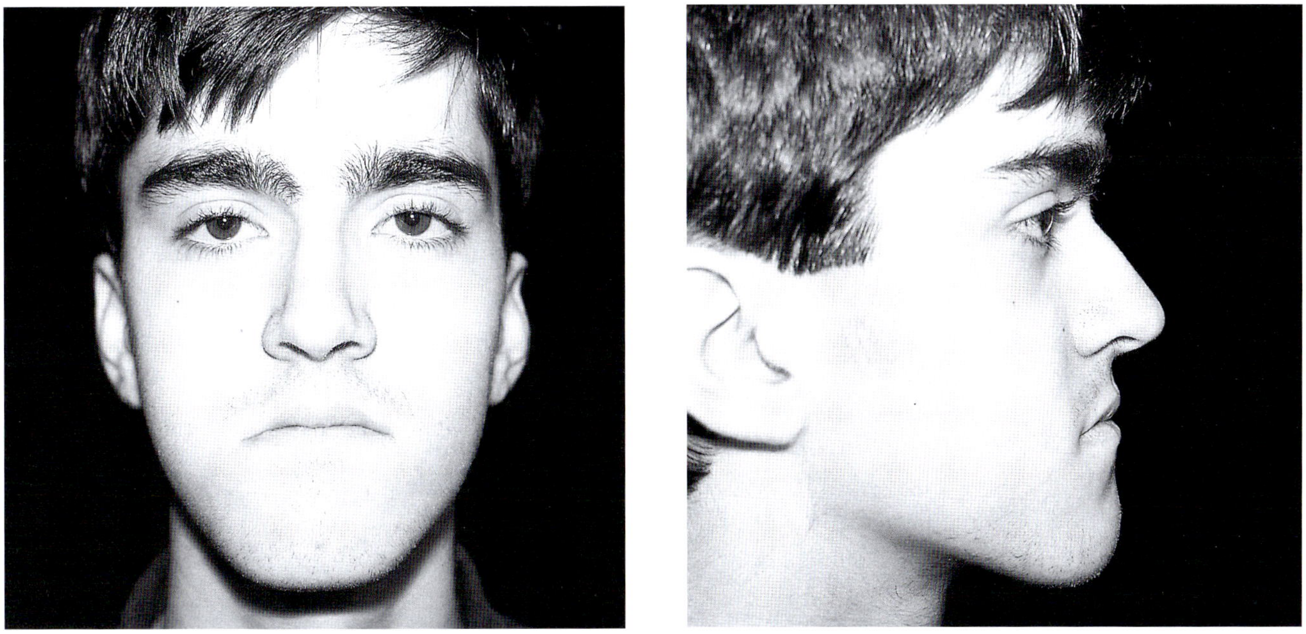

A B

FIG. 5-6 A Class III patient following orthodontic attempts to correct his malocclusion. Secondary to mandibular incisor retraction, there is loss of lower lip support making the chin appear even more prominent.

For all these aesthetic reasons, orthodontic attempts to camouflage the Class III malocclusion should be reserved for only modestly affected individuals. Those that have the greatest possible chance of achieving an adequate aesthetic result should have the following facial characteristics prior to commencing orthodontic correction: a relatively prominent nose, a lower lip that is sagittally prominent relative to the chin, and a deep labiomental fold. These facial characteristics will allow for some orthodontic retraction of the lower incisor teeth as well as maxillary incisal flaring without further sacrificing facial aesthetics. For instance, if the lower lip is prominent relative to the chin and there is a deep labiomental fold, the retraction of the incisor teeth will not necessarily add to the prominence of the chin or make the labiomental fold too shallow, a characteristic that would only add to the overall prominence of the chin. In addition, a rather prominent nose allows for significant surgical augmentation of the midface as well as the nasolabial angle area. This would provide for improved facial balance and possible correction of the overall facial concavity. At the same time the relatively prominent nose would not appear too small.

Aesthetic Considerations Involving Presurgical Orthodontics

Once a decision has been made to perform jaw surgery, a presurgical orthodontic plan must be formulated to prepare the patient for surgery. This is undertaken to eliminate most, if not all, dental compensations that have been built into the occlusion to achieve the desired degree of intra-arch leveling, to eliminate any dental interferences that would prevent ideal repositioning of the jaws, to allow for maximum interdigitation of the maxillary and mandibular teeth at the time of surgery, and to provide the surgeon with orthodontic appliances that can be used to secure intermaxillary fixation at the time of surgery.

The most basic and fundamental component of presurgical orthodontic care involves the elimination of dental compensations (Fig. 5-7). These may have occurred spontaneously and/or secondary to prior orthodontic attempts to correct a skeletally based malocclusion. Classic dental compensations involve excessive uprighting of maxillary incisors in the Class II patient and excessive retroinclination of mandibular incisors in the Class III patient. Presurgical orthodontic reversal of these compensations have profound aesthetic consequences that can be both favorable and detrimental to the appearance of the patient in the presurgical period. As the uprighted maxillary incisors are flared in the Class II patient, the angular relationship between the nose and the upper lip improves. However, the apparent sagittal deficiency of the mandible appears to be exaggerated because the overjet may be increased. As the retroinclined mandibular incisors are flared in the Class III patient, labiomental relations may improve, but the apparent

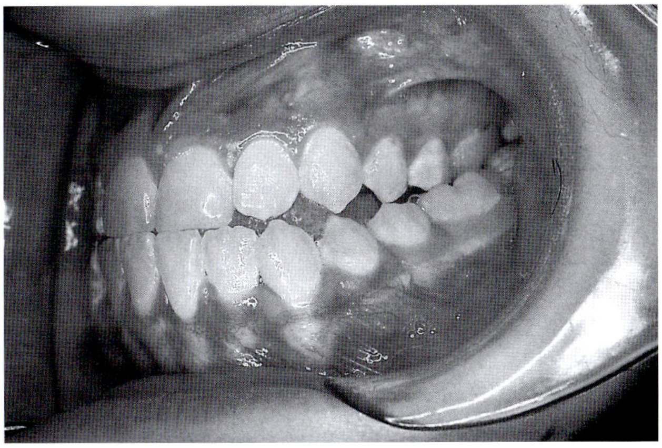

A

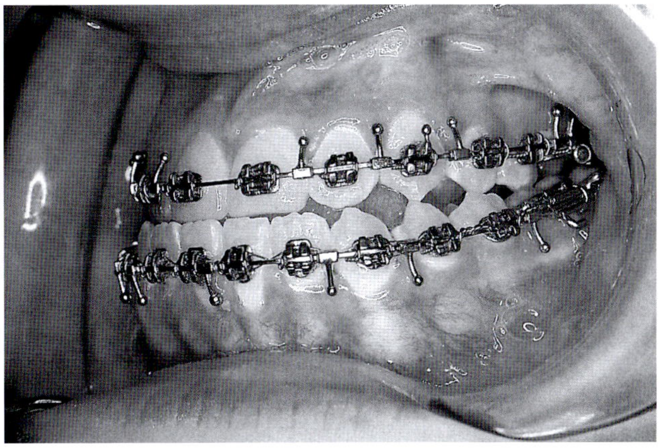

B

FIG. 5-7 Classic orthodontic presurgical decompensation of a Class III patient. A: An edge to edge anterior bite due to retroinclined mandibular incisors. B: As the mandibular incisors are proclined, the Class III discrepancy increases.

sagittal discrepancy between the maxilla and mandible becomes worse. As dental compensations are corrected and the occlusal discrepancy becomes greater, the extent to which the jaw or jaws are repositioned also becomes greater. These presurgical orthodontic manipulations have obvious treatment planning implications.

Dental compensations may have also taken place in the vertical dimension. This involves the prior extrusion of maxillary incisors in an attempt to close skeletal-based anterior open-bite malocclusions. It is incumbent for the surgeon to understand that such tooth movement may be unstable and that it is likely to relapse following any surgical correction. Accordingly, it is usually advisable for the orthodontist to presurgically reverse such compensations. As a result, the anterior open bite is made worse (Fig. 5-8). This again has important treatment planning consequences. Aesthetically, it can change the extent to which the vertical position of the anterior maxilla is surgically altered. In addition, it may indicate the need for segmental orthodontic preparation and segmental maxillary surgery.

There are some treatment planning decisions that must be made prior to commencing presurgical orthodontics that will have profound effects on the final aesthetic outcome. Perhaps the most important of these involves the extent to which presurgical leveling of the mandibular dentition with an exaggerated curve of Spee should take place (Fig. 5-9). This is typically seen in the Class II, deep anterior bite patient because of supraeruption of the mandibular incisor teeth. Presurgical leveling by mandibular incisor intrusion will create a very different aesthetic result following surgery as compared to the result in which postsurgical leveling of the mandibular dentition is accomplished by premolar extrusion. Presurgical leveling minimizes the extent of overbite and therefore minimizes the amount of clockwise ro-

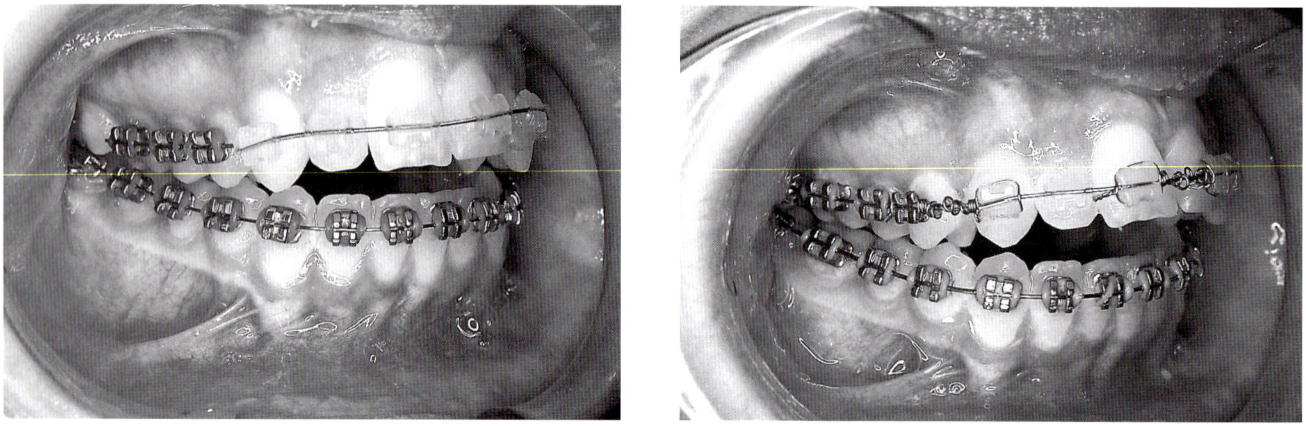

A B

FIG. 5-8 Classic segmental orthodontic presurgical decompensation of a Class II, open-bite patient. A: Prior orthodontic attempts to close the bite have involved retraction and extrusion of incisors. B: As the incisor teeth are decompensated, that is, as they are intruded and flared, the open bite is exaggerated. This patient will undergo a four-piece Le Fort I procedure to surgically level the maxilla.

tation of the mandibular body as it is surgically advanced into a Class I relationship. As clockwise rotation is minimized, pogonion will primarily translate anteriorly. On the other hand, postsurgical leveling maximizes the overbite and, therefore, increases the extent to which the mandibular body will rotate clockwise with surgical advancement. As a result, pogonion will translate inferiorly (lower face height increases) at the expense of anterior advancement. Accordingly, in a Class II anterior deep-bite patient who has a relatively prominent chin and short lower face, presurgical orthodontic leveling should be kept to a minimum. The majority of orthodontic level-

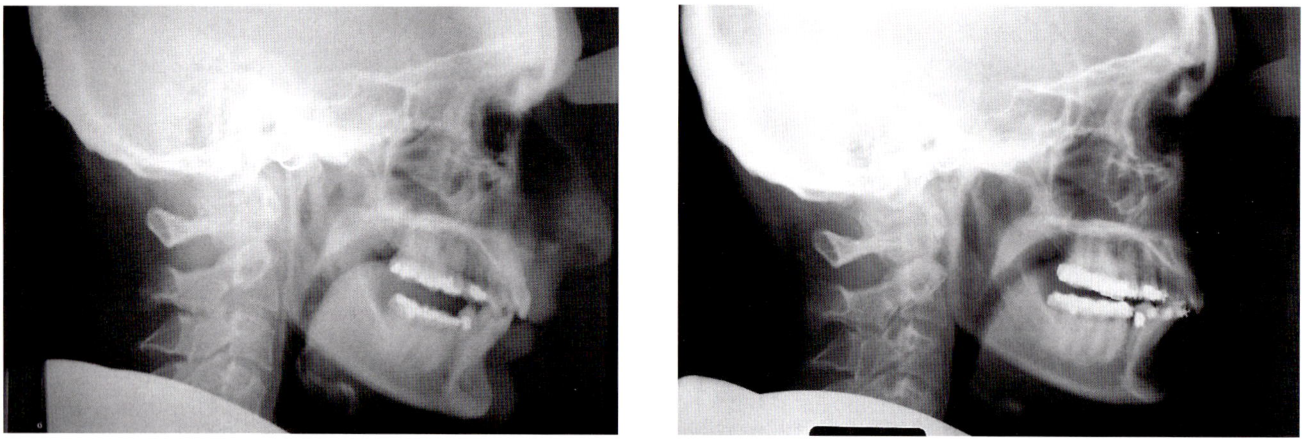

A B

FIG. 5-9 A: Lateral cephalometric X-ray film demonstrating exaggerated mandibular curve of Spee. B: Presurgical leveling of the mandibular dentition secondary to incisor intrusion.

ing should take place in the postopertative period. This consideration becomes even more critical because the results of reduction genioplasty are usually suboptimal. Reduction genioplasty frequently becomes necessary in the Class II, anterior deep-bite patent who undergoes presurgical and not postsurgical orthodontic leveling of the mandibular dentition.

Another aesthetic consideration regarding presurgical orthodontics involves the axial inclination of the maxillary incisor teeth when surgical rotation of the maxilla is planned. This is most frequently a concern in the Class II, long-face patient in which surgical correction involves counterclockwise rotation of both the maxilla and mandible. This type of rotational movement causes flaring of the maxillary anterior dentition, the extent of which depends on the point of rotation in the maxilla (Fig. 5-10). If the maxillary incisor teeth are already flared, uprighting in the presurgical phase of orthodontics may be indicated.

The final aesthetic, presurgical orthodontic decision in which the surgeon should be involved concerns the advisability of extracting teeth. There are two aesthetic reasons for this. The first involves the severe Class II patient in which the extent of overjet is inadequate to allow for significant mandibular advancement. In such instances, mandibular premolar extractions and subsequent orthodontic uprighting of mandibular incisors significantly increase the available overjet. Other surgical options to increase the extent of mandibular advancement would be maxillary advancement and/or maxillomandibular counterclockwise rotation. These latter two options must, of course, be aesthetically appropriate for the individual patient.

The second aesthetic indication for extraction therapy involves the patient with excessive flaring of the maxillary incisor teeth. In such instances, maxillary premolar extractions provide the orthodontist with adequate arch length for orthodontic retraction of the maxillary anterior dentition. It is this author's opinion that the excessive morbidity that is associated with surgical closure of such premolar extraction sites make this a compelling aesthetic orthodontic consideration.

Less common, but of equal aesthetic importance, is the advisability of orthodontic opening of previously closed extraction spaces. This concept has assumed greater clinical appeal with the advent of dental implants, which eliminate the need for fixed crown and bridge prostheses. This approach can become an aesthetic option in the Class II patient who has previously undergone orthodontic correction involving premolar extraction therapy and is now in a Class I dental relationship but unhappy with his or her appearance. Overjet can be created in this fashion to allow for mandibular advancement. Nevertheless, many of these patients require counterclockwise rotation of both the maxilla and mandible to achieve adequate projection of the lower face (Fig. 5-10).

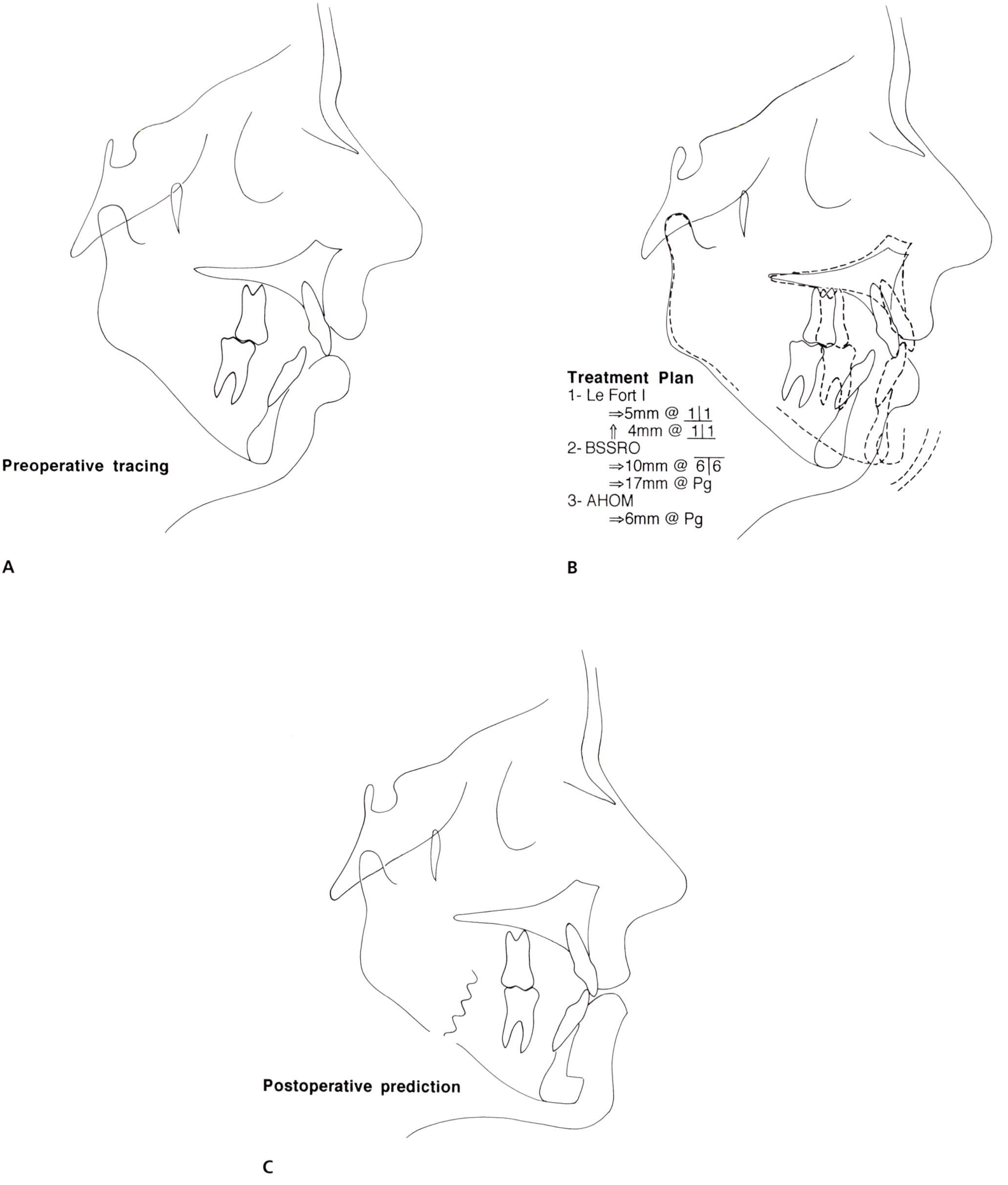

FIG. 5-10 A: Cephalometric tracing of a patient with a Class II malocclusion and a long face. B: Counterclockwise rotation of the maxilla and mandible are necessary to achieve adequate projection of the lower face. As the maxilla is rotated, the incisors are flared. C: Final prediction tracing depicting 23 mm of advancement of soft tissue pogonion.

Suggested Reading

1. Merrifield, L.: Differential Diagnosis. *Seminars in Orthodontics.* 2:241, 1996.
2. Proffit, W.R. and Ackerman, J.L.: Diagnosis and Treatment Planning in Orthodontics. In T.M. Graber and R.L. Vanarsdall (Eds.): *Orthodontics: Principles and Techniques.* St. Louis: Mosby, 1994.
3. Proffit, W.R. and White, R.P. (Eds.): *Surgical Orthodontic Treatment.* St. Louis: Mosby, 1991.
4. Sarver, D.M.: Video cephalometric diagnosis (VCD): A new concept in treatment planning. *Am J Orthod Dentofacial Orthop* 110:128, 1996.

Technical Considerations

CHAPTER 6

Intraoperative technical considerations will be discussed only to the extent that they may make an aesthetic difference. These include factors that have a direct aesthetic impact and those that influence the aesthetic result in a more indirect fashion. For instance, an example of a technical variation that may directly influence the aesthetic result is an alar cinch suture. An example of an indirect factor would be technical variations that affect stability, which may have a more gradual yet longer term effect on the aesthetic outcome.

Maxillary Surgery

Direct Aesthetic Factors

1. Anterior Vertical Maxillary Position

When performing Le Fort I surgery, only one intraoperative measurement need be determined that will affect the aesthetic outcome (this assumes that an intermediate splint is employed if two-jaw surgery is being performed). This determination is the vertical position of the maxillary incisor teeth. Depending on the aesthetic goals of surgery, one may want to let this remain the same, to superiorly reposition the incisor teeth to achieve less incisor exposure, or to inferiorly reposition the incisal edge for greater incisor exposure. This quantitative determination becomes particularly problematic when the maxilla is being moved in the sagittal as well as in the vertical dimension. Measurements of osteotomized segments or osteotomy gaps along the piriform aperture are notoriously inaccurate for many reasons: (1) an error of only 1 mm may represent a disproportionately large percentage of the total movement desired at the maxillary incisor edge, (2) measurements along the piriform aperture correlate more accurately with canine position rather than with incisal position (particularly if there is any planned alteration of the occlusal plane), and (3) the lack of parallelism between the piriform aperture and true vertical causes poor correlation between these two anatomic areas.

A better alternative, although not totally infallible, is the vertical distance from a fixed, extraoral anatomic point to the maxillary arch bar at the level of the central incisors. The incisal edge itself is not chosen as a point of reference because an occlusal splint may be covering the incisal edge. This author has chosen to use the right medial canthus as the extraoral point. It represents a much greater vertical distance than any intraoral measurement could provide, and therefore, the percentage of potential error is less. This greater distance becomes particularly important when vertical maxillary displacement is accompanied by movement in the sagittal dimension. The vertical measurement is initially taken following all soft tissue dissection but prior to any osteotomies. It is then remeasured following maxillary repositioning but prior to any application of fixation devices. In this way, one can easily adjust the vertical position of the anterior maxilla without having to remove and reapply fixation devices (Fig. 6-1).

2. Vertical Level of Maxillary Osteotomy

Much attention has been focused on altering the level of the Le Fort I osteotomy in an effort to achieve better aesthetic results. This primarily re-

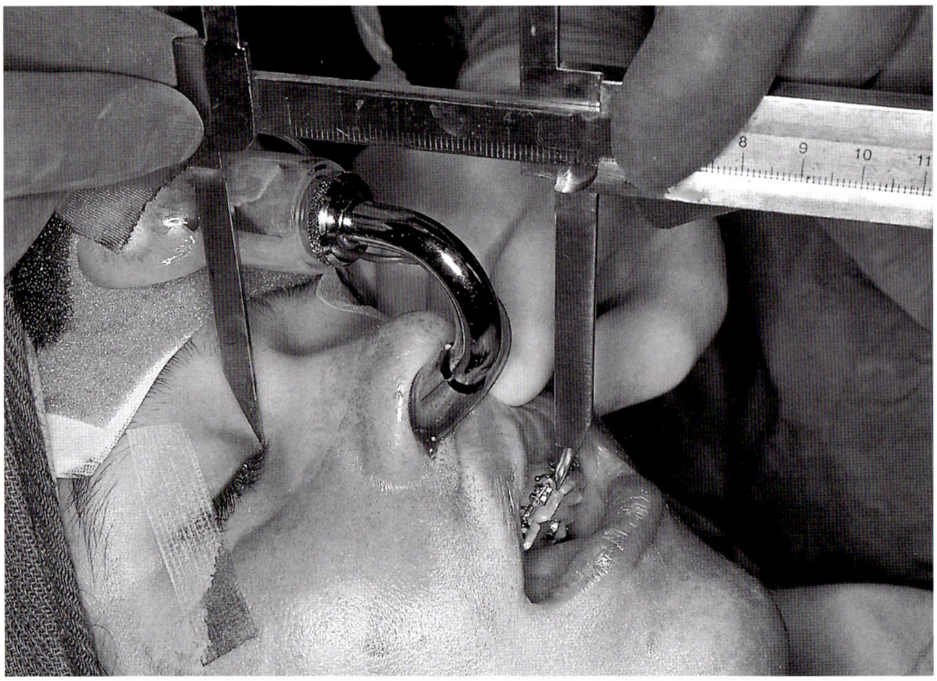

FIG. 6-1 Intraoperative measurement of vertical distance from right medial canthus to maxillary arch bar at central incisors.

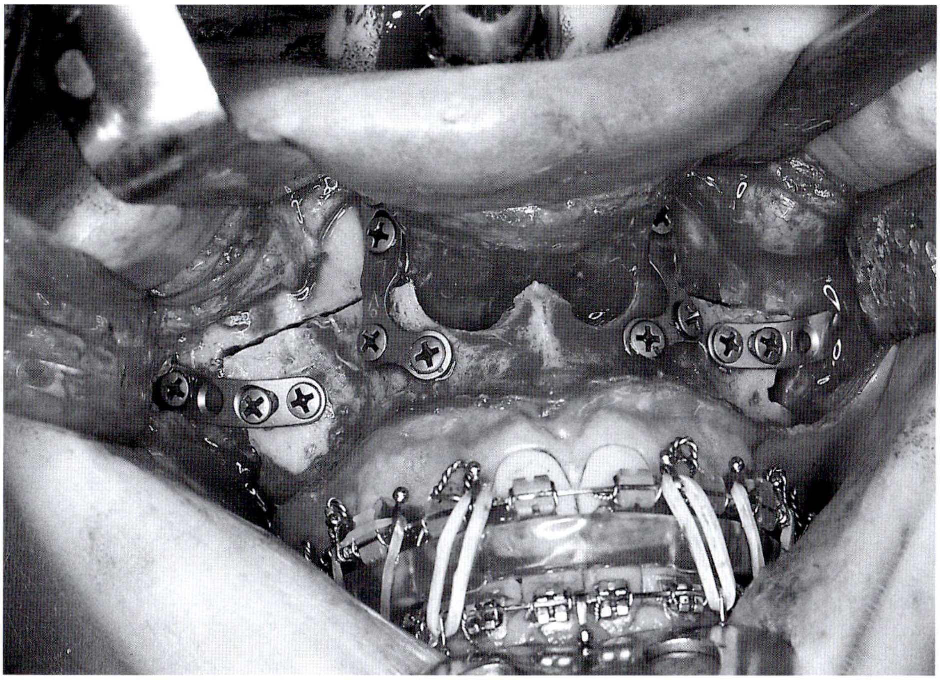

FIG. 6-2 The high Le Fort I osteotomy. Note level of osteotomy is well above the nasal floor and 3 mm below the infraorbital foramen.

lates to the patient with maxillary hypoplasia who has sagittal deficiency extending into the infraorbital region. By placing the osteotomy at a higher level than in the traditional Le Fort I procedure, the infraorbital region would presumably be augmented. Unfortunately, placing this bone cut at a higher level limits the versatility of the Le Fort I procedure. For instance, it minimizes the extent to which one can rotate the maxilla about the vertical axis because of asymmetric augmentation that would occur in the infraorbital regions. It also limits the extent to which one can alter the maxillary occlusal plane because variable degrees of infraorbital augmentation would occur relative to the maxillary dentition. Multiple maxillary segmentation also becomes more problematic from a technical point of view. Therefore, high Le Fort I osteotomy (Fig. 6-2) is performed only with nonsegmental maxillary advancement unaccompanied by rotation about the vertical and/or transverse planes. Alternatively, the osteotomy may be placed well below the infraorbital foramen and the facial skeleton superior to the osteotomy augmented with block hydroxyapatite. Fixation can be achieved either by wedging the block into the osteotomy gap (Fig. 6-3) or by using a single lag screw (Fig. 6-4). This provides adequate augmentation in the infraorbital area without limiting the versatility of the Le Fort I osteotomy (Fig. 6-5). In addition, a higher level of augmentation relative to the infraorbital foramen can be easily achieved.

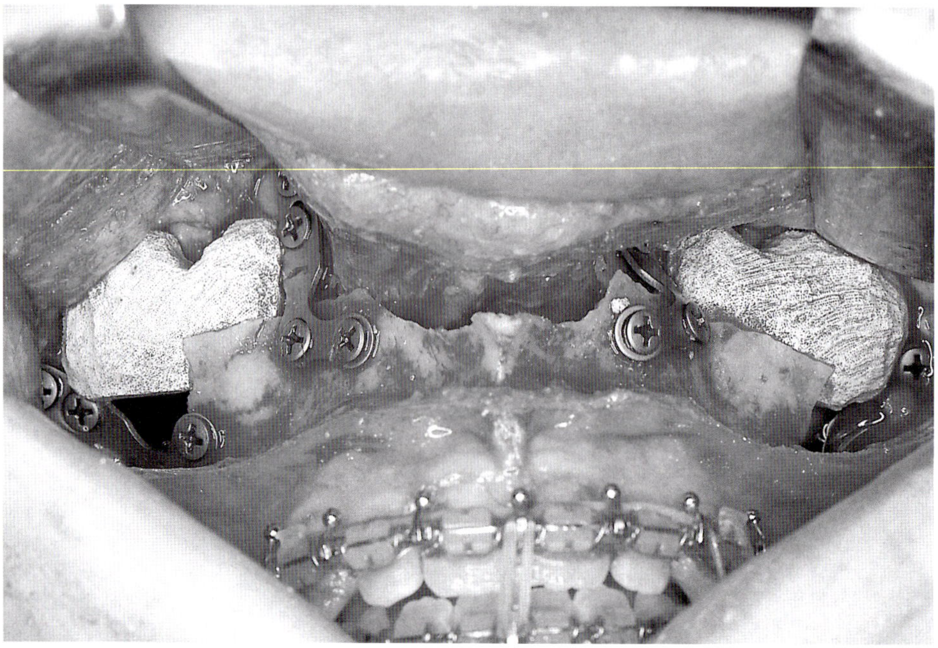

FIG. 6-3 Blocks of porous hydroxyapatite wedged into Le Fort I osteotomy to further augment the midface. Note notching to accommodate the infraorbital nerve. Compared to the high Le Fort I osteotomy, superior levels of augmentation are more easily achieved.

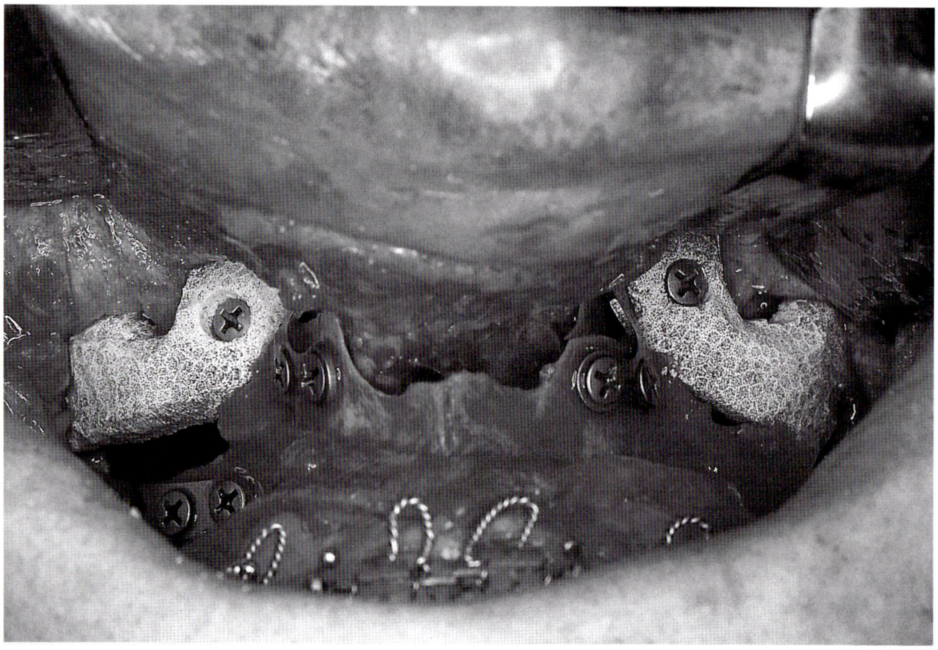

FIG. 6-4 Single lag screw used to secure hydroxyapatite blocks for infraorbital augmentation.

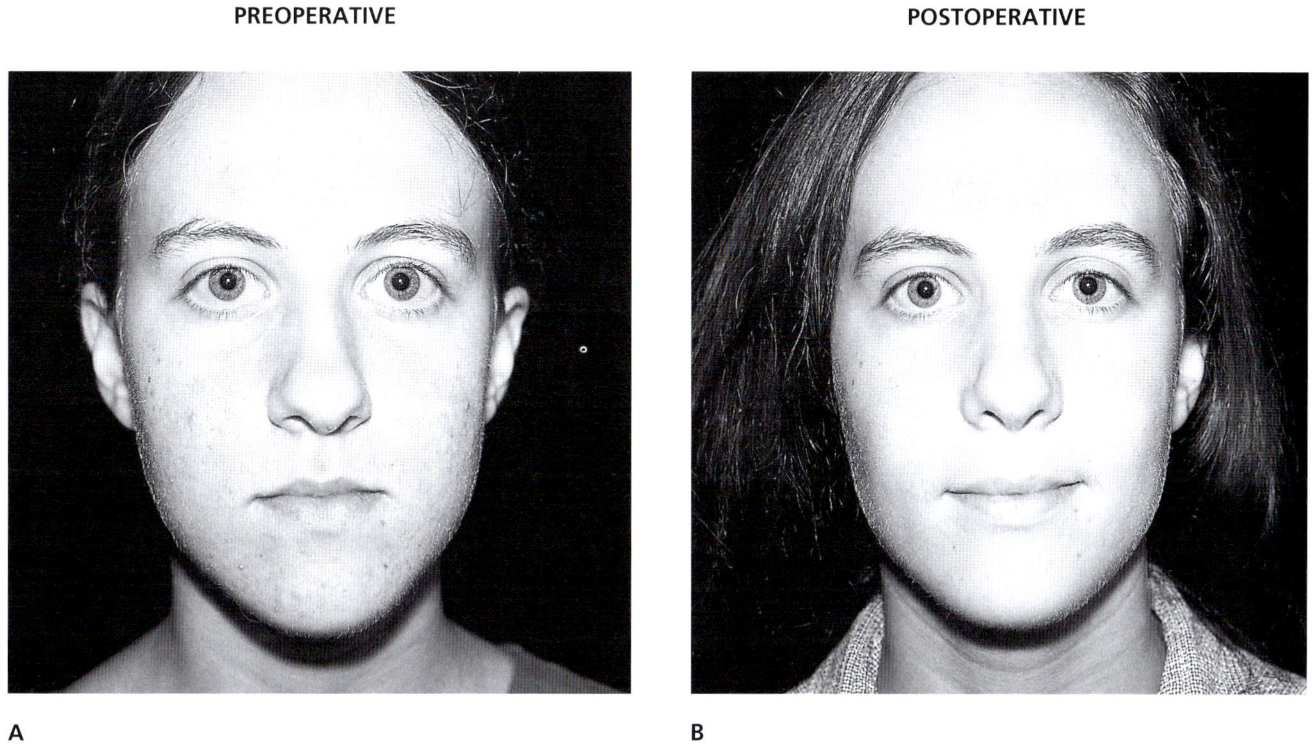

FIG. 6-5 A: Patient with maxillary deficiency extending to the infraorbital region with inferior scleral show. B: Following Le Fort I advancement osteotomy and hydroxyapatite augmentation of infraorbital region, the scleral show has been dramatically reduced.

3. Maxillary Segmentation

The only aesthetic indication for segmentalizing the maxilla is to alter the axial inclination of the maxillary incisor teeth. When performed for this reason, it is clearly preferable to place the interdental osteotomy between the lateral incisor and canine teeth so that sufficient rotation of the anterior segment can be achieved without altering the incisal vertical dimension.

4. Alar Cinch Closure

Without question, unattractive widening of the nasal base can occur following maxillary surgery. This can significantly detract from an otherwise good aesthetic result. Although nasal base widening may be beneficial to patients with narrow nasal bases, nasal base widening in patients with normal to excessive preoperative nasal base width is clearly undesirable. In an effort to avoid this, the alar cinch suture was devised. The physiologic rationale for this was based on the assumption that nasal widening was due to the subperiosteal release of the transverse nasalis muscle, which causes excessive alar flaring once the muscle is no longer inserted into bone. However, the lack of nasal widening observed in patients having undergone maxillary surgery in which the maxilla has been repositioned either posteriorly or inferiorly discounts this reasoning. Alar base widening following maxillary surgery appears to be a mechanical response to enhanced skeletal support of the nasal base; that is, it occurs only with maxillary superior and/or anterior repositioning. Therefore, alar base cinching merely represents a mechanical attempt to narrow the alar base width. This author presently utilizes a 2-0 nonresorbable suture placed in figure-eight fashion through the intraoral incision into the tissues in the vicinity of the alar base insertion. It includes all of the soft tissues just deep to the dermis. Tightening of the suture affects a decrease in the distance between the right and left alar lobules (Fig. 6-6).

One needs to be well aware of several factors regarding this technique. First and foremost, it is unreliable. It works reasonably well in some patients and not at all in others (Fig. 6-7). Given its lack of predictability, one can hardly expect to control the degree to which one affects alar base width, as has been suggested by other surgeons; it is neither qualitatively nor quantitatively predictable! Lastly, if it is effective in narrowing alar base width, one must expect the perinasal and perioral soft tissues to change in some other dimension. Consider this technique to be similar to a Z-plasty procedure. One gains in length in one direction by giving it up in another direction. If alar base width decreases, it must be accompanied by increased nasal tip projection, increased upper lip projection, or increased length of the upper lip. This then becomes a variable to consider as it relates to degree of incisal show following maxillary impaction surgery.

In those patients in whom this technique does not work, postoperative Weir excisions of the nasal lobules is usually effective in achieving nasal base narrowing. This is usually performed no sooner than several months following maxillary surgery (Fig. 6-7).

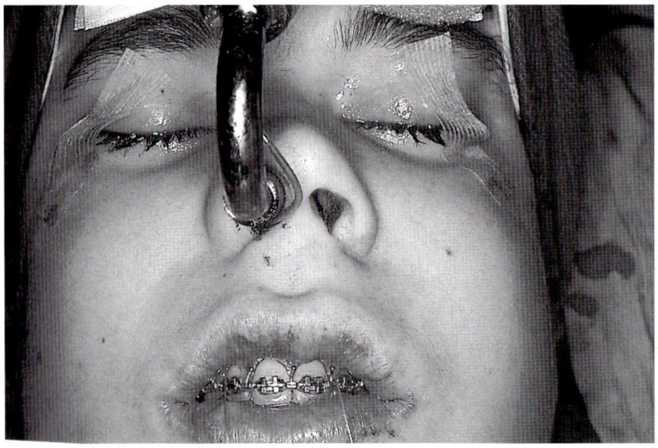

A

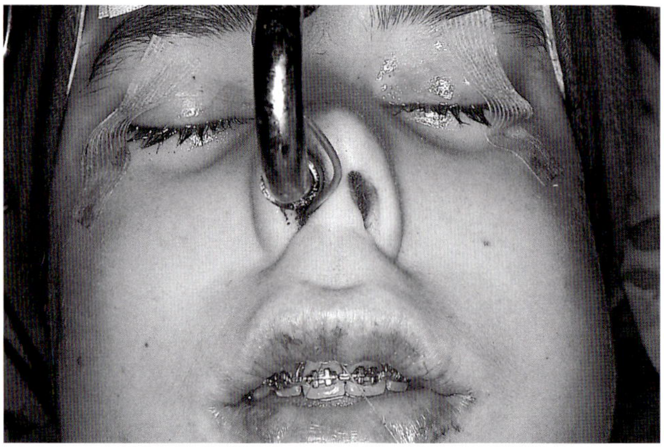

B

FIG. 6-6 A: Intraoperative view of nasal base following Le Fort I osteotomy and prior to alar cinching. B: Following alar cinching, there is narrowing of the alar base width. Notice increased nasal projection and upper lip lengthening.

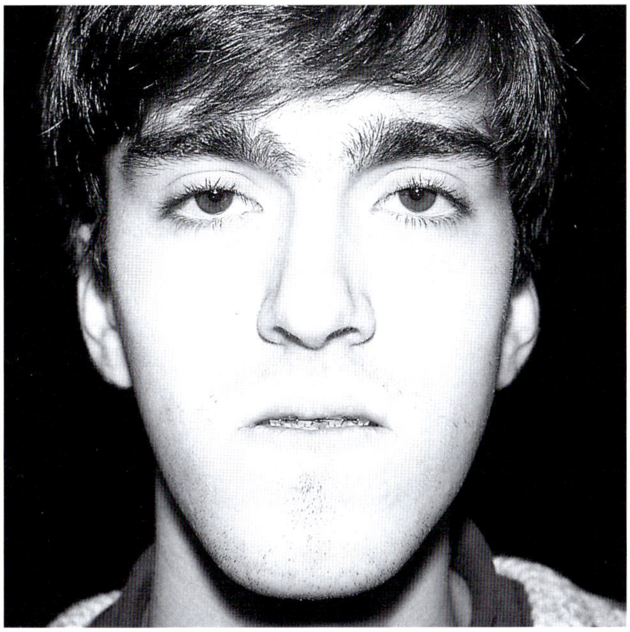

A

B

C

FIG. 6-7 A: A patient prior to double-jaw surgery involving impaction and advancement of the maxilla and placement of an alar cinch suture. B: Following surgery, there is obvious unattractive widening of the nasal base. Alar cinching did not work in this patient. C: The same patient after having had Weir excisions performed 6 months following jaw surgery.

Indirect Factors

The only consideration to be discussed here are the technical details involved in inferior maxillary repositioning with the use of hydroxyapatite placed interpositionally. This is important because it has significantly enhanced the stability of vertical maxillary lengthening procedures and, as a result, has had a long-term effect on the aesthetic outcome.

Following maxillary downfracture, the maxilla is temporarily wired to the mandible in the desired occlusal relationship. The condyles are then seated and the maxillomandibular complex is rotated about the condyles until the desired anterior maxillary vertical dimension is obtained. Six millimeter thick, porous, hydroxyapatite blocks are then carved to the desired dimensions and placed from the piriform buttress posteriorly along the lateral nasal wall. Piriform aperture bone plates are then applied. If double-jaw surgery is being performed, one can utilize eccentric screw placement to achieve compression to ensure that the hydroxyapatite blocks will not dislodge (this represents the only indication in all of orthognathic surgery for using compression plating). The remainder of the anterior lateral maxillary osteotomy gap is then filled with hydroxyapatite block, and then the zygomatic buttress plates are applied (Fig. 6-8). Here again, compression may be utilized if double-jaw surgery is being performed. What is the reasoning for double-jaw surgery being a prerequisite for the utilization of plate

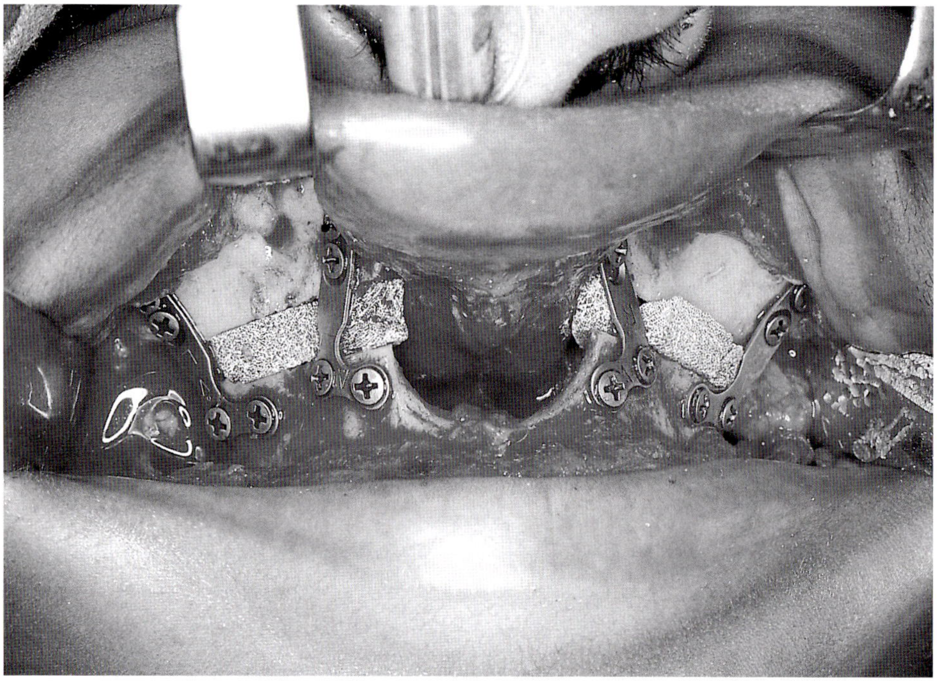

FIG. 6-8 Intraoperative view demonstrating placement of hydroxyapatite blocks into osteotomy gap following inferior repositioning of maxilla.

compression in the maxilla? One does not need to be concerned about the theoretic risks of condylar displacement or uneven application of compression that may alter the occlusal relationship because the mandibular surgery that immediately follows will negate any of these disadvantages.

Mandibular Surgery

Direct Factors

1. Proximal Segment Position

Correct positioning of the proximal segment during bilateral sagittal split ramus osteotomy is aesthetically important because it recreates the appropriate skeletal highlighting in the region of the gonial angle. This not only defines the posterior border of the ramus but also creates the visual impression of posterior face height. Provided that rigid fixation is correctly utilized, overrotation of the proximal segment is now rarely seen. Several techniques may be employed to insure proper position of the proximal segment as it relates to its under- or overrotation and, hence, the appearance of the gonial angle region. Although proximal segment positioners have been championed by many, this author finds their use time consuming, burdensome, and without any guarantee that it will fulfill its primary purpose, which is correct positioning of the condyle in the glenoid fossa. I prefer to use a much simpler technique based on the geometry of the osteotomy gap created in the sagittal dimension between the proximal and distal segments. In straightforward mandibular advancements with little rotation, one should basically observe parallelism between the proximal and distal bone cut. When advancement is accompanied by rotation, the osteotomy gap created becomes characterized by a divergent relationship depending on the type of rotation. Once the correct shape of the ostoetomy gap is created by proximal segment rotation, a bone clamp is applied and the fixation screws placed. Though this is a rough guide to proximal segment rotation, it appears to be all that is clinically necessary to achieve good gonial aesthetics.

2. Percutaneous versus Intraoral Screw Placement

Elaborate right-angle instrumentation has been designed to allow intraoral placement of hardware to stabilize the sagittal split ramus osteotomy. The obvious alleged advantage of this is to avoid a visible cheek scar. I continue to use a percutaneous technique through an approximate 6-mm cheek incision. It is oriented in a vertical direction, parallel to the relaxed skin tension lines, and placed in the most convenient location to afford easy screw placement. It is a mistake to place this incision below the inferior border of the mandible and attempt to retract the soft tissues superiorly. This incision is closed with a running 6-0 fast-absorbing catgut suture. The resulting scars are unnoticeable, and I have yet to revise one (Figs. 6-9, 6-10).

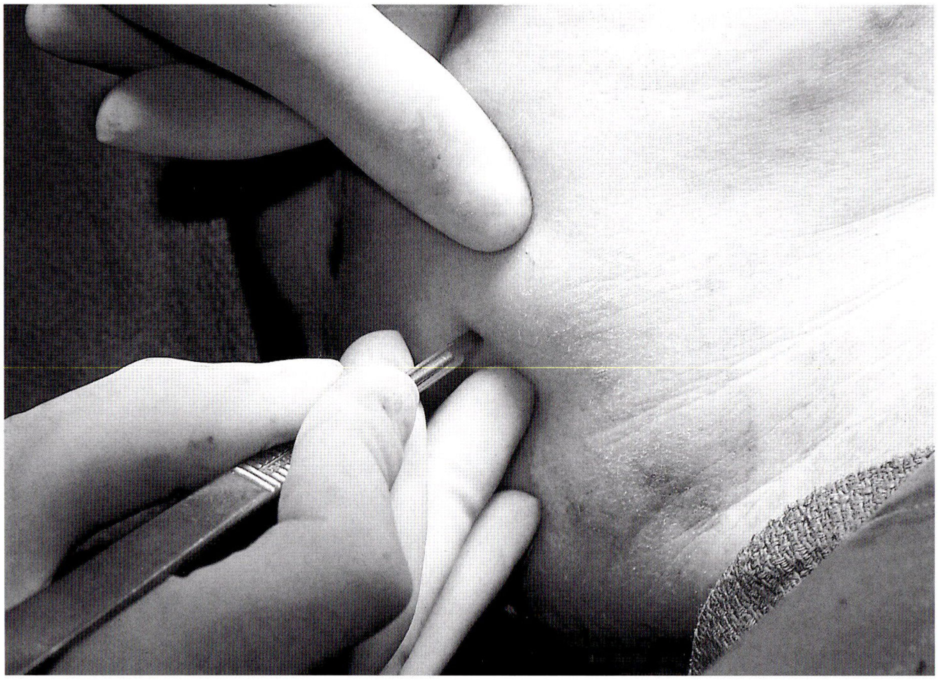

FIG. 6-9 Placement of percutaneous stab in cheek to allow rigid fixation to be performed in patient undergoing sagittal split ramus osteotomy.

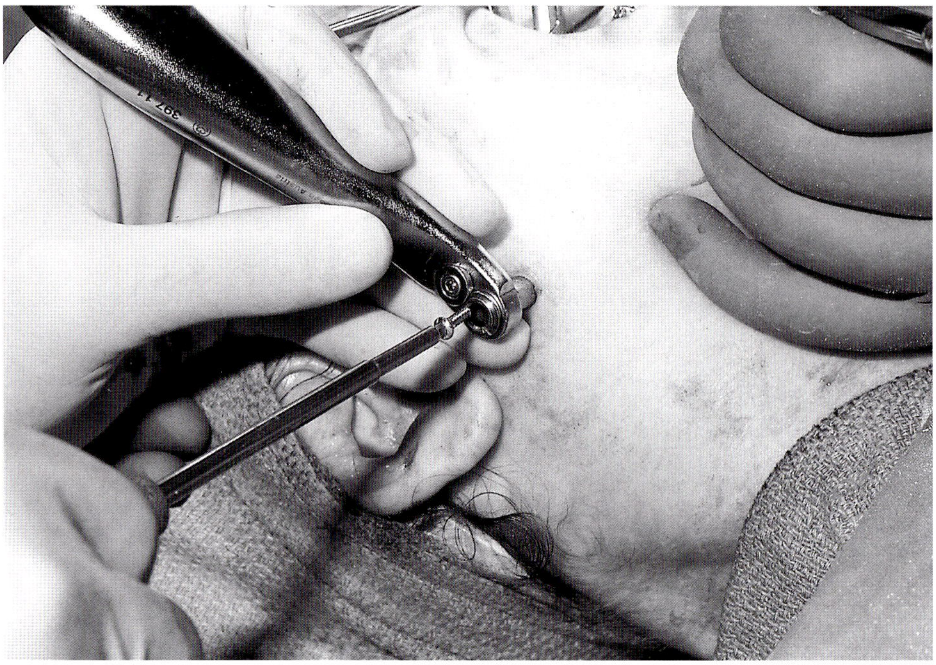

FIG. 6-10 Placement of instruments through trocar into oral cavity.

Indirect Factors

The indirect technical considerations that relate to the long-term aesthetic outcome following sagittal split ramus osteotomy are all concerned with enhancing skeletal stability when extensive mandibular advancements and/or counterclockwise rotation of the distal segment is required.

1. Extended Osteotomy

The osteotomy along the ramus and onto the body of the mandible is extended to the first molar, second premolar region. This allows for a large interface of bone contact and, in turn, a greater healing interface. It also allows for multiple options in terms of hardware placement (Fig. 6-11).

2. Fixation Hardware

For fixation, 2.7 mm screws are utilized. The outer cortex or proximal segment is overdrilled with a 2.7 mm diameter drill bit, and the inner cortex or distal segment is drilled with a 2 mm drill bit. The inner cortex is then tapped. Screw tightening achieves compression between the proximal and distal segments.

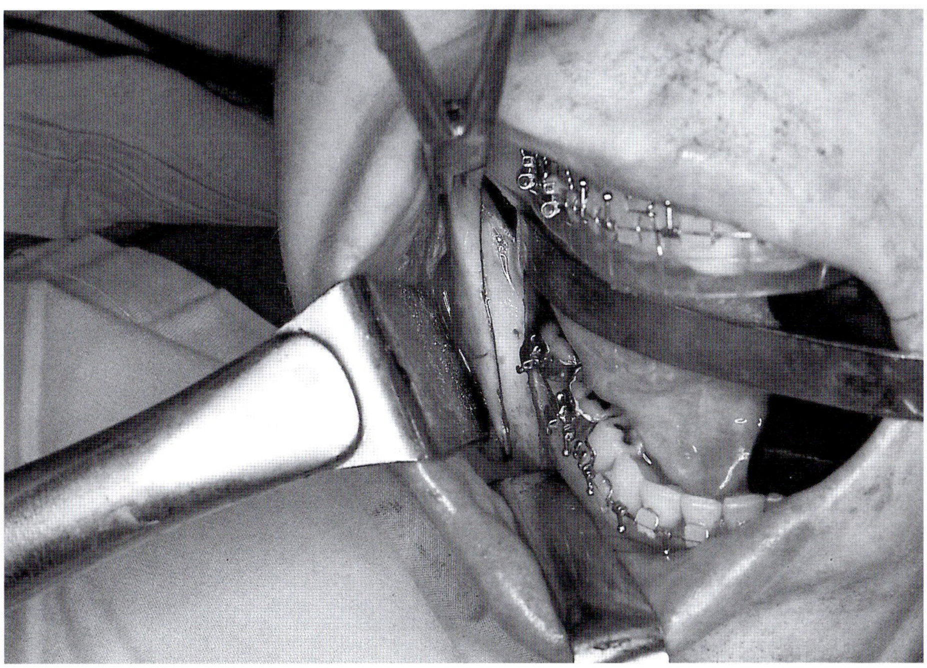

FIG. 6-11 Extended sagittal split ramus osteotomy along buccal cortex into region of first molar.

3. Screw Placement

All screws are placed along the superior border of the ramus so that all are positioned superior to the inferior alveolar canal. This mitigates any nerve compression (Fig. 6-12). To avoid lateral condylar displacement with compression of the segments, it is occasionally necessary to remove bone in one or more areas along the osteotomy interface or to place a bone shim within the osteotomy.

4. Suprahyoid Myotomies

It is occasionally necessary to perform suprahyoid myotomies to allow for a passive anterior repositioning of the distal mandibular segment. This is done through the osteotomy cut that is utilized for an advancement genioplasty. Through this approach, one can release the anterior belly of the digastric and geniohyoid muscles, if necessary.

5. Third Molar Removal

All impacted mandibular third molars should be removed at least 6 to 9 months prior to sagittal split ramus osteotomy. This allows adequate time for mature bone to fill the extraction site so that screw placement is not compromised.

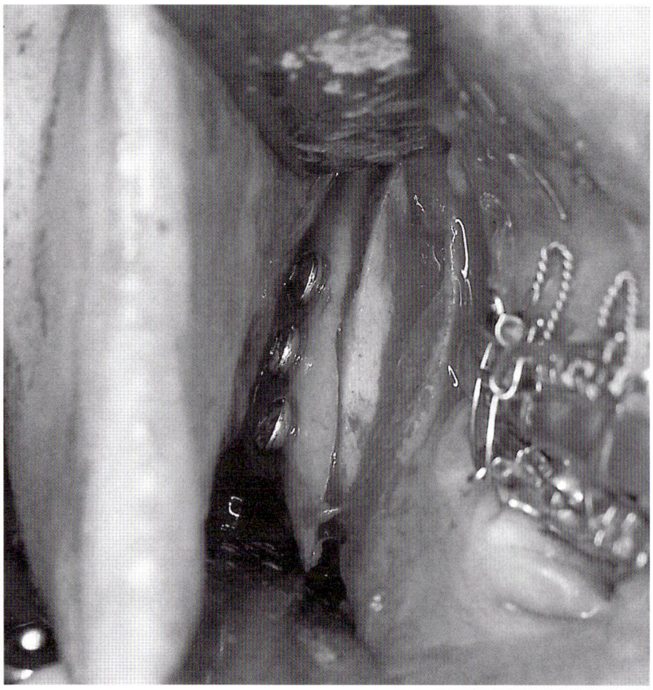

FIG. 6-12 Preferred placement of lag screws along superior border of proximal segment and superior to inferior alveolar nerve. This avoids compression of nerve.

Chin Surgery

All technical considerations related to the aesthetic outcome following chin surgery will be discussed separately in Chapter 11.

Suggested Reading

1. Bell, W.H., Proffit, W.P., and White, R.P. (Eds) *Surgical Correction of Dentofacial Deformities*. Philadelphia, Saunders, 1992.
2. Epker, B.N. and Fish, L.C.: *Dentofacial Deformities: Integrated Orthodontic and Surgical Correction*. St. Louis: Mosby, 1986.
3. Rosen, H.: Definitive surgical correction of vertical maxillary deficiency. *Plast Reconstr Surg* 85:215, 1990.
4. Rosen, H.: Occlusal plane rotation: Aesthetic enhancement in mandibular micrognathia. *Plast Reconstr Surg* 91:1231, 1993.

Surgical Corrections

PART 2

This section deals with utilizing those aesthetic principles developed in Part 1 to achieve the desired aesthetic outcome. I believe that these principles are applicable in the overwhelming majority of patients undergoing surgical corrections. However, one should not misconstrue statements made as sacrosanct. Exceptions will always be encountered. It is the thoughtful surgeon, unbound by rigid dogma, who can aesthetically adapt.

Because treatment plans are based on describing the morphologic appearance of the individual and deemphasizing the skeletal diagnosis, this section is divided into four basic, "bread-and-butter" morphologic facial patterns typically treated. Admittedly, this format is arbitrary, because there is a wide degree of overlap among the different morphologic types. For instance, the facially convex, posteriorly divergent patient may have a long face as well. The separation into four major morphologic types was done for organizational purposes only. However, it does assist in encouraging one to think in terms of what the patient looks like and not what the X-ray film looks like.

It should be emphasized that only the more commonly encountered developmental deformities are addressed. Congenital problems, particularly clefts and major lower face asymmetries such as hemifacial microsomia, were intentionally omitted.

Minimal follow-up periods of 12 months are provided in patient presentations. If no specific length of follow-up is stated, one should assume it is at least 12 months. Longer follow-up times are mentioned when they add further information regarding the stability of a given procedure.

Facial Concavity and Anterior Divergence: The Class III Deformity

CHAPTER

7

Regardless of the skeletal etiology for a Class III malocclusion, the common morphologic denominator among these patients is the visual impression of a flat to concave profile. Whatever its severity may be, facial flatness and/or concavity is aesthetically displeasing and it must be corrected. On the other hand, excessive degrees of anterior facial divergence may be an attractive quality provided it is found in association with overall facial convexity. Accordingly, the common surgical denominator in the overwhelming majority of Class III patients should be maxillary advancement surgery, while the treatment planning variable is the necessity for simultaneous mandibular surgery and possibly chin surgery as well. This statement is applicable whether the Class III malocclusion is caused by maxillary deficiency or mandibular prognathism, or a combination of both.

This seemingly controversial approach is predicated on sound rationale. First and most importantly, the aesthetic results will be enhanced. This is primarily due to the restoration of overall facial convexity without undue reliance on contraction of facial skeletal volume. Although an anteriorly divergent profile may result, overall facial convexity will have been restored without sacrificing the skeletal support of the soft tissues in the perimandibular and submental regions. It is worthwhile here to reiterate the aesthetic principles established regarding facial disproportion in the sagittal dimension: excessive degrees of anterior facial divergence can be attractive provided (1) that it is accompanied by overall facial convexity in the midsagittal and parasagittal planes, (2) that there are normal maxillomandibular relationships as evidenced by the sagittal posturing of the upper lip relative to the lower lip, and (3) that there is adequate projection of the nasal dorsum and nasal tip.

There are other compelling reasons to espouse maxillary advancement surgery as the workhorse for correcting the Class III malocclusion regardless of skeletal etiology. Of importance is the fact that maxillary advancement surgery is a technically easier operation to perform compared to the sagittal split ramus osteotomy. Secondly, the risk of operative complications is less with maxillary advancement. The primary morbidity of maxillary surgery relates to the level of the osteotomy in the posterior maxilla at its junction with the pterygoid plates. As long as this is kept as low as possible, the risks of excessive blood loss and fracture extending superiorly into the orbit are rare. On the other hand, sagittal split ramus osteotomy, even in the best of hands, carries a significant risk of permanent injury to the inferior alveolar nerve with subsequent sensory loss in the lower lip and chin region.

An additional reason for favoring maxillary surgery as opposed to mandibular surgery is its extreme versatility. Because one can easily change the vertical position of the maxilla as well as its sagittal position, enhanced aesthetic results can be obtained. This is particularly applicable as it relates to our ability to elongate the maxilla to provide greater degrees of incisal show and increases in skeletal volume in the vertical as well as the sagittal dimensions. In addition to changes in vertical dimension, maxillary surgery also affords the surgeon the opportunity to segmentalize the maxilla not only to achieve the best occlusal result, but also to provide for the most ideal axial inclination of the maxillary incisor teeth.

The role of osseous genioplasty in the aesthetic correction of the Class III skeletal deformity must not be underestimated. Mandibular prognathism is frequently accompanied by anterior vertical mandibular excess. Although restoring a normal occlusion (regardless of which jaw is moved) will reduce the vertical dimension of the lower face by improving intercuspation, further reduction in the height of the anterior mandible is frequently desirable. Osseous genioplasty, performed by making two parallel horizontal osteotomies with an appropriate ostectomy of intervening bone, will not only achieve a reduction in vertical dimension but also create the visual illusion of flattening an excessively steepened mandibular plane.

Osseous genioplasty is also frequently indicated to advance the symphyseal chin segment in many Class III individuals. This is particularly important when employing mandibular setback surgery because chin advancement will minimize any reduction in skeletal volume in the sagittal dimension. The aesthetic sine qua non for chin advancement in patients undergoing mandibular setback is a shallow labiomental fold as well as soft tissue pogonion being posterior to the lower lip. These latter two aesthetic elements are not infrequently seen in the Class III individual.

Surgical Options

The choice of which surgical procedure is ultimately chosen for correction of a Class III individual must be dictated by his or her facial morphology and not by skeletal diagnoses. To reiterate, the morphologic feature that must be corrected is the facial flattening or concavity. Accordingly, there are three basic surgical options from which to choose. The first option is mandibular setback surgery. This option should be chosen least frequently because of the reduction in skeletal volume with subsequent loss of soft tissue support that necessarily occurs in this procedure. Regardless of the skeletal etiology responsible for the Class III abnormality, this surgical approach is indicated in a relatively small, select group of individuals.

The second option is maxillary advancement surgery. Its great advantage over mandibular setback surgery is that it augments skeletal volume as opposed to reducing it; in addition, it is safer and easier to perform and provides one with tremendous versatility in terms of the vertical dimension of the lower face as well as the option of maxillary segmentation.

The third surgical option is double-jaw surgery. This most frequently involves maxillary advancement with some type of mandibular rotation or modest mandibular setback. On occasion, double-jaw surgery may involve mandibular advancement as well as maxillary advancement in those selected Class III patients in whom adequate projection of the mid- and lower face could not be achieved with maxillary advancement alone. This is frequently considered necessary as it relates to the soft tissue goals of the individual patient. The choice of a double-jaw procedure provides the surgeon with the opportunity to achieve the best possible aesthetic result independent of any occlusal constraints. This remains the single greatest advantage of double-jaw surgery over isolated maxillary or mandibular surgery.

Morphologic Features of Class III Patients

It bears repeating that the decision of which surgical option to exercise is dictated by the facial morphology and the underlying relationship of the skeletal foundation to the facial mask. It is not a skeletal diagnosis that determines the surgical treatment plan. Qualitative, passive changes in soft tissue drape secondary to skeletal repositioning is a critical determining factor. There are, additionally, six basic morphologic features that one needs to consider in determining the most aesthetically appropriate surgical plan. These are sagittal or midline aesthetics of the midface, parasagittal aesthetics of the midface, a soft tissue qualitative assessment of the mid- and lower face, the extent of the sagittal discrepancy between the maxilla and the mandible, the existence of any lower face asymmetry, and the existence of any vertical dysplasia of the maxilla.

1. Sagittal Aesthetics of the Midface

This primarily relates to the size of the nose, the morphology of the nasolabial angle, and the prominence of the upper lip particularly at its base. A nose with excessive projection would certainly favor a maxillary advancement procedure because the apparent size of the nose diminishes as the maxilla is advanced. Mandibular setback surgery in such an individual would only make the nose appear larger. On the other hand, a nose with inadequate projection favors posterior repositioning of the mandible because maxillary advancement surgery will make the nose appear smaller. The morphology of the nasolabial angle and the prominence of the upper lip are interrelated inasmuch as the nasolabial angle will appear overly obtuse and poorly defined secondary to maxillary deficiency, with subsequent lack of lip support. In this situation, an increase in definition of the nasolabial angle as it becomes more acute is rather predictable as one advances the maxilla. Accordingly, this morphologic factor strongly favors maxillary advancement surgery. An additional predictable change in nasolabial angle morphology occurs with inferior repositioning of the maxilla as it is advanced. In this situation, the nasolabial angle will typically widen and become softer in appearance. Unfortunately, these are the only two instances in which one can reliably predict the change in nasolabial angle morphology; I have seen nasolabial angles become both more acute and more obtuse following sagittal advancement of the maxilla.

2. Parasagittal Midface Aesthetics

This relates to the skeletal support of the soft tissues in both the infraorbital and paranasal regions. Class III individuals who lack support of the upper cheek should clearly be treated with maxillary advancement surgery. Such individuals may even have minimal degrees of inferior scleral show, which can be improved with maxillary advancement even at the Le Fort I level. Attempting to camouflage this by isolated onlay augmentation in the infraorbital region without accompanying Le Fort I advancement provides inconsistent results. Therefore, sagittal deficiency in this area is best treated with skeletal displacement rather than isolated augmentation. Paranasal flattening also favors maxillary advancement surgery to correct Class III malocclusions. However, unlike the infraorbital region, this area is much more amenable to onlay augmentation only using either cranial bone or block hydroxyapatite. My preference is to use block hydroxyapatite because of the unpredictable degree of resorption of cranial bone.

3. Soft Tissue Support of the Mid- and Lower Face and Neck

The adequacy of skeletal support of the soft tissues must be taken into consideration. In the midface, this includes a well-defined zygomatic complex with a suggestion of a submalar hollow and a well-supported malar fat pad in the absence of excessive depth to the nasolabial fold. Adequate support of the soft tissues in the lower face is indicated by a well-defined inferior border of the mandible and the absence of any jowl formation. Inadequate submental soft tissue support may be due to absolute or relative redundancy of skin, fat, and platysma muscle. In the absence of well-supported tissue in these areas, the cervical mandibular angle is poorly defined and excessively obtuse.

Maxillary advancement surgery to correct a Class III malocclusion is strongly indicated in patients who lack soft tissue support in the midface so that skeletal highlighting in this area can be created. In addition, it is also obviously indicated in individuals who lack support in the lower face and neck. Further deterioration of the soft tissue support in these areas would occur with mandibular setback surgery. On the other hand, posterior repositioning of the mandible can be considered in those individuals who have a well-supported facial mask in combination with a Class III malocclusion secondary to a large mandible.

4. Extent of Sagittal Discrepancy

As the extent of the sagittal discrepancy between the maxilla and the mandible becomes greater, the chances of a good aesthetic result from single-jaw surgery become less. This is true of both large maxillary advancements as well as large mandibular setbacks. As a general rule, skeletal displacements of 10 mm or more in either the maxilla or the mandible should be rarely undertaken in Class III malocclusions secondary to developmental deformities. (This statement does not apply to congenital problems such as in the cleft lip and palate patient.) Extensive maxillary advancements will usually result in undesirable nasal aesthetics. More often than not, the nasal base will widen excessively, and the projection of the nose may appear to be inadequate. Large mandibular setbacks should be avoided because of the undesirable consequences of reductions in skeletal volume with secondary lack of soft tissue support. Accordingly, double-jaw surgery, that is, maxillary advancement with modest mandibular setback, is usually indicated in such individuals.

5. Lower Face Asymmetry

Class III malocclusion accompanied by lower face asymmetry will frequently require double-jaw surgery for the best aesthetic correction. This is most frequently seen in two clinical situations: (1) when an occlusal cant or tilt exists and (2) when there is mandibular asymmetry with a dental midline shift and one wishes to correct the sagittal discrepancy with maxillary advancement surgery. In this latter instance, double-jaw surgery to advance the maxilla and derotate the mandible is required to prevent shifting of the maxilla off of the midline as it is advanced.

6. Vertical Maxillary Dysplasia

A Class III malocclusion may exist in the presence of vertical maxillary excess with or without an anterior open bite. In such instances, double-jaw surgery is almost always required because the real sagittal discrepancy between the maxilla and the mandible becomes greater secondary to mandibular autorotation after the maxilla is intruded. Under this circumstance, isolated correction in the maxilla usually requires an extensive maxillary advancement, which is undesirable for the reasons already stated. Mandibular setback surgery of some degree is usually advisable. Although less common, mandibular prognathism can also be seen with vertical maxillary deficiency. This, too, ordinarily requires double-jaw surgery.

Patient Presentations

The following patient presentations were chosen to help illustrate the aforementioned aesthetic principles for correction of the Class III patient. Skeletal cephalometric values are included only to substantiate for the reader the skeletal diagnosis and to illustrate how this information may be at variance with the clinical impression of the patient's facial morphology. It also serves to emphasize the lack of clinical relevance of these time-honored parameters and how the ultimate treatment plan differed from what the cephalometric evaluation would have suggested.

All patients treated with maxillary surgery underwent alar base cinching.

Case Study 7-1

This patient represents an appropriate candidate for mandibular setback surgery. In the frontal view, there is good vertical proportionality and lower face asymmetry with the chin point deviated to the left of the midline. In the lateral view, there is an anteriorly divergent profile with modest overall flattening in the midsagittal plane. There is good infraorbital and paranasal support. There is rather modest projection of the nasal tip. There is good soft tissue support throughout the mid- and lower face. There is a modest Class III sagittal discrepancy with a mandibular dental midline shift to the left of 3.5 mm.

The patient underwent sagittal split ramus osteotomy to set the mandible back and rotate it to the right to restore lower face symmetry. Postoperatively there is a good occlusion. There is a modest Class II canine relationship due to a tooth mass discrepancy secondary to small maxillary lateral incisors. A more attractive result would have been obtained with augmentation of the nasal spine region to create a greater degree of facial convexity and some modest elevation of the nasal tip.

The primary indications for mandibular setback surgery in this patient are the rather small size of the nose, mandibular asymmetry, adequate parasagittal midface aesthetics, well-supported soft tissues in the lower face and neck, and no vertical maxillary dysplasia. If maxillary surgery had been used to correct the sagittal discrepancy, a double-jaw procedure would have been necessary because of the mandibular asymmetry. For all of these reasons, mandibular setback seemed appropriate. This combination of morphologic factors in this patient is somewhat unusual in my surgical practice; this accounts for the small number of patients treated with mandibular setback surgery alone.

PREOPERATIVE POSTOPERATIVE

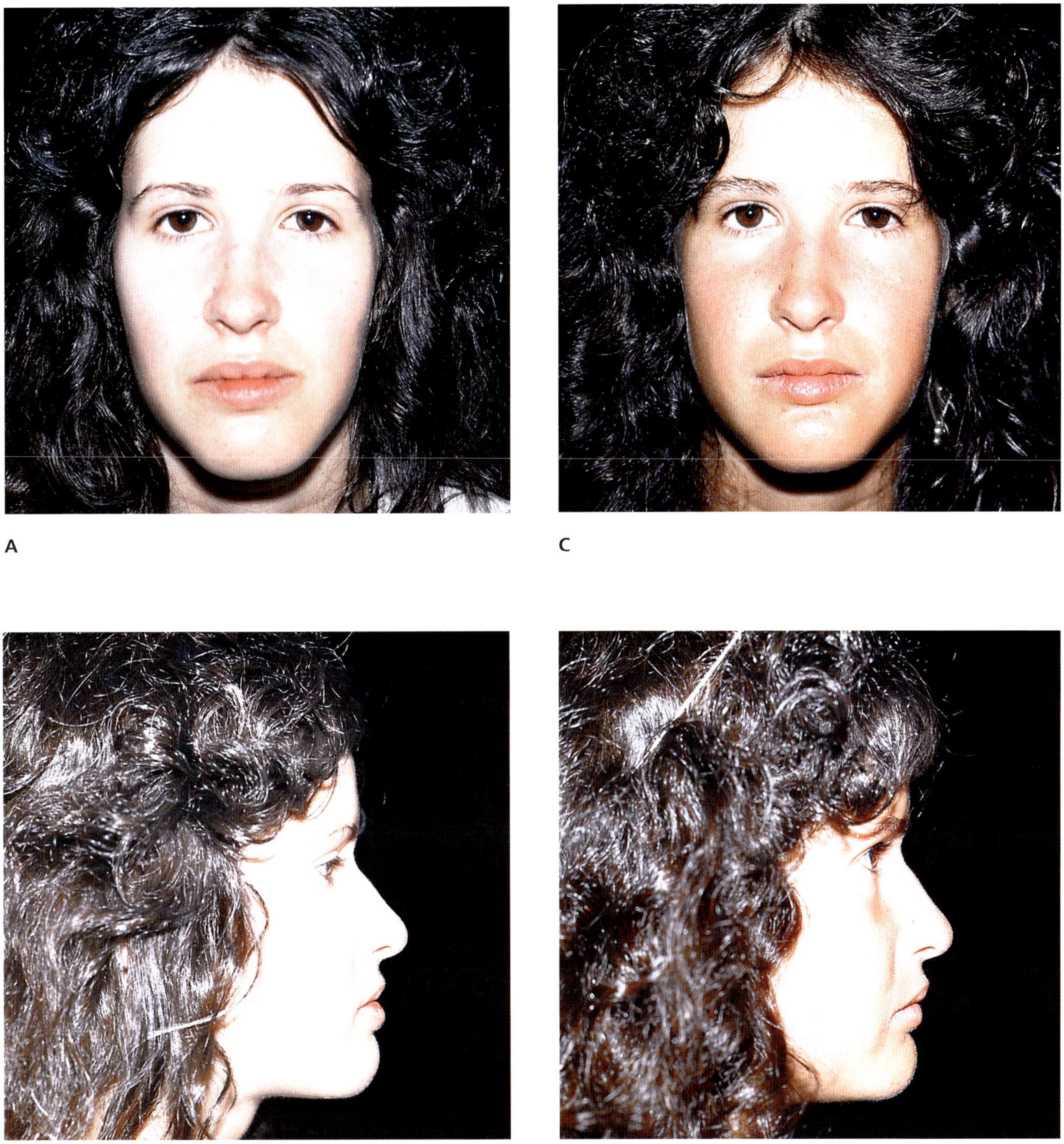

A C B D

CASE STUDY 7-1 A & B: A Class III patient whose morphologic features indicated mandibular setback surgery. Those features included adequate midface projection parasagittally, a relatively small nose, good soft tissue support in the lower face, and mandibular asymmetry.

PREOPERATIVE **POSTOPERATIVE**

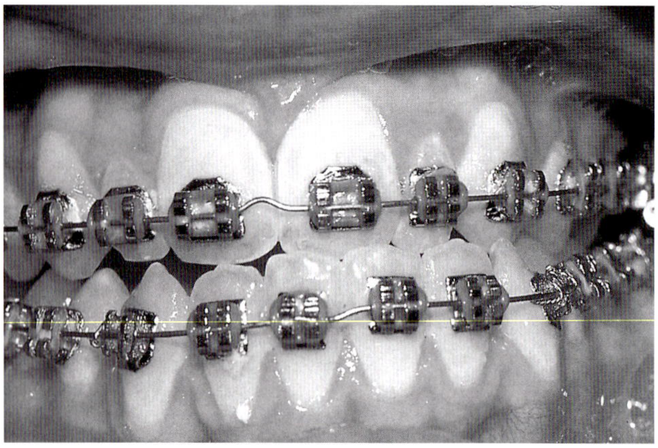

E

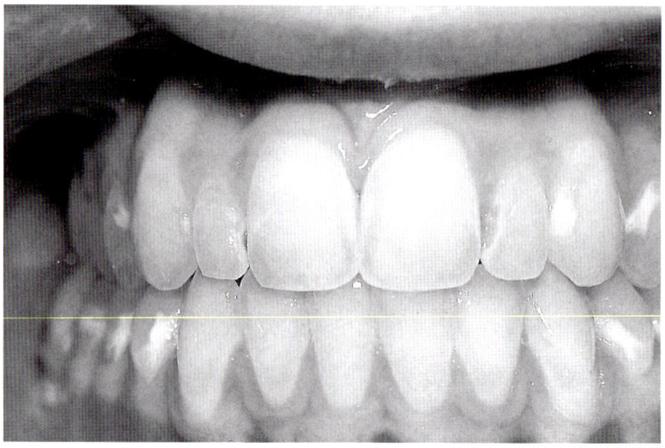

G

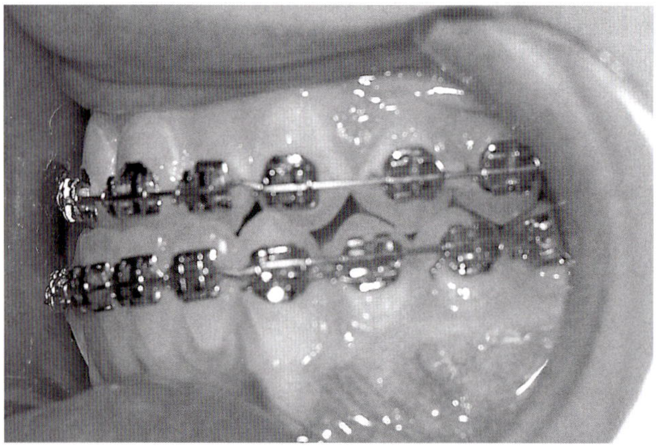

F

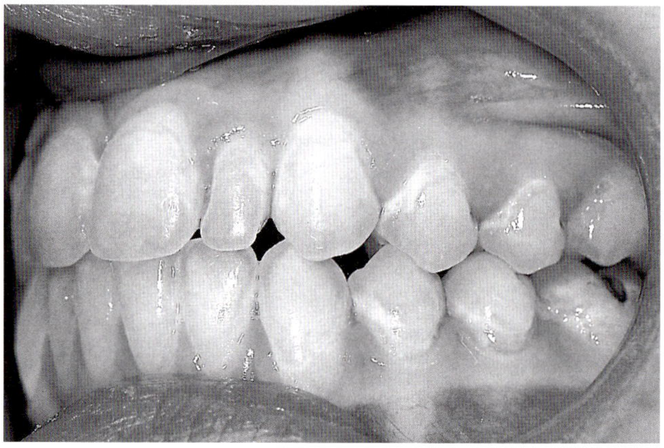

H

Fig. 7-1 (*Continued*) E & F: The preoperative occlusion is characterized by a 3.5-mm midline discrepancy and a tooth mass discrepancy due to small lateral incisions. Orthodontist: Dr. James Ackerman

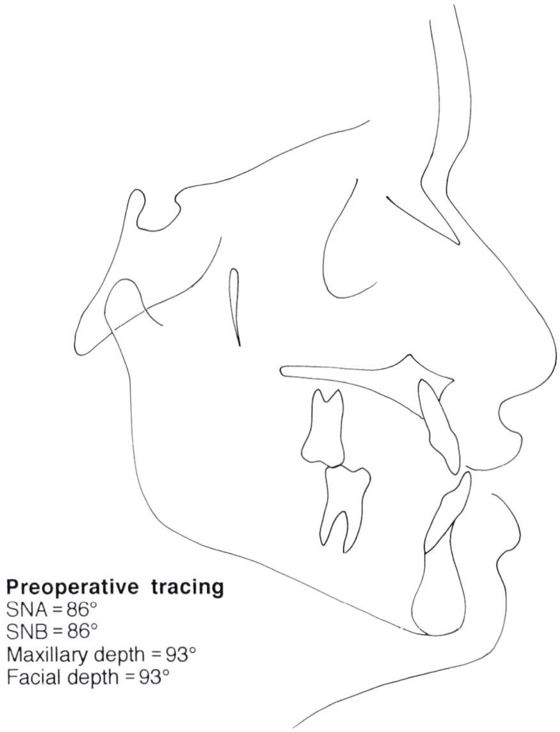

I

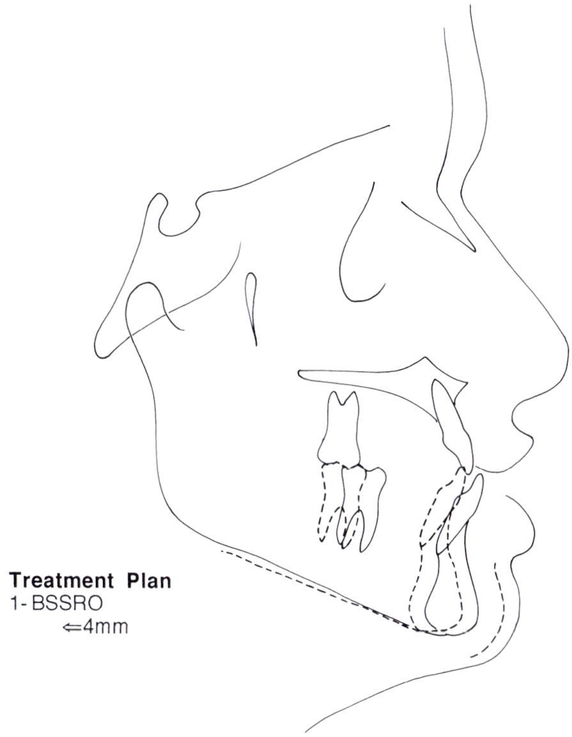

J

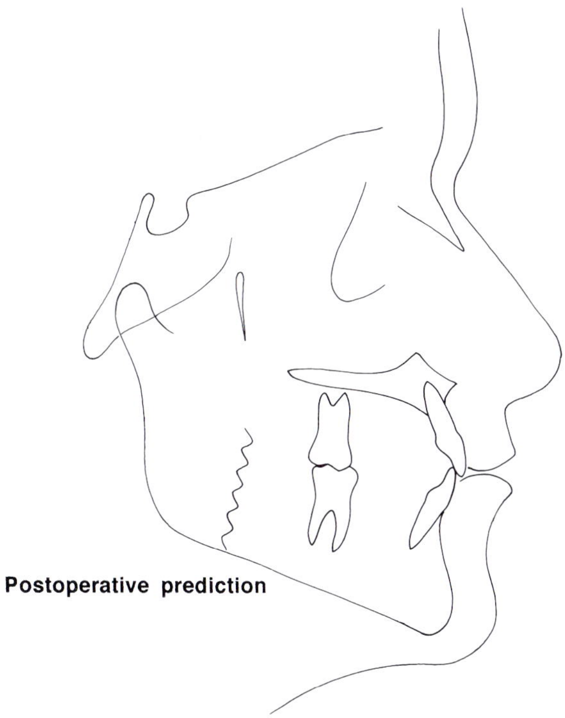

K

FIG. 7-1 *(Continued)*

Case Study 7-2

The patient in this figure has a skeletal diagnosis of mandibular prognathism. The patient's morphology in the frontal view is characterized by a modest degree of excessive lower face height with minimal mentalis hyperfunction. There is good lower face symmetry. In the lateral view, there is overall anterior divergence with midsagittal convexity. However, there is parasagittal concavity with poor infraorbital and paranasal support. The mandibular plane angle is somewhat steepened. There is adequate soft tissue support throughout the lower face. There is a Class III occlusion and a midline dental discrepancy in the maxillary dentition of 2 mm to the right of the facial midline. Cephalometric skeletal and soft tissue analyses support a skeletal diagnosis of mandibular prognathism.

Maxillary advancement surgery was undertaken in this patient because of the significant parasagittal flattening. This provided the opportunity to reposition the maxilla inferiorly to achieve a bit more incisal exposure and to correct the midline dental discrepancy with maxillary rotation. A vertical reduction genioplasty was undertaken to further reduce the height of the lower face. However, sagittal movement of the symphyseal bone segment was avoided.

The postoperative result is characterized by an overall softening of the appearance and enhancement of facial balance. In the frontal view, this is due to a reduction of the overall dominance of the mandible as a result of the vertical reduction of the chin. Note the improved relations between the upper lip and incisor teeth due to maxillary vertical elongation. The profile remains anteriorly divergent to an excessive degree; however, there is now facial convexity both midsagittally and parasagittally. The soft tissue support of the lower face has not been sacrificed.

This patient represents a rather classic example of the aesthetic rationale for treating mandibular prognathism with maxillary advancement. This result could have been improved with soft tissue augmentation of the upper lip. This would have provided better balance between the upper and lower lips and would have further softened the patient's appearance.

PREOPERATIVE | **POSTOPERATIVE**

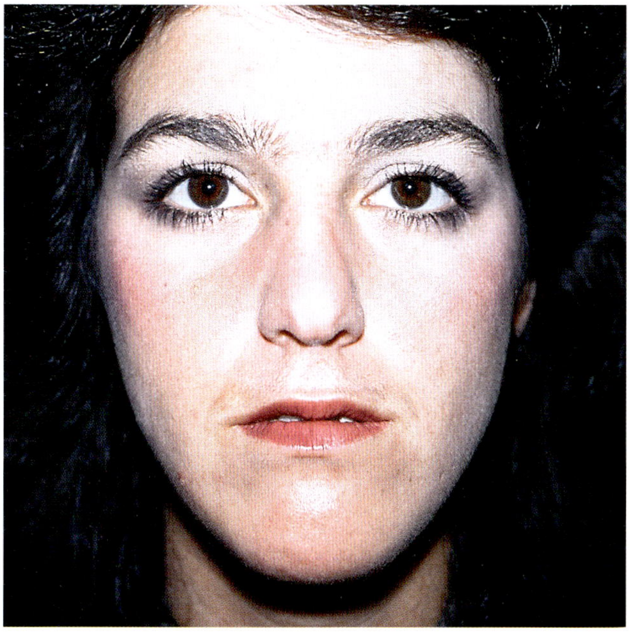

A

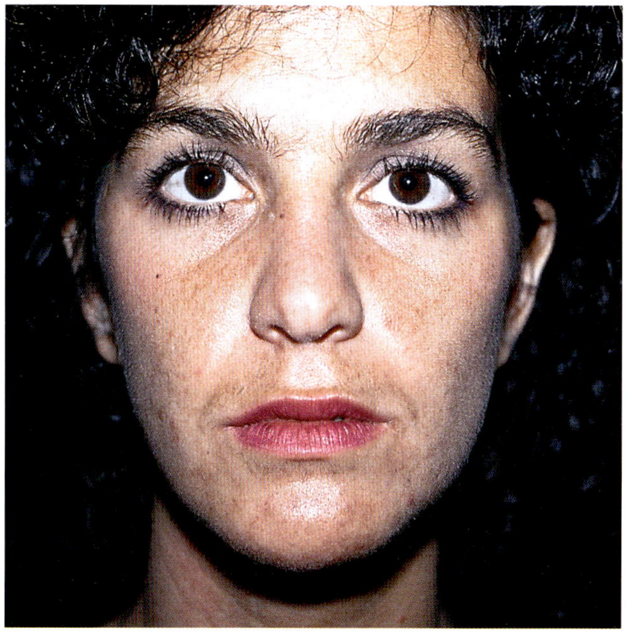

C

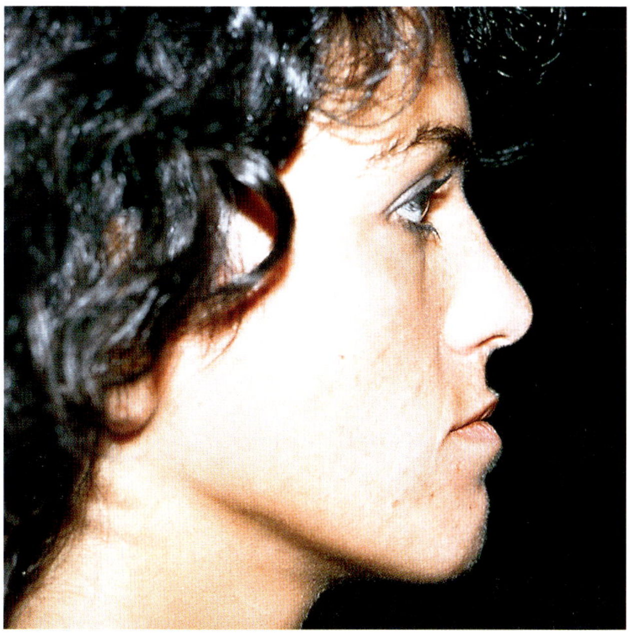

B

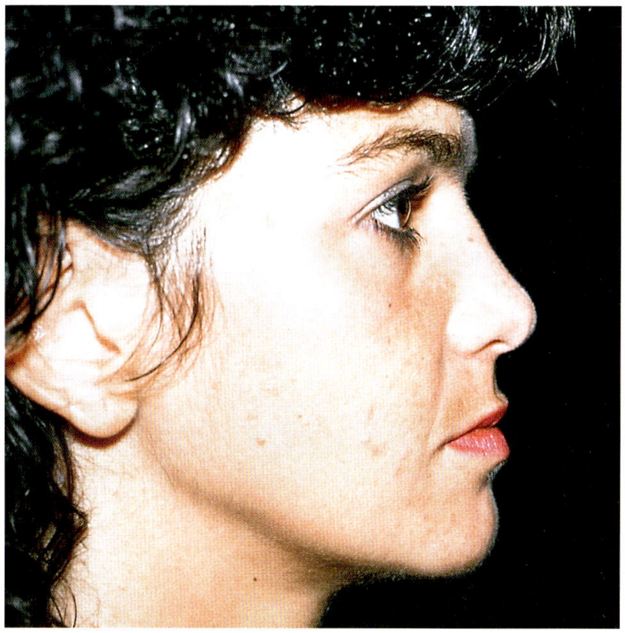

D

CASE STUDY 7-2 A & B: A patient with a skeletal diagnosis of mandibular prognathism. Morphologic features including parasagittal flattening and minimal incisal exposure indicated a maxillary correction. The symphyseal vertical excess and steep mandibular border were corrected with vertical reduction genioplasty.

PREOPERATIVE | POSTOPERATIVE

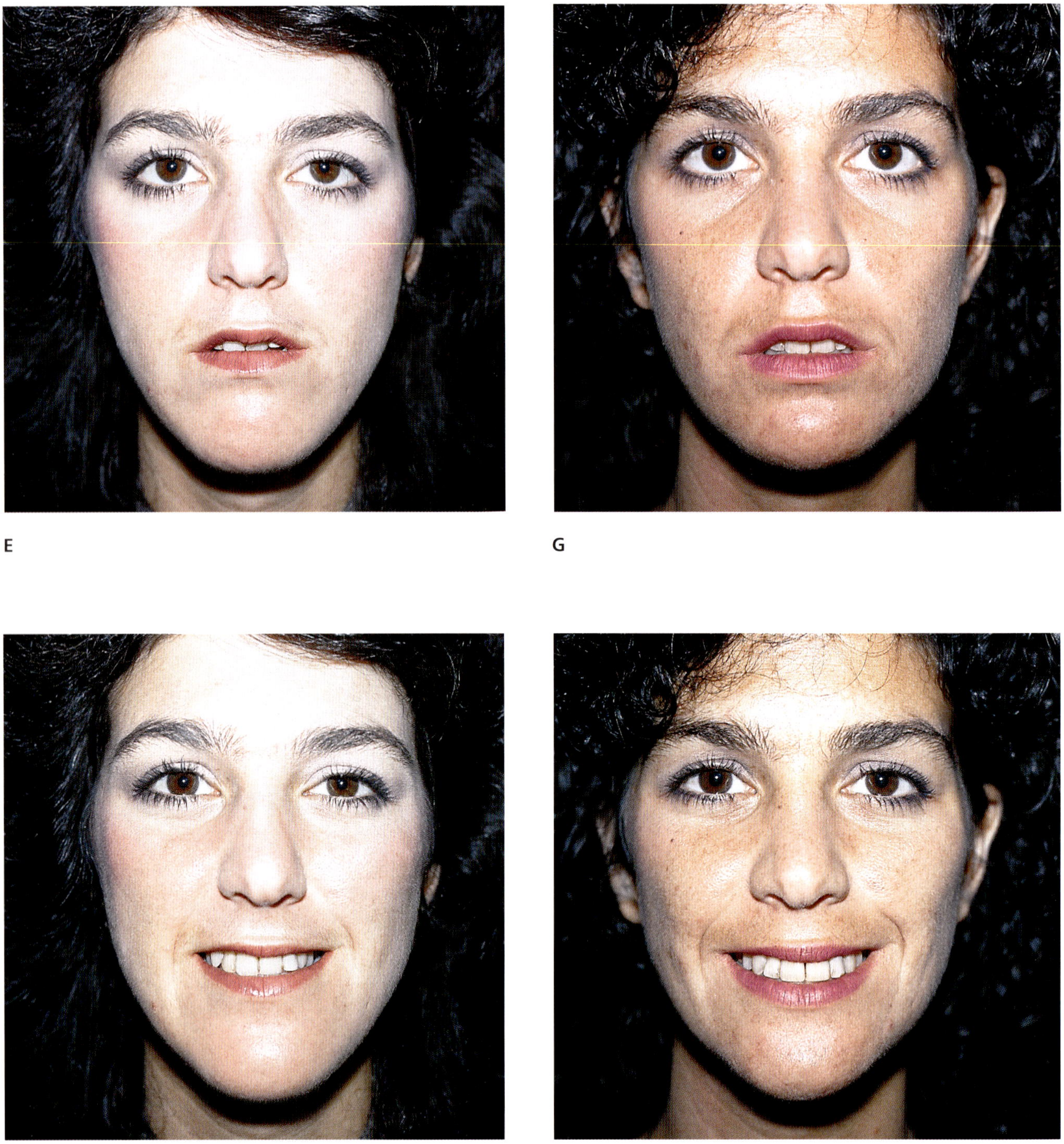

E G

F H

Fig. 7-2 (*Continued*) E & F: Note the maxillary midline deviation to the right. G & H: Following surgery, there is greater maxillary incisor exposure due to a 3-mm vertical displacement. From: Rosen, H.: Maxillary advancement for mandibular prognathism: Indications and rationale. *Plast Reconstr Surg* 87:823, 1991. Reprinted with permission.

PREOPERATIVE | **POSTOPERATIVE**

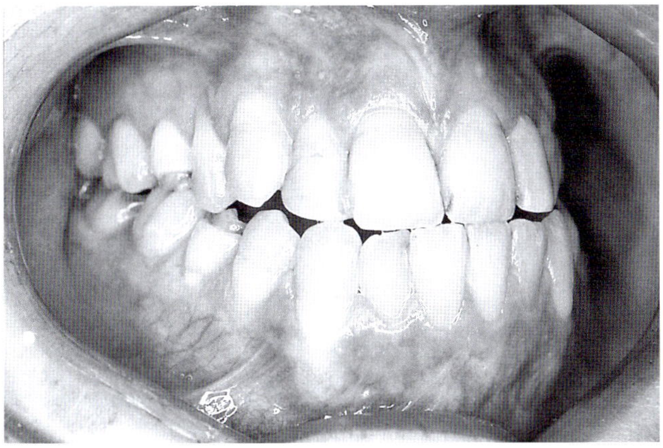

I

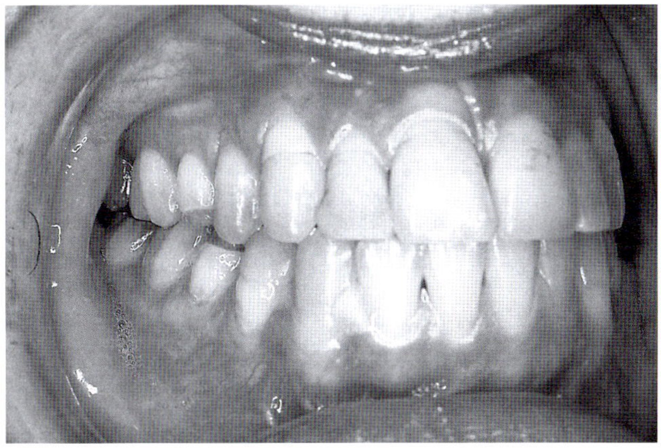

J

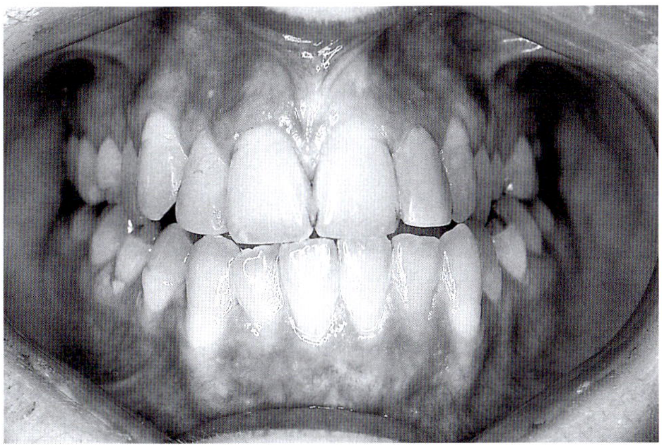

K

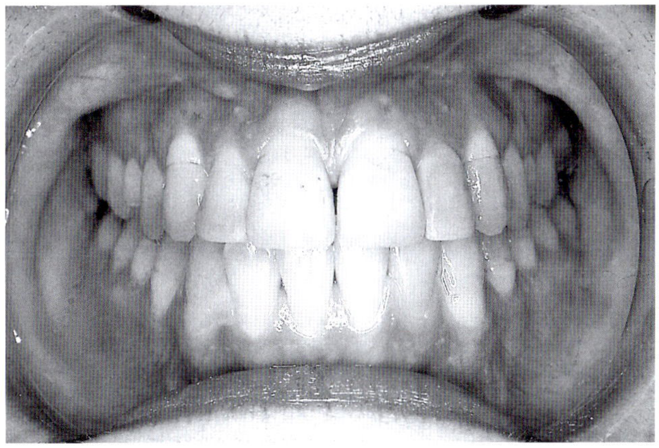

L

FIG. 7-2 (*Continued*) I & J: Note midline discrepancy of 2mm. This is due to the maxilla being to the right of the craniofacial midline. Orthodontist: Dr. James Ackerman

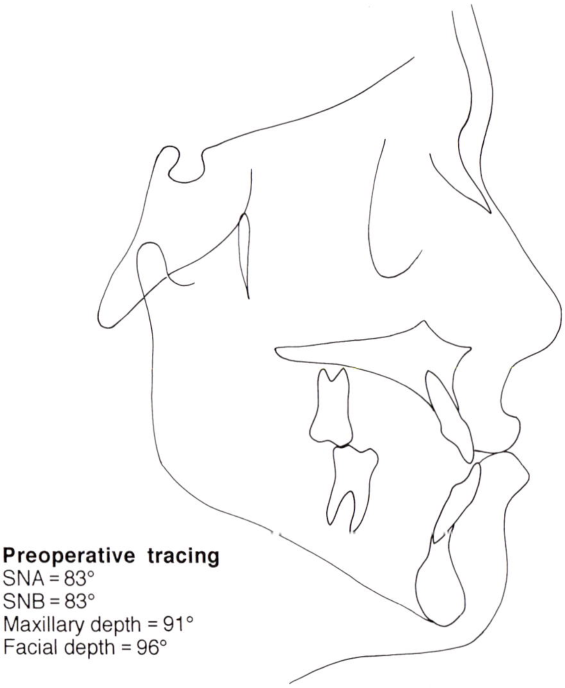

M

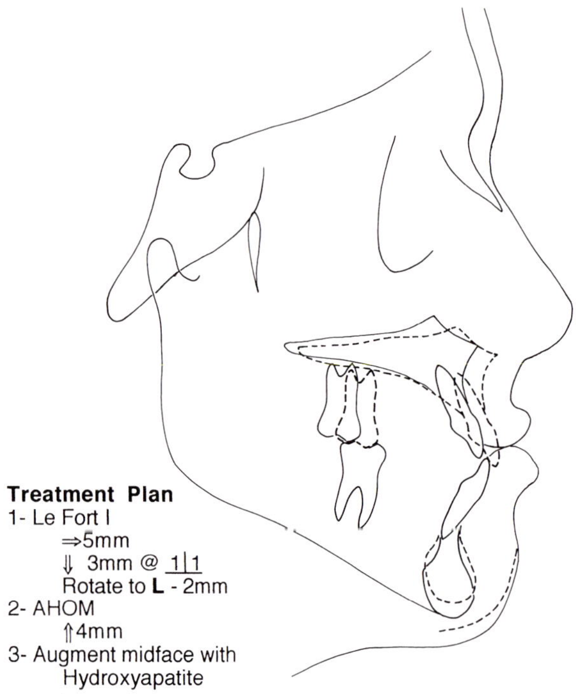

N

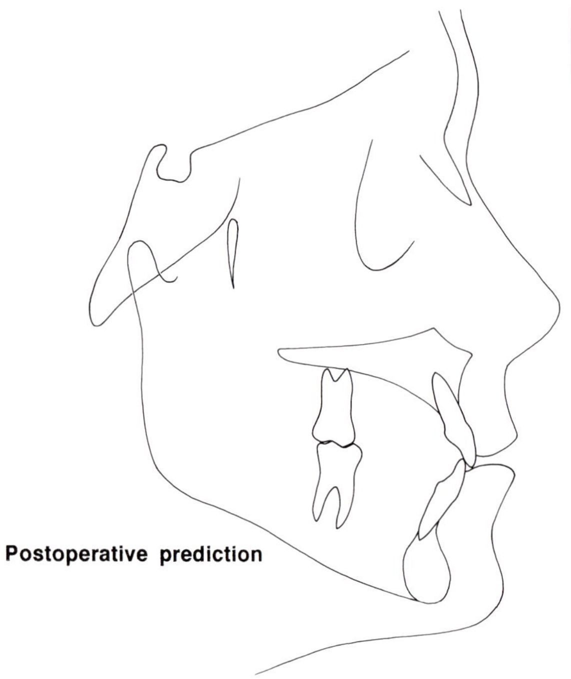

O

FIG. 7-2 *(Continued)* M: Cephalometric analysis confirms a diagnosis of mandibular prognathism.

Case Study 7-3

This patient represents another example of an individual with mandibular prognathism treated with maxillary advancement surgery. In the frontal view, there is good vertical proportionality and facial symmetry. There is an apparent inverse relationship between the upper and lower lips. There is poor soft tissue support throughout the perimandibular region. There is minimal tooth exposure in both repose and with full smiling. In the lateral view, there is an anteriorly divergent profile with midsagittal flattening. However, the midface structures all have adequate sagittal projection, particularly in the infraorbital and paranasal areas. There is normal projection of the nasal dorsum and tip. Again, note the poor soft tissue support throughout the parasymphyseal area with jowl formation. The lower lip is well in advance of the upper lip. There is a Class III occlusal relationship with an approximate 8-mm sagittal discrepancy. Cephalometric analysis supports the diagnosis of mandibular prognathism.

Maxillary advancement surgery was chosen in this patient in order to preserve skeletal volume in view of the soft tissue excess in the perimandibular region. In addition, vertical lengthening of the maxilla was treatment planned to provide for a greater degree of incisal exposure. The maxilla was advanced 6 mm and lengthened 3 mm. Postoperatively, there is an acceptable aesthetic result with an enhanced profile due to greater midsagittal and parasagittal facial convexity. There has been no further deterioration of the skeletal support of the soft tissues of the lower face.

The aesthetic results could have been markedly improved with greater vertical displacement of the maxilla. Note there is little improvement in the extent of incisal exposure due to lengthening of the upper lip. Greater vertical displacement would have also minimized the alar widening. Lastly, modest liposuctioning of the jowls should have been performed.

PREOPERATIVE

POSTOPERATIVE

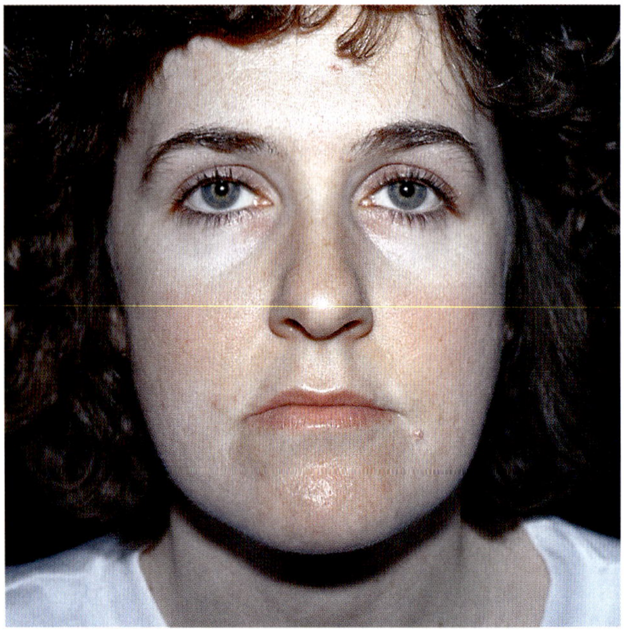

A

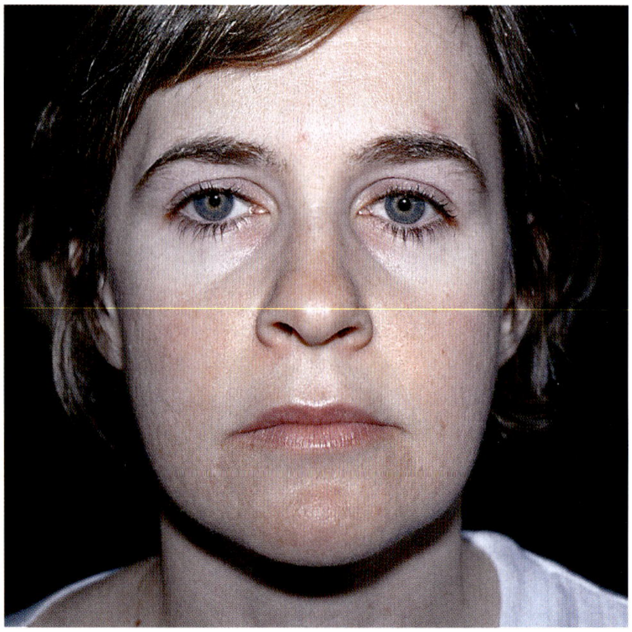

C

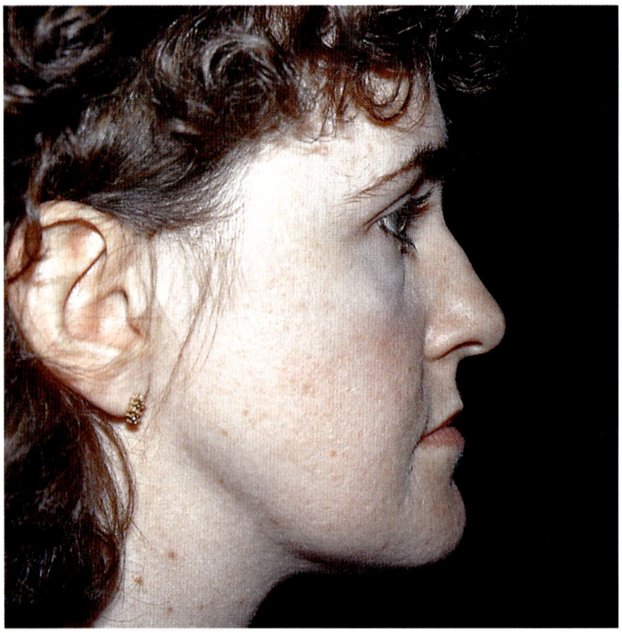

B

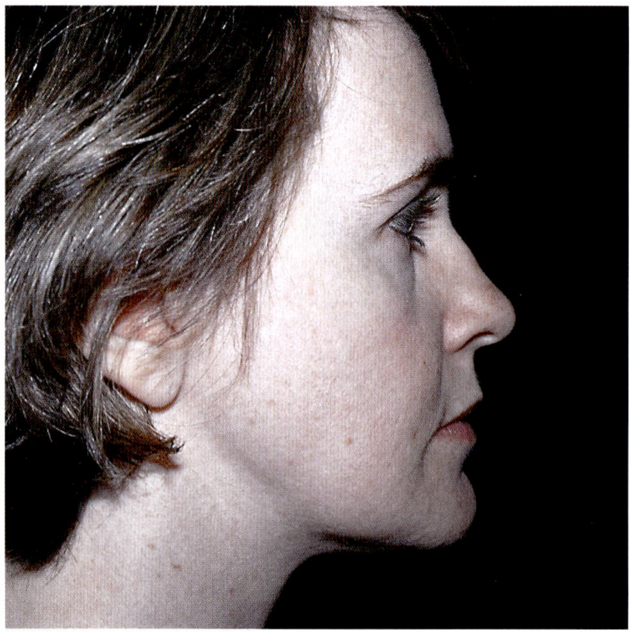

D

Case Study 7-3 A & B: A patient with a Class III maloclusion treated with maxillary advancement because of the desire to preserve skeletal volume in the lower face. Note the perimandibular soft tissue fullness. The skeletal diagnosis is mandibular prognathism.

PREOPERATIVE **POSTOPERATIVE**

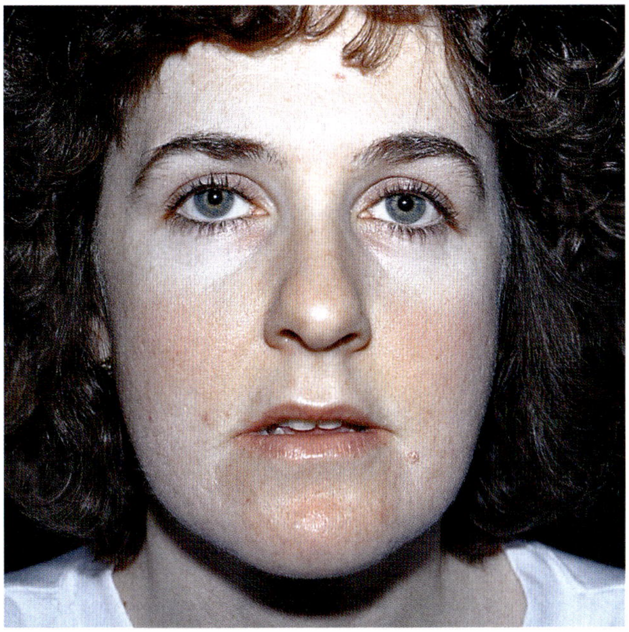

E

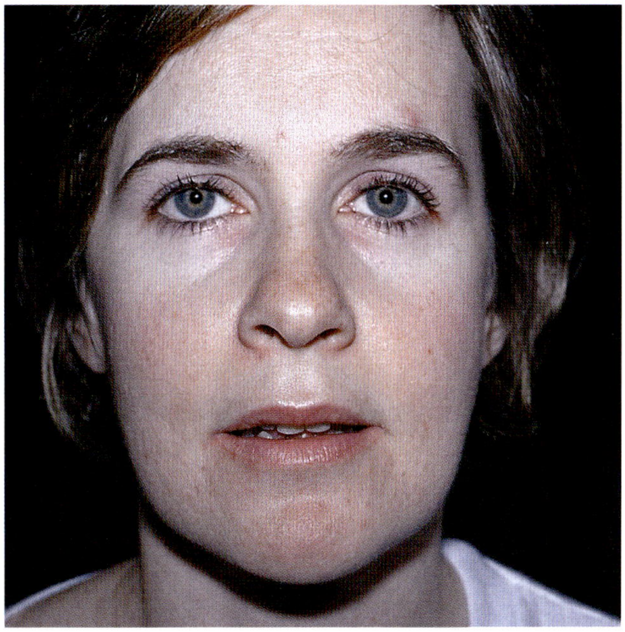

G

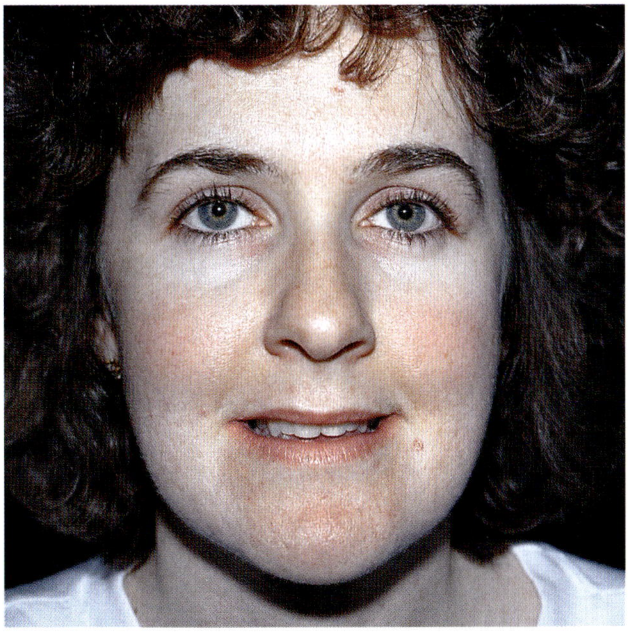

F

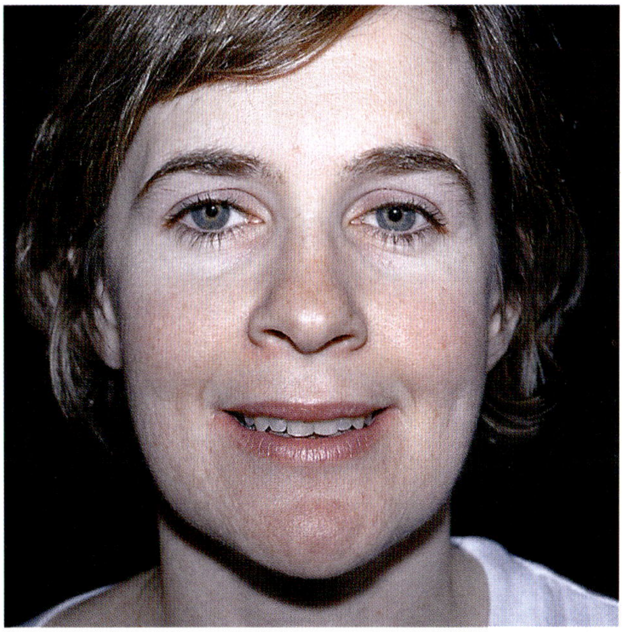

H

FIG. 7-3 (*Continued*) G & H: Despite maxillary lengthening as well as advancement, there is little improvement in incisal exposure due to elongation of the upper lip. Note the excessive alar base width despite alar cinching.

PREOPERATIVE **POSTOPERATIVE**

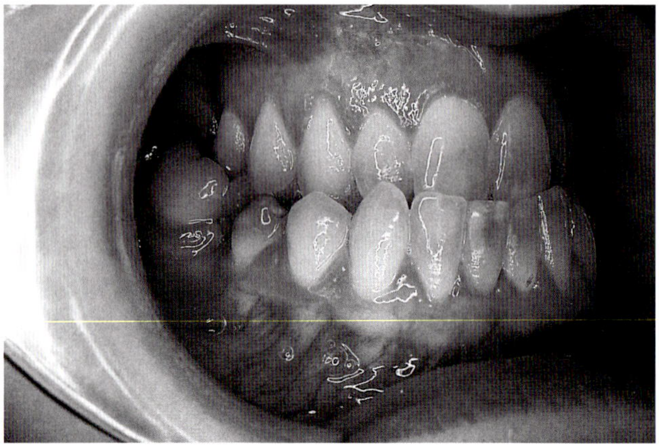

I

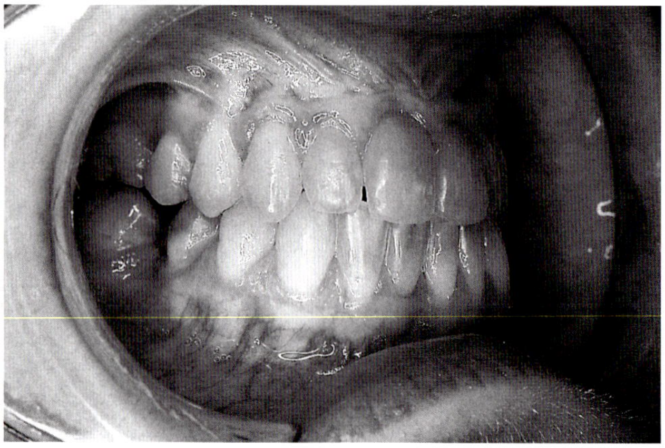

K

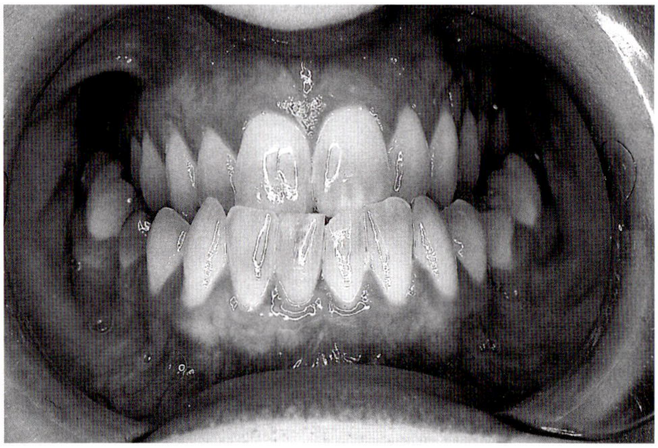

J

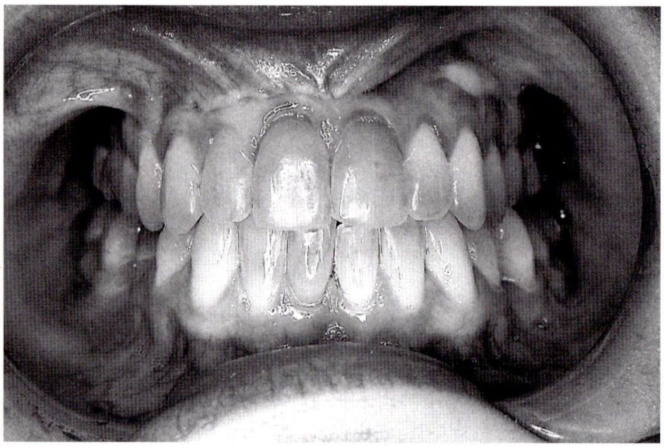

L

FIG. 7-3 (*Continued*) K & L: A six-year occlusal result. Orthodontist: Dr. Julian Freeman

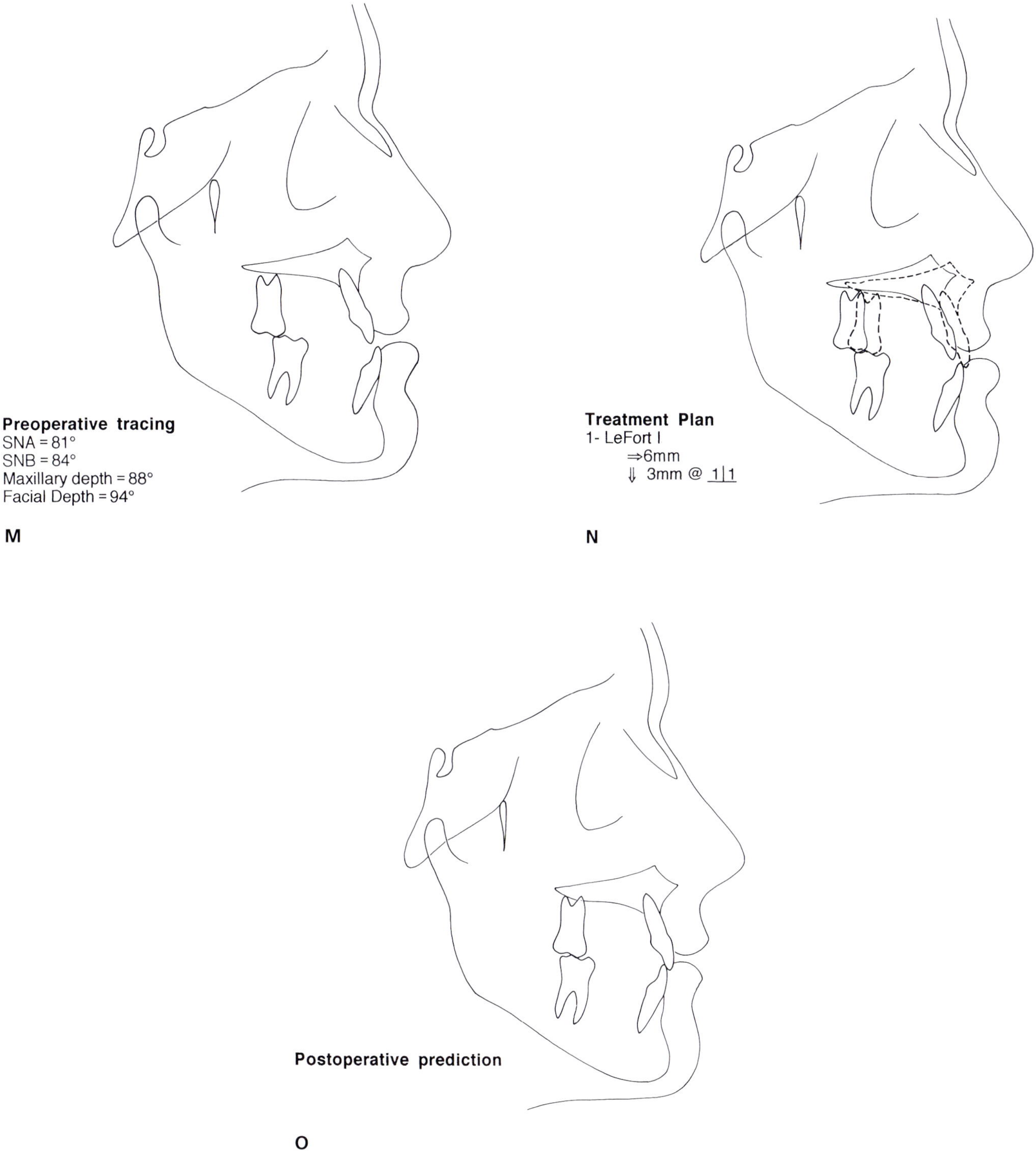

FIG. 7-3 (*Continued*) M: The cephalometric diagnosis is mandibular prognathism.

Case Study 7-4

This patient exhibits clinical maxillary deficiency in that there is significant infraorbital deficiency with scleral show and a lack of support to the soft tissues of the paranasal region and upper lip. There is a tapering of the lower face secondary to a vertically elongated chin, a modest dorsal nasal hump, and good soft tissue support throughout the lower face and neck. There is a Class III malocclusion. Cephalometric analysis confirms maxillary deficiency.

The patient is an excellent candidate for maxillary advancement surgery because the infraorbital and midface morphology represent the most severe aesthetic deficiencies. The surgical correction involved maxillary advancement at the Le Fort I level of 7 mm; hydroxyapatite augmentation of the infraorbital region; vertical reduction and advancement genioplasty of 4 mm and 4 mm, respectively; and a secondary rhinoplasty.

Postoperatively, there is a significant softening of the appearance in the frontal view due to the elimination of the scleral show, a fuller midface, and less tapering of the lower face. The profile is now anteriorly divergent secondary to the advancement component of the genioplasty, with excellent overall facial convexity and good labial relationships. There is excellent support of the soft tissues throughout the face.

This patient illustrates the effectiveness of Le Fort I surgery in improving ocular-adnexal relations and the applicability of advancement genioplasty provided the labiomental fold is shallow and the lower lip is well in advance of pogonion.

PREOPERATIVE | **POSTOPERATIVE**

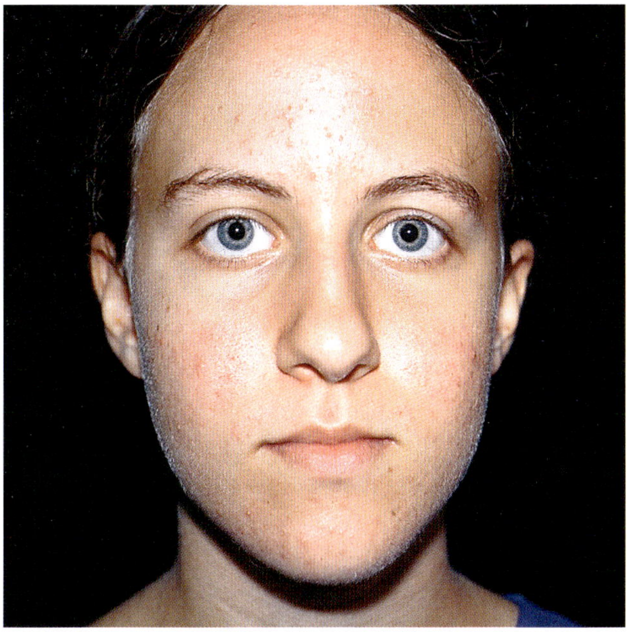

A

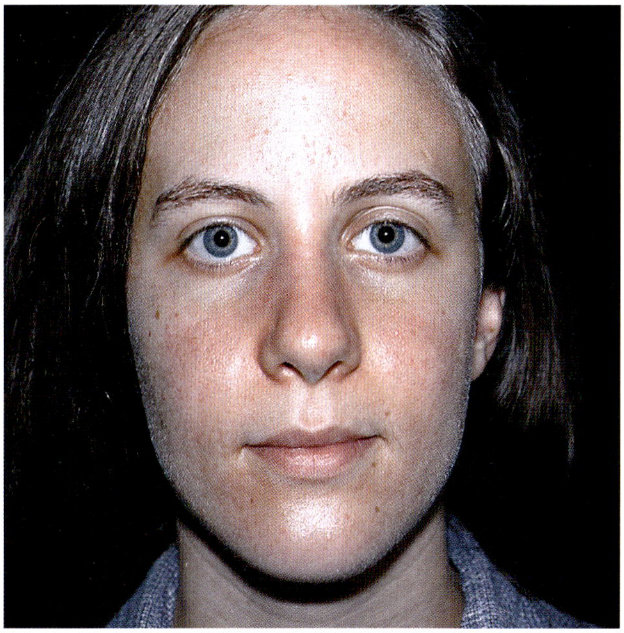

C

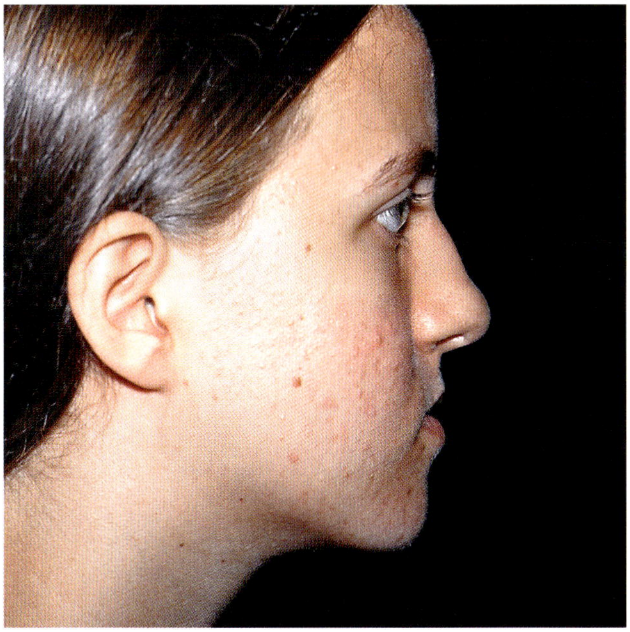

B

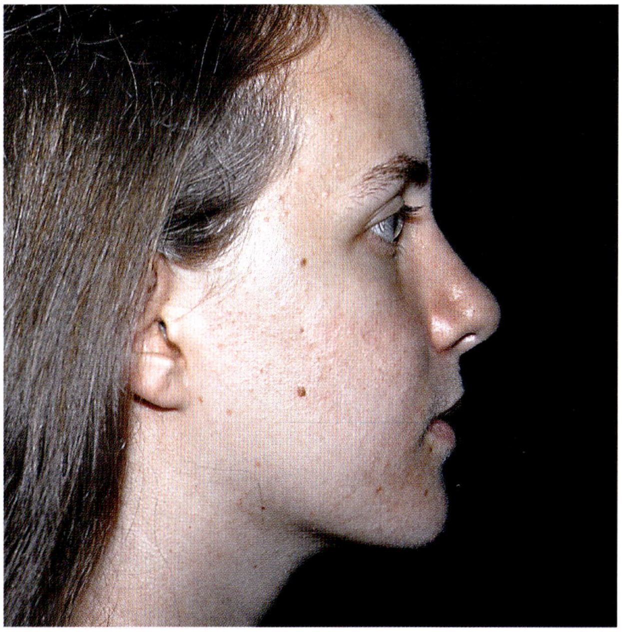

D

CASE STUDY 7-4 A & B: A patient with maxillary deficiency. Note the parasagittal flattening extending into the infraorbital region and the long, tapered chin. C & D: The postoperative appearance is characterized by enhanced skeletal support of the midface soft tissues and anterior divergence with overall convexity.

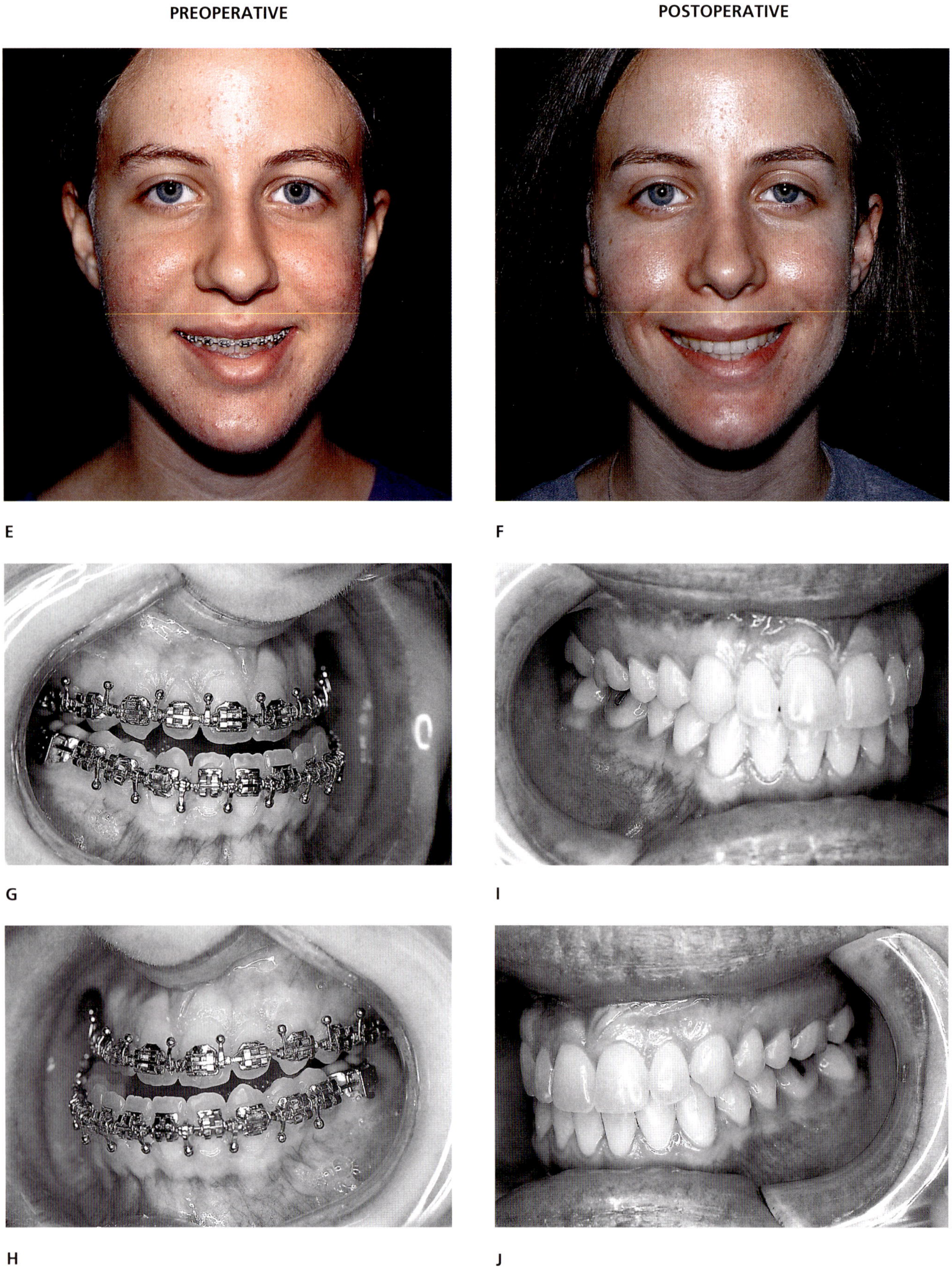

FIG. 7-4 (*Continued*) Orthodontist: Dr. Richard Drummond

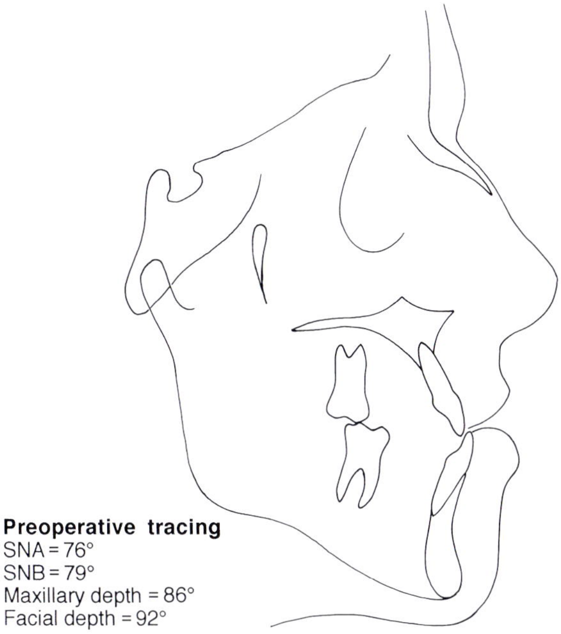

K

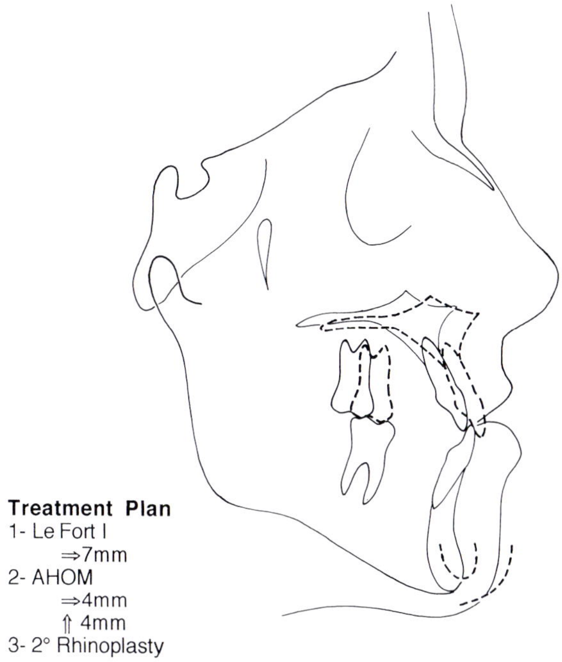

L

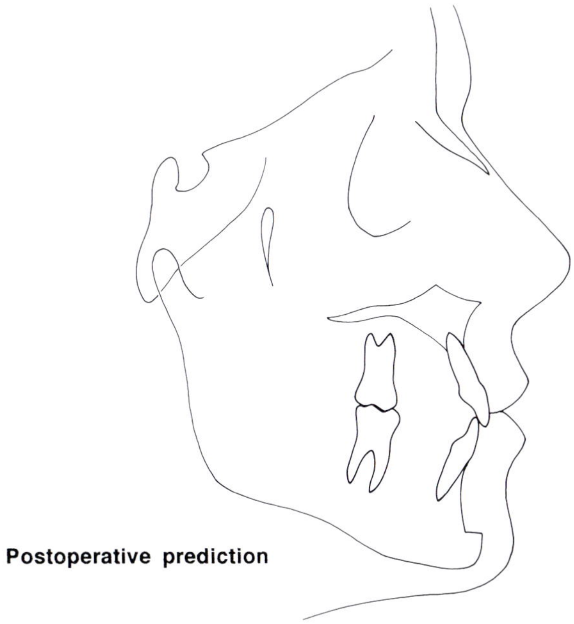

M

FIG. 7-4 *(Continued)* K: The skeletal diagnosis is maxillary deficiency.

Case Study 7-5

This patient also has sagittal deficiency of the midface. This is manifested by lack of support of the upper lip with its secondary thinning, minimal vermillion exposure, paranasal flatness, obtuse nasolabial angle, and an overall flat profile. There is a lack of incisal exposure with the lips at rest. There is a Class III malocclusion. There is a subtle submental fullness.

The patient is a perfect candidate for maxillary advancement surgery with appropriate midface aesthetics, appropriate nasal aesthetics, and a rather modest sagittal discrepancy between the maxilla and the mandible. Surgical correction required a 6-mm maxillary advancement with a 4-mm maxillary lengthening. In addition, submental liposuctioning was performed. There is slight downward and backward rotation of the mandible due to maxillary lengthening.

Postoperatively, there is greater fullness of the vermillion and improved upper lip support. The paranasal deficiency has been entirely eliminated, and the nasolabial angle has become more acute. There is the visual illusion of decreased nasal projection due to enhanced paranasal and upper lip support. There is now approximately 2 mm of incisal exposure with the lips at rest. The cervical mandibular angle has been improved due to liposuctioning.

PREOPERATIVE

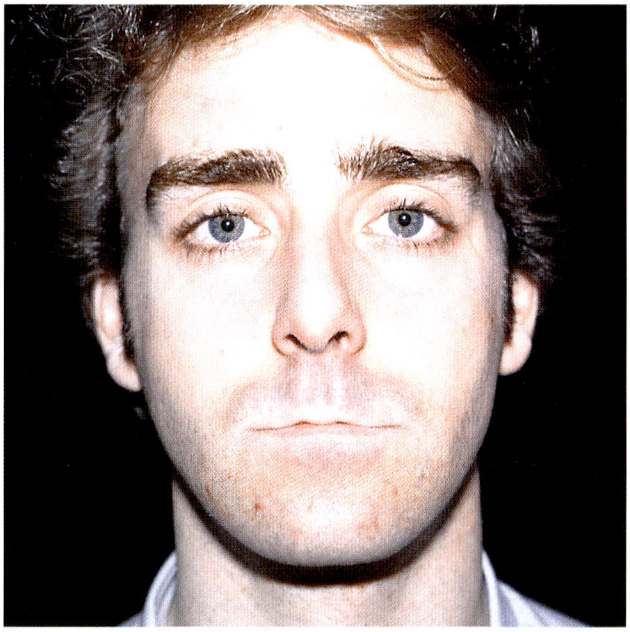

A

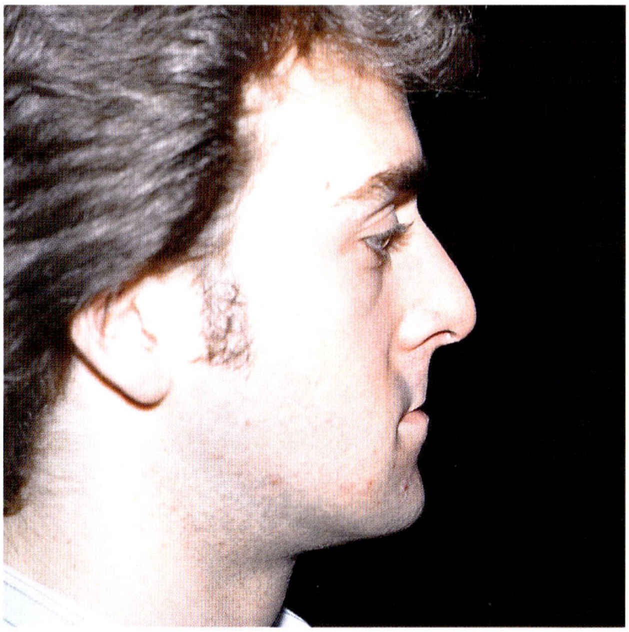

B

POSTOPERATIVE

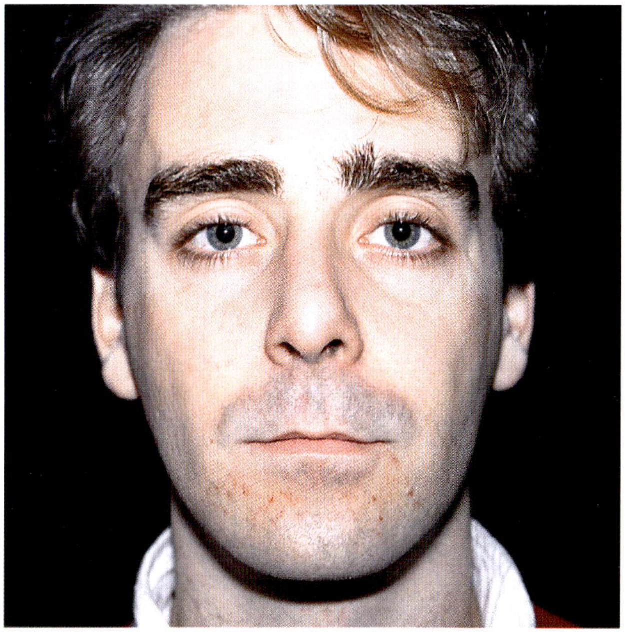

C

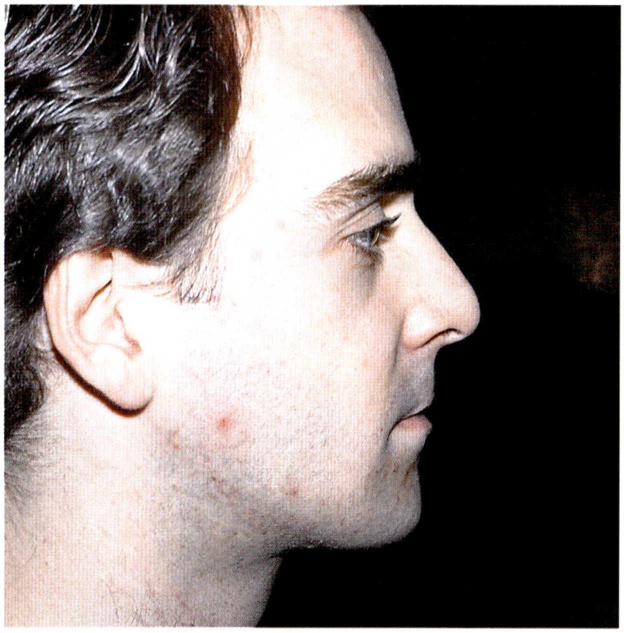

D

CASE STUDY 7-5 A & B: A patient with maxillary deficiency treated with maxillary advancement and lengthening.

PREOPERATIVE | **POSTOPERATIVE**

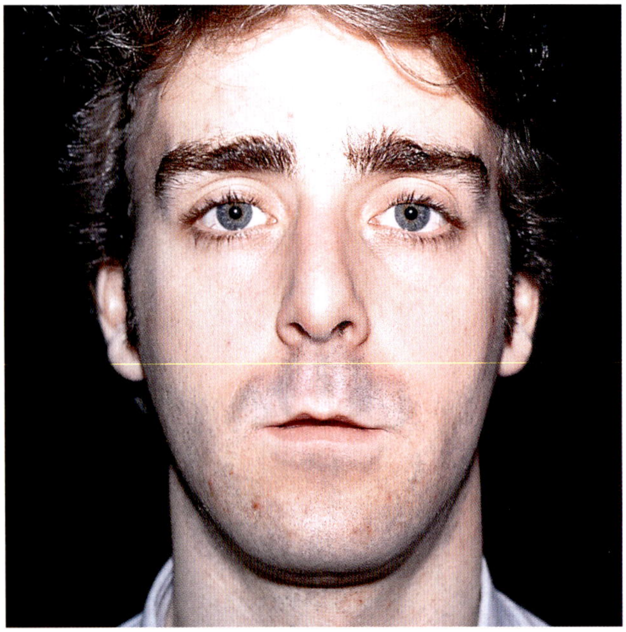

E

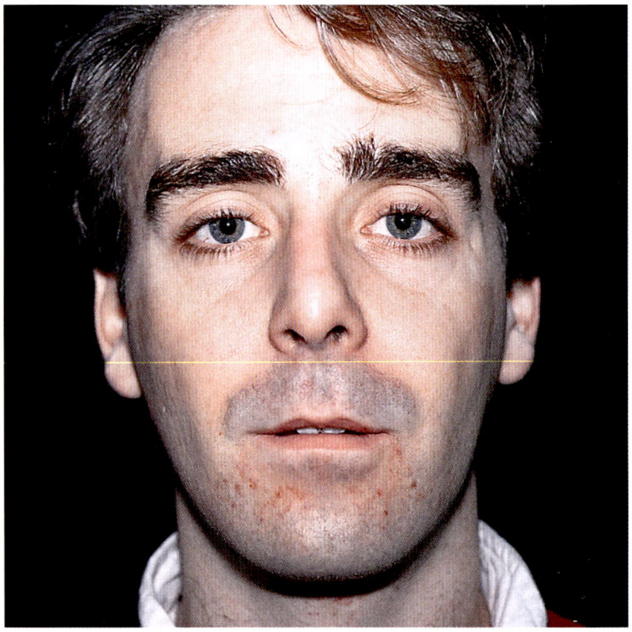

F

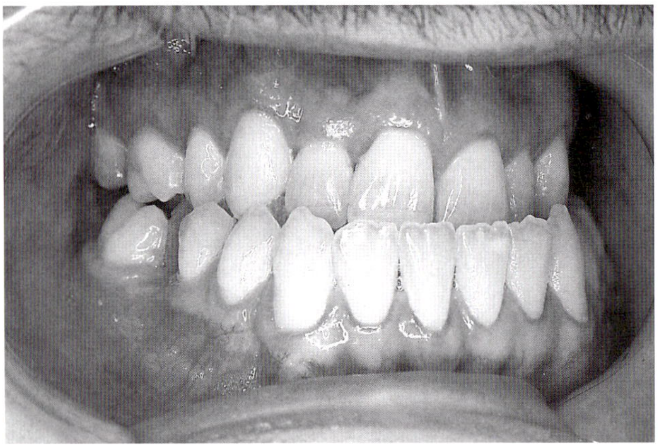

G

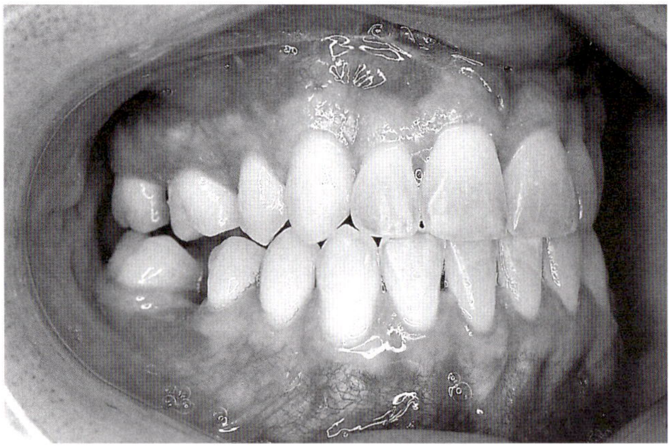

I

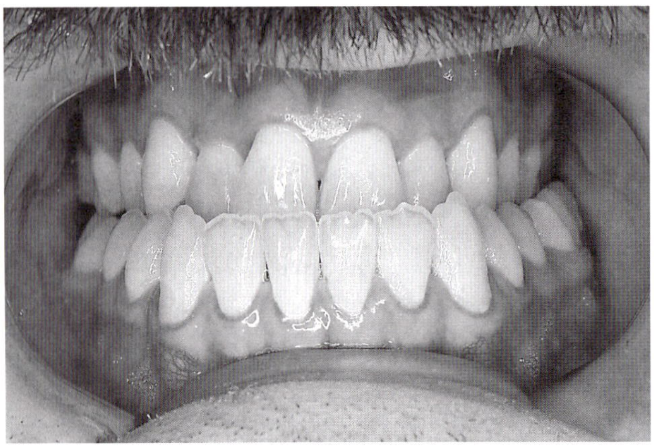

H

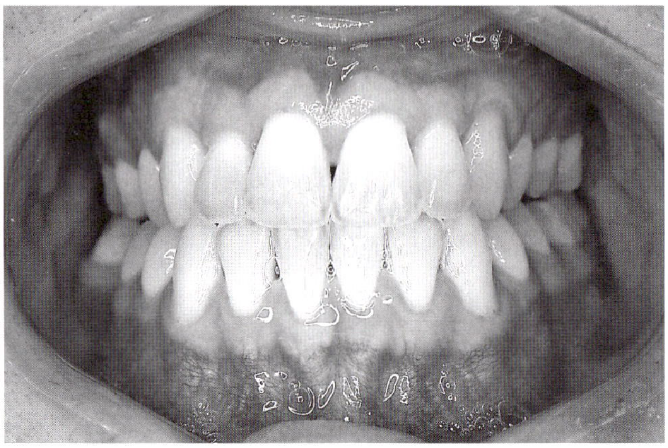

J

Fig. 7-5 (*Continued*) F: Notice enhanced upper lip support, fuller vermillion and increased incisal exposure. Orthodontist: Dr. William Roberts

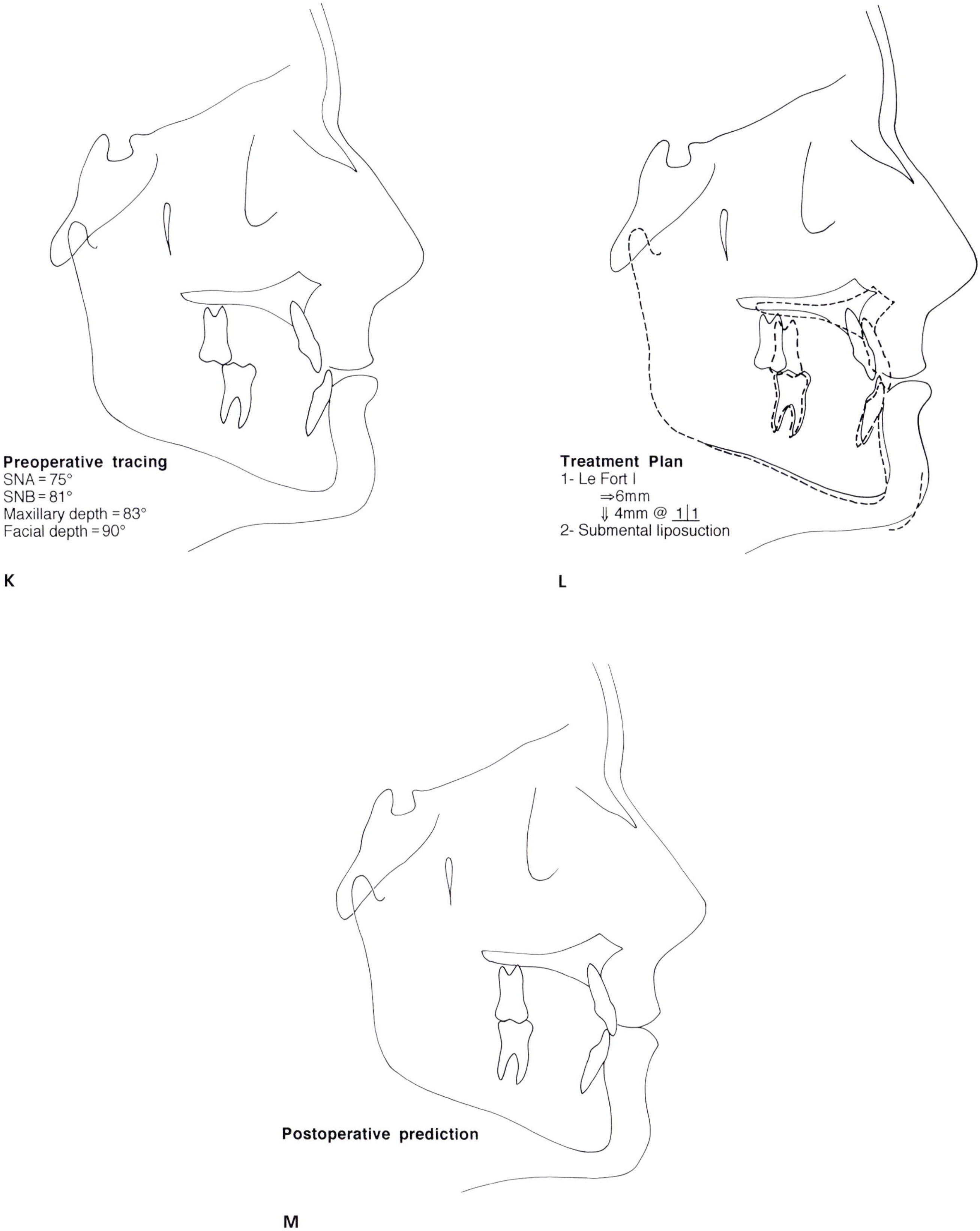

FIG. 7-5 *(Continued)*

Case Study 7-6

This patient obviously has mandibular prognathism. Cephalometric analysis confirms this and suggests the maxilla is in normal position. In the frontal view, there is good vertical proportionality and facial symmetry. The soft tissues are adequately supported. In profile, there is marked anterior divergence with midsagittal and parasagittal flattening. The lower lip is prominent relative to the upper lip and chin. Examination of the occlusion as well as the lateral cephalometric radiograph demonstrates a rather severe sagittal discrepancy between the maxilla and the mandible of approximately 12 mm. Single-jaw surgery, regardless of which jaw is repositioned, would achieve a less than adequate result because of the magnitude of movement required. Therefore, double-jaw surgery is indicated. Despite the skeletal diagnosis, the surgical correction will rely primarily on skeletal expansion in an effort to preserve soft tissue support. The maxilla is moved anteriorly 8 mm, and the mandible is set back 4 mm. In addition, the chin was advanced 4 mm so that there was no overall sacrifice in the sagittal projection of the lower face.

Postoperatively, there is a marked improvement in the sagittal relations seen in both the frontal and profile views. The profile is still markedly divergent anteriorly. There is now overall facial convexity both in the sagittal and parasagittal planes. The nose appears to project less due to the maxillary advancement. However, given the ethnic background of the patient, this does not detract from the aesthetic result. There is maintenance of excellent soft tissue support.

PREOPERATIVE POSTOPERATIVE

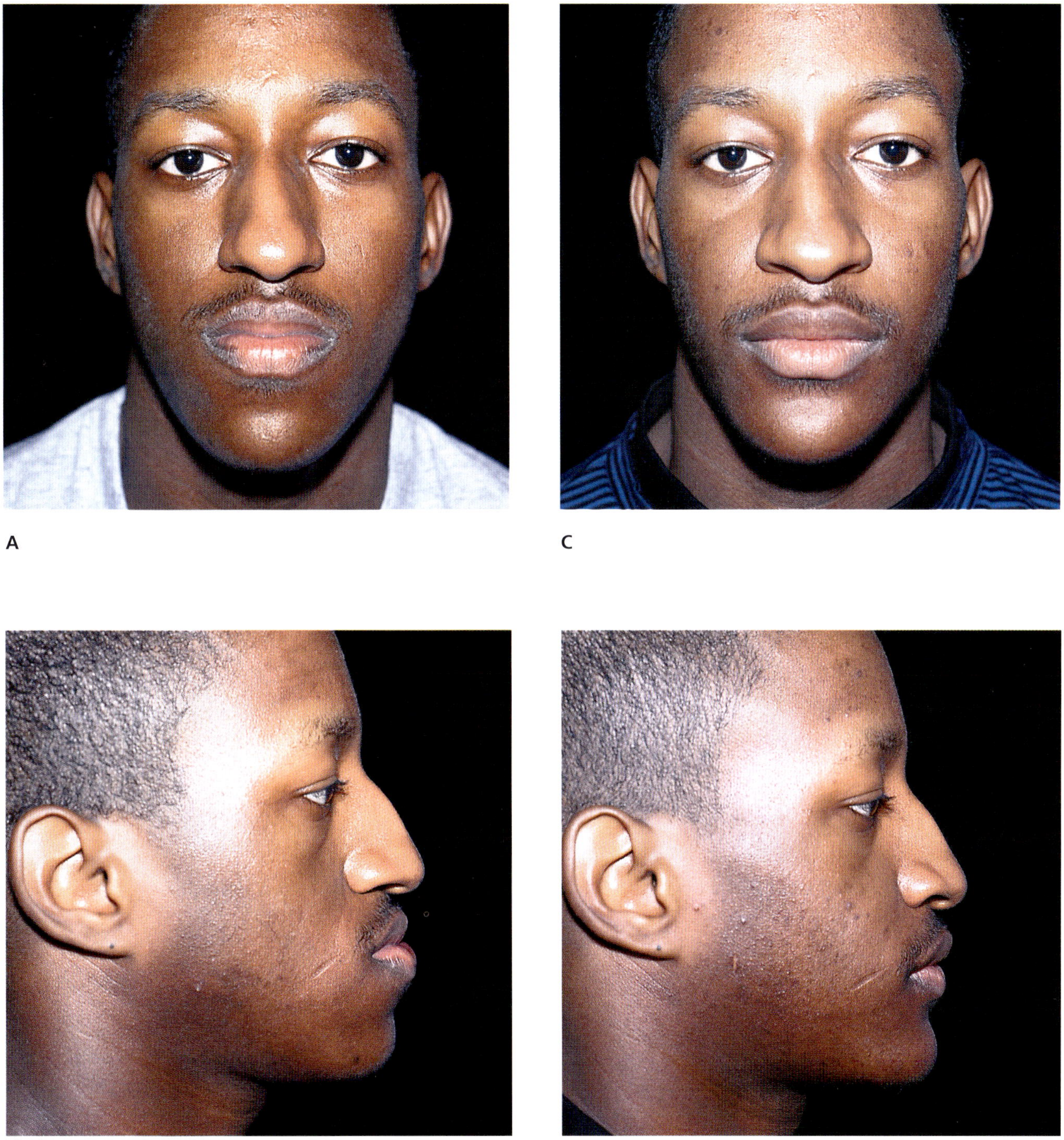

CASE STUDY 7-6 A & B: A patient with 12-mm Class III discrepancy due to mandibular prognathism treated with 8-mm maxillary advancement, 4-mm mandibular setback, and 4-mm advancement genioplasty. C & D: There is no loss of skeletal support to the soft tissues.

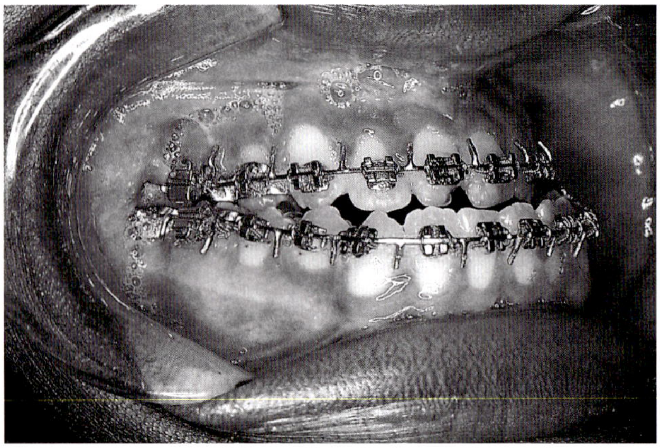

E

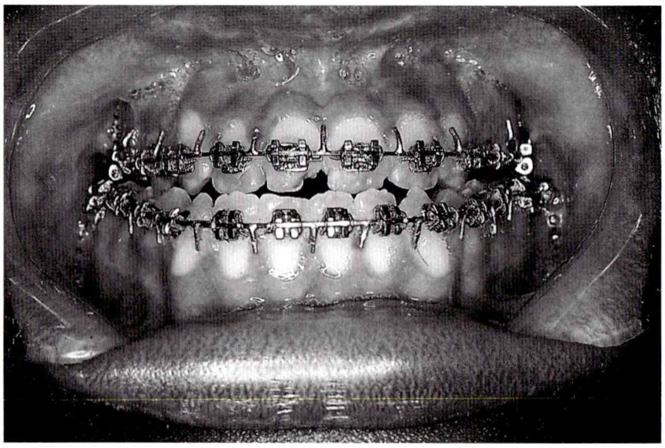

F

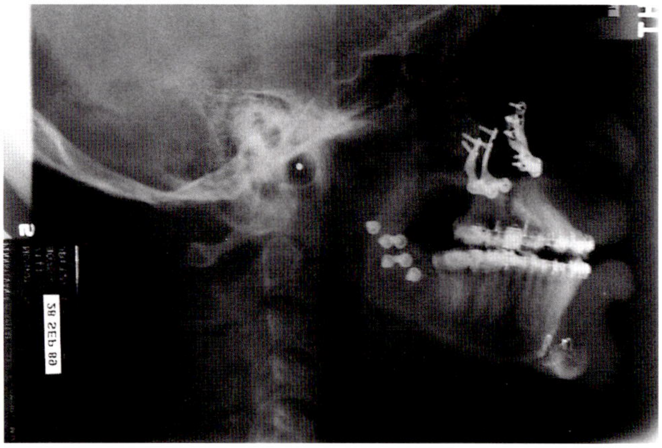

G

FIG. 7-6 (*Continued*) E & F: Preoperative occlusion. G: Postoperative lateral cephalogram demonstrating surgical correction. (No postoperative occlusal views available.) Orthodontist: Dr. Joseph Damone

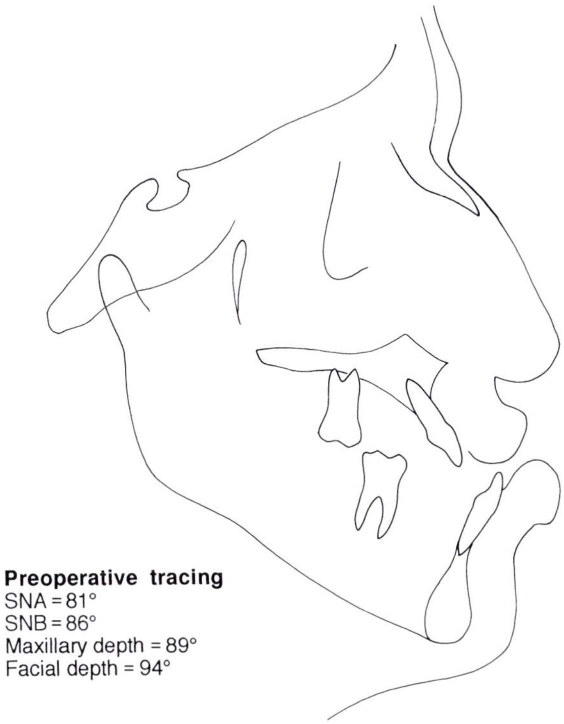

H

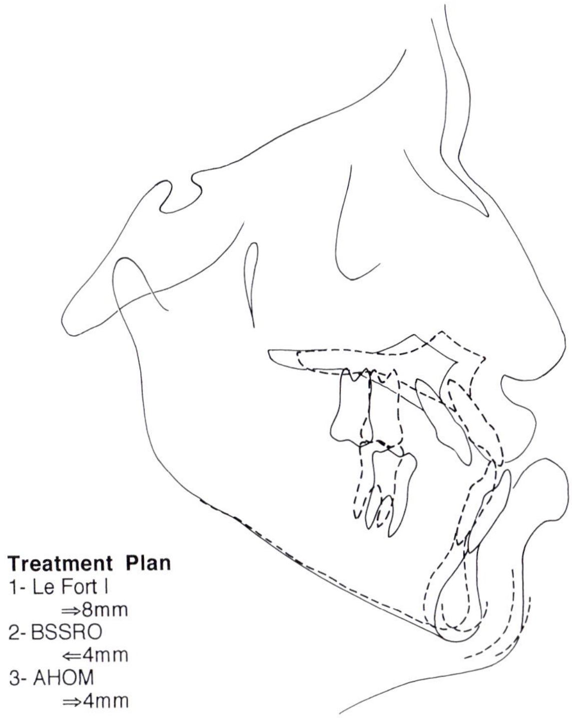

I

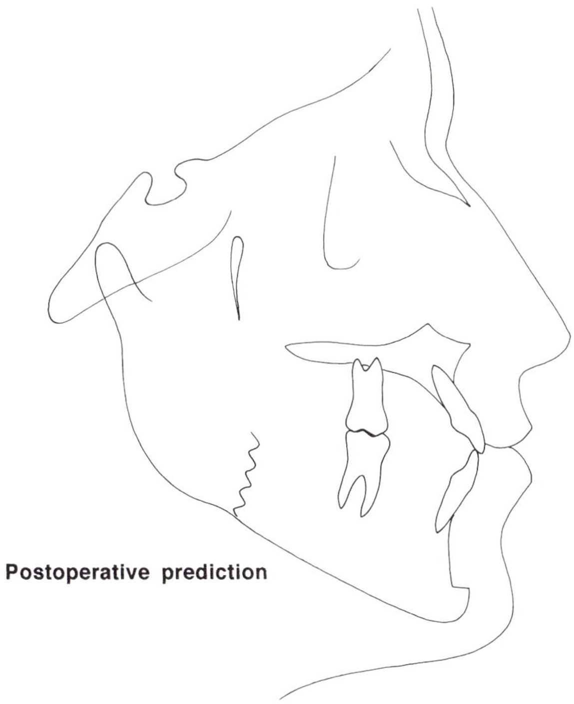

J

FIG. 7-6 (*Continued*) I: Note that there is no loss of projection of the soft tissue pogonion. Anterior divergence has been preserved and facial convexity has been achieved. From: Rosen, H.: Maxillary advancement for mandibular prognathism: Indications and rationale. *Plast Reconstr Surg* 87:823, 1991. Reprinted with permission.

Case Study 7-7

This patient underwent double-jaw surgery to correct mandibular prognathism because of the asymmetric component of the deformity. The asymmetry is clearly demonstrated in the frontal view with the chin point deviated to the left of the craniofacial midline. The profile view reveals good projection of the midfacial structures in the midsagittal plane, with an anteriorly divergent profile clearly due to an enlarged mandible, and parasagittal flattening. This clinical impression is confirmed cephalometrically; that is, the maxilla is normally positioned and the mandible is large. The occlusion is Class III, and the mandibular midline is deviated to the left by the width of a lower incisor.

Double-jaw surgery was chosen because of the patient's desire to maintain a strong lower jaw and the surgeon's desire to maintain soft tissue support. Accordingly, two-piece Le Fort I surgery was carried out to advance the maxilla 6 mm and to widen it 4 mm. Bilateral sagittal split ramus osteotomy rotated the mandible to the right but with no posterior shift. In addition, a genioplasty was carried out to further shift the chin 3 mm to the right.

In the frontal view, postoperatively, excellent lower face symmetry has been achieved. However, there is some unattractive alar base flaring, which could be adequately corrected with secondary alar base excisions. The profile view is characterized by maintenance of excessive anterior divergence but with good overall facial convexity both sagittally and parasagittally. There has been maintenance of excellent soft tissue support throughout the lower face and neck and improvement in soft tissue support in the midfacial region, particularly in the area of the nasolabial fold. The occlusal result is a 5-year follow-up.

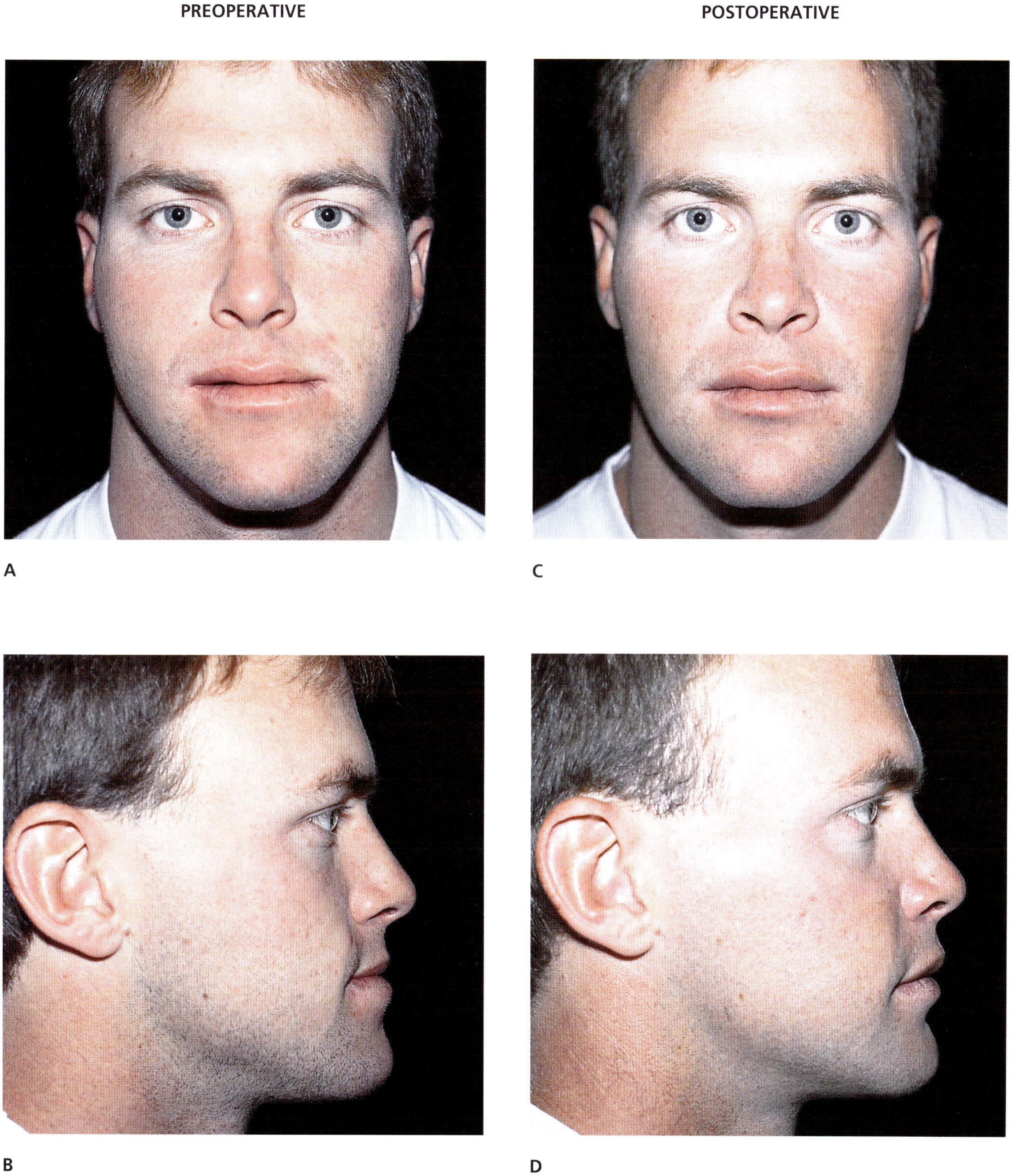

CASE STUDY 7-7 A & B: A patient with asymmetric mandibular prognathism treated with double jaw surgery. There is slight parasagittal flattening and vertical facial excess. C & D: There is no reduction of anterior divergence and enhancement of facial convexity. The soft tissues remain well supported in the lower face. Note the reduction in facial height secondary to improved occlusal interdigitation.

PREOPERATIVE POSTOPERATIVE

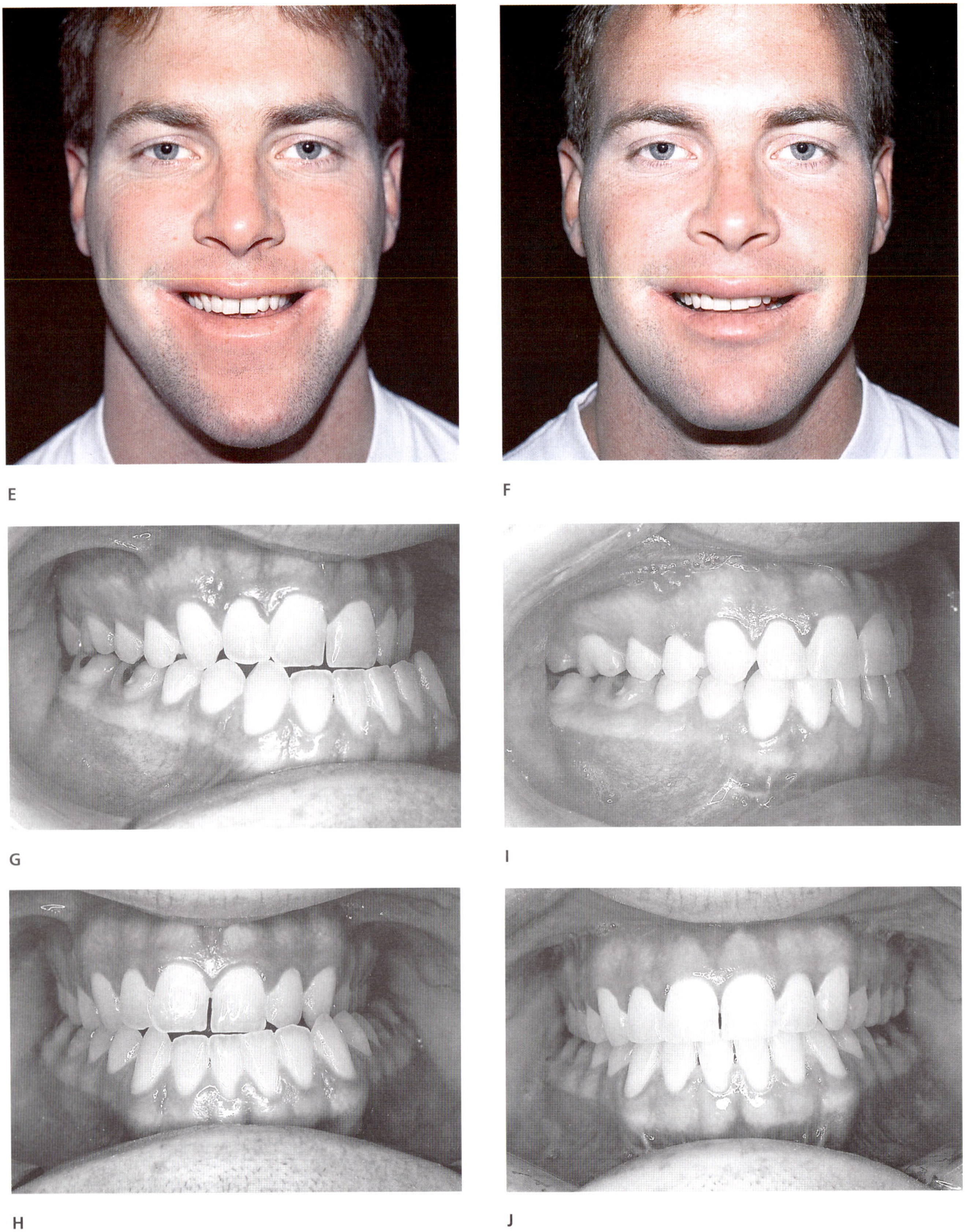

FIG. 7-7 *(Continued)* I & J: A five-year occlusal follow-up. From Rosen, H: Maxillary advancement for mandibular prognathism: Indications and rationale. *Plast Reconstr Surg* 87:823, 1991. Reprinted with permission.

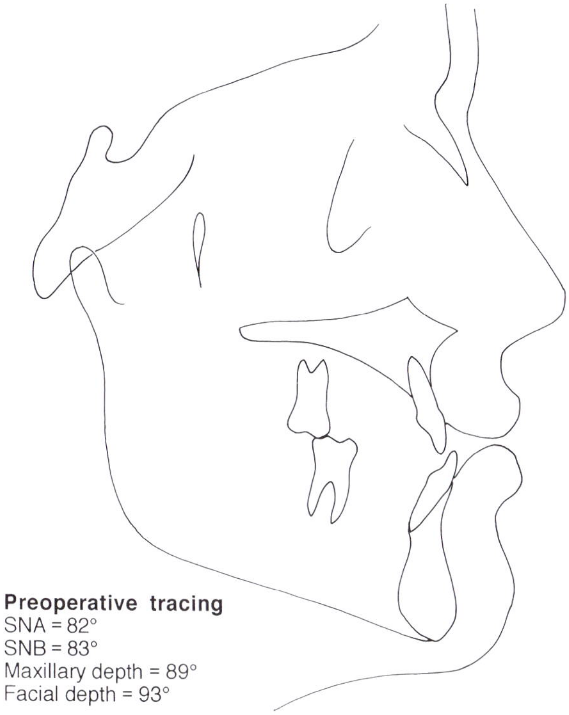

K

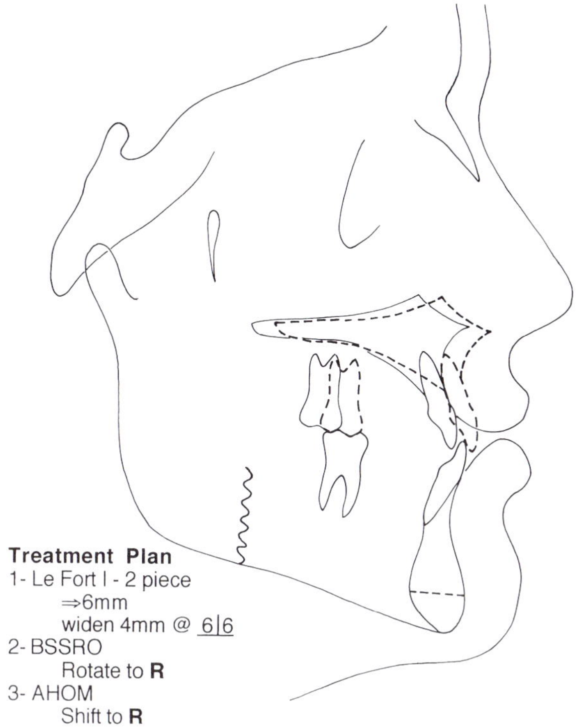

L

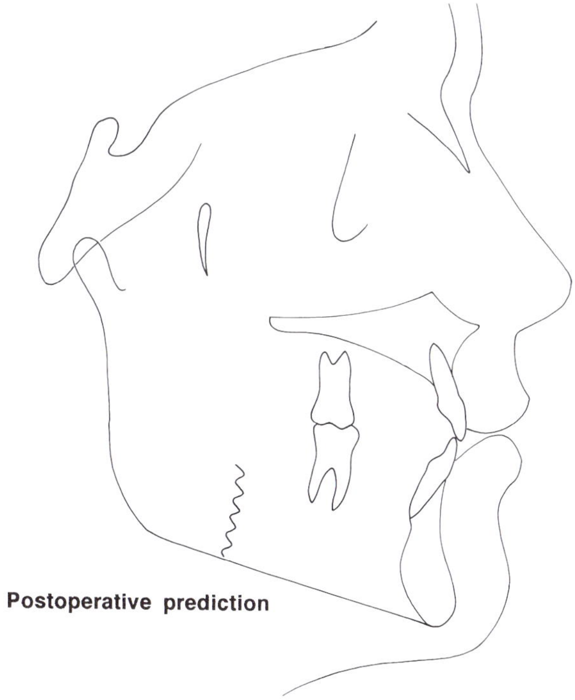

M

FIG. 7-7 *(Continued)* L. Note that there is no loss of skeletal projection of the lower face. Orthodontist: Dr. James Ackerman

Case Study 7-8

This patient also has significant lower face asymmetry due to asymmetric mandibular prognathism. Cephalometric analysis again confirms a large mandible and normal maxilla. There is definite flattening of the facial profile, and there is overall anterior facial divergence. There is significant lower face asymmetry with the chin being deviated to the left of the craniofacial midline. There is moderate submental fullness, and the soft tissues are rather thick. The occlusion is Class III with the mandibular midline deviated to the left by the width of a lower incisor.

Double-jaw surgery is indicated to preserve and enhance skeletal support of the soft tissues and simultaneously correct the facial asymmetry. The treatment plan involves advancing the maxilla and vertically elongating the maxilla, 6 mm and 3 mm, respectively; sagittal split ramus osteotomy to derotate the mandible to the right but with no posterior shift; and osseous genioplasty to shift the chin further to the right.

Postoperatively, excellent facial symmetry has been achieved. There is an increase in exposure of the four incisors. There is now overall facial convexity in the presence of an anteriorly divergent profile. The soft tissue support of the lower face has been preserved, and the soft tissue support of the midface has been enhanced. A much better aesthetic result could have been achieved if the submental area had been treated with liposuction. Secondary alar base excisions to narrow the alar base was declined by the patient.

PREOPERATIVE | **POSTOPERATIVE**

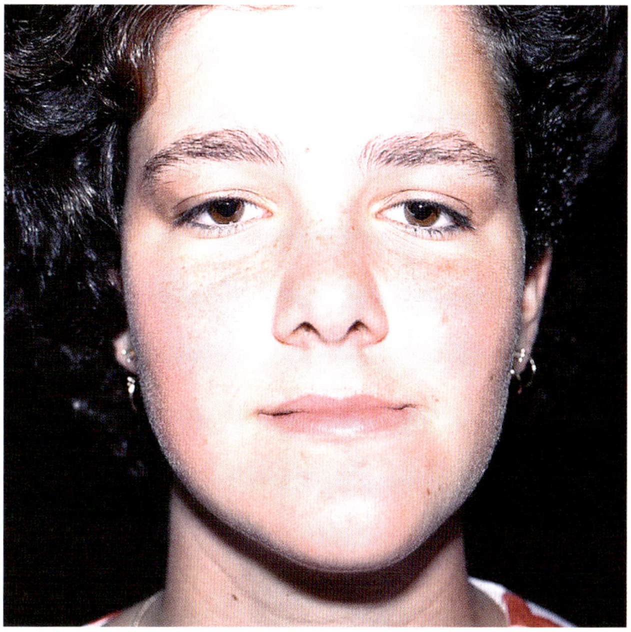

A

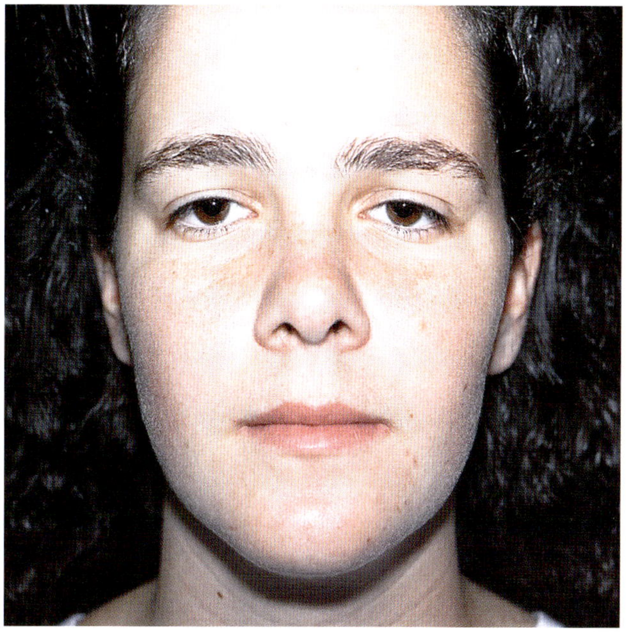

C

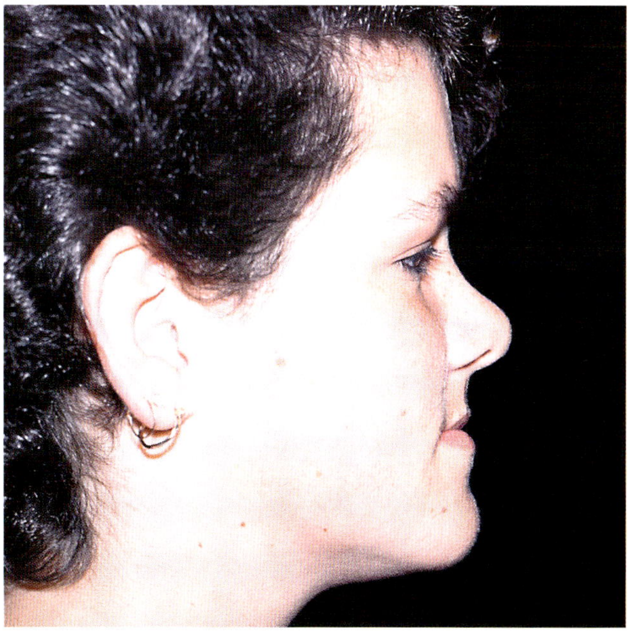

B

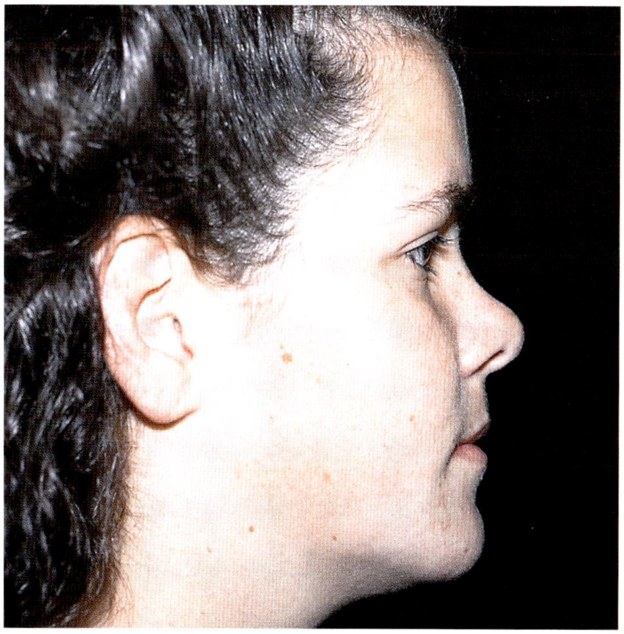

D

CASE STUDY 7-8 A & B: A patient with asymmetric mandibular prognathism treated with maxillary advancement to preserve soft tissue support in the lower face and mandibular rotation. C & D: There is no loss of skeletal volume in the lower face. Facial convexity has been enhanced.

PREOPERATIVE **POSTOPERATIVE**

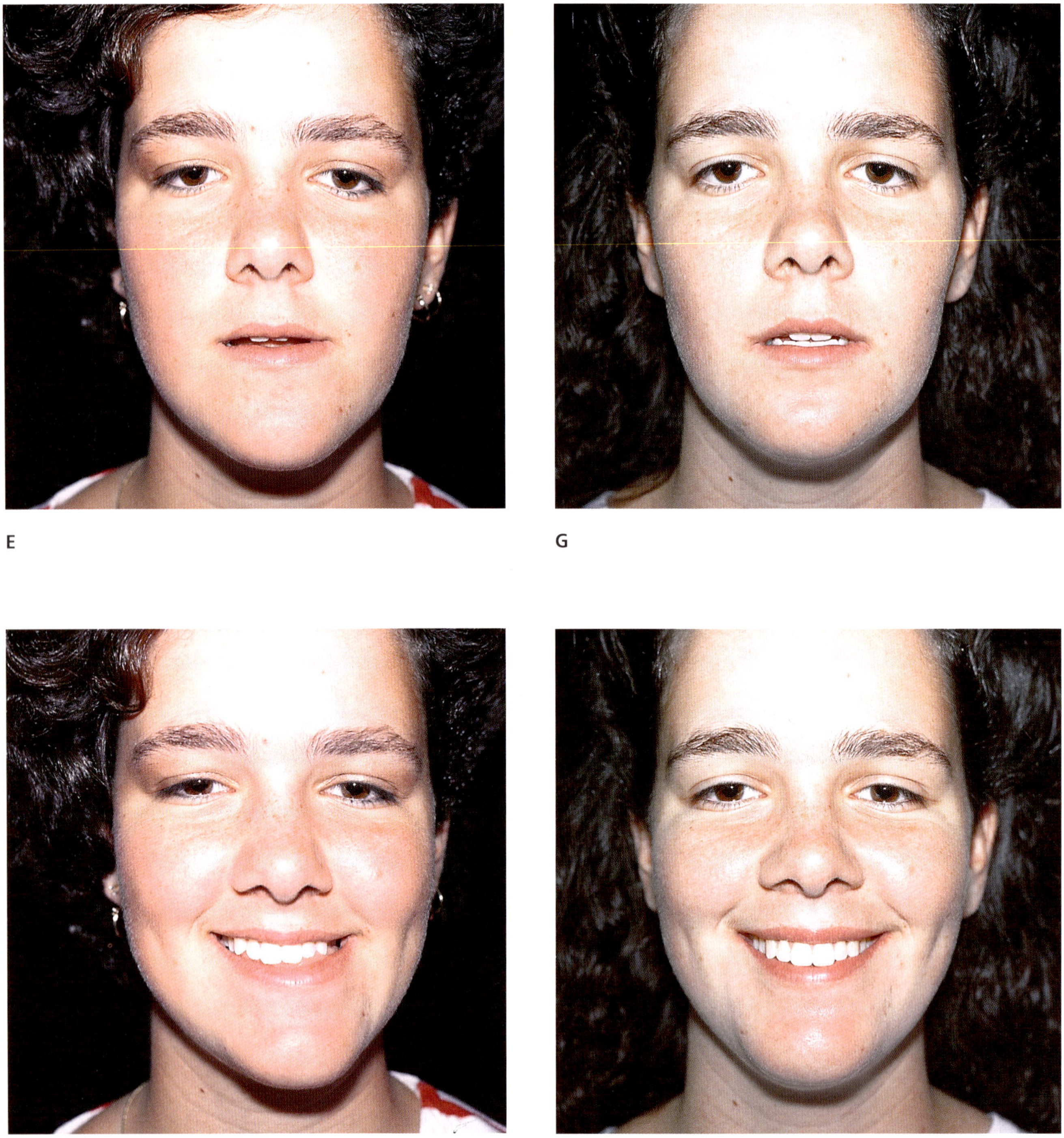

E G

F H

FIG. 7-8 (*Continued*) E & G: Notice increased maxillary incisal exposure due to maxillary lengthening. From Rosen, H.: Bimaxillary surgery. In M. Habal (Ed.): *Advances in Plastic and Reconstructive Surgery*. St. Louis: Mosby, Year Book, Chicago, 10:93, 1993. Reprinted with permission.

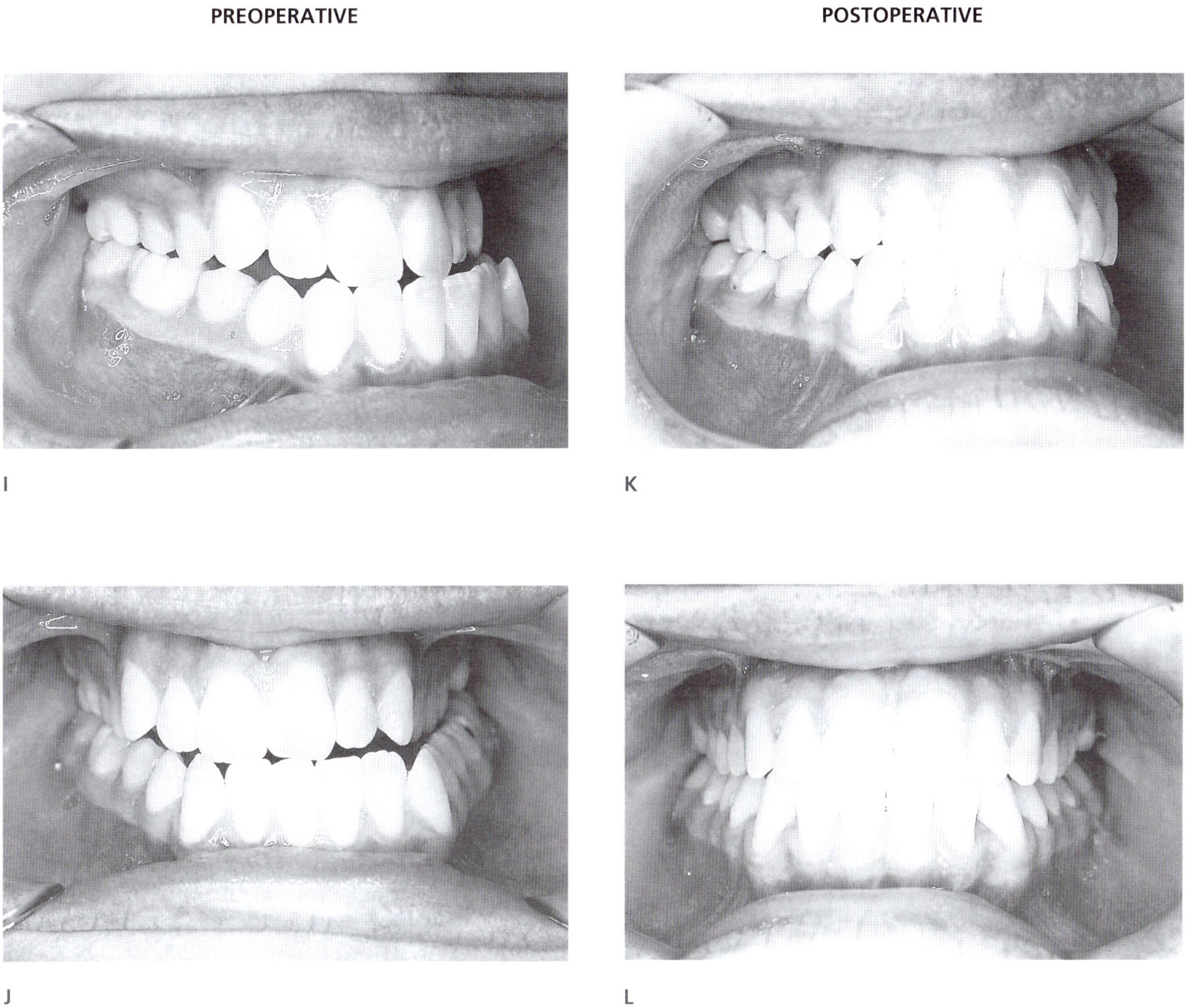

FIG. 7-8 (*Continued*) K & L: The occlusal result at three years. Orthodontist: Dr. James Ackerman

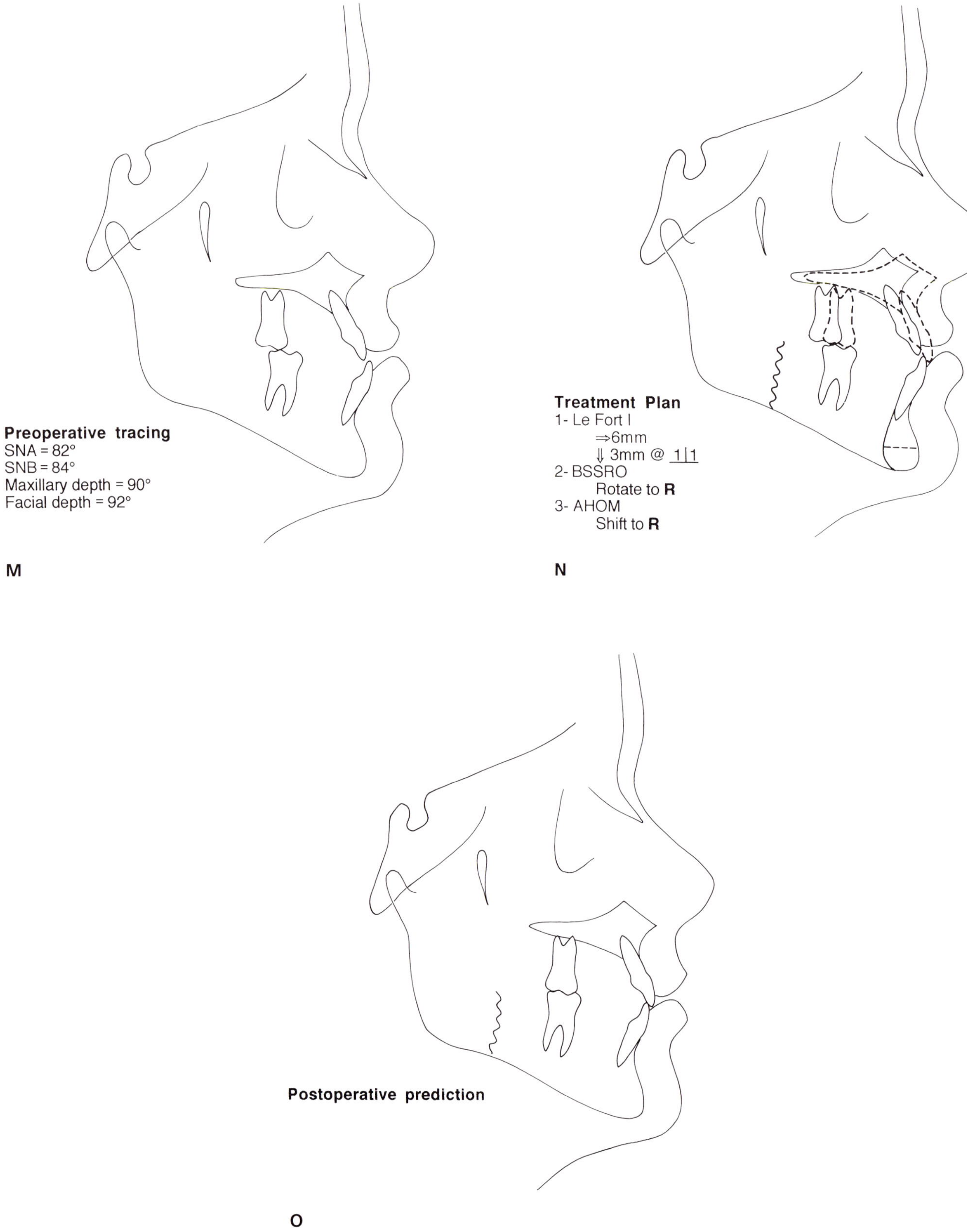

Fig. 7-8 (*Continued*) N: There is no planned posterior movement of pogonion.

Case Study 7-9

This patient demonstrates still another indication for double-jaw surgery, that is, the correction of mandibular prognathism when it coexists with a long face secondary to vertical maxillary excess. A double-jaw procedure is indicated in this patient because maxillary impaction surgery to correct the vertical maxillary excess causes the sagittal discrepancy between the maxilla and the mandible to significantly worsen as the mandible autorotates upward and forward.

This particular patient's facial morphology is characterized by a marked increase in the vertical dimension of the lower face secondary to vertical maxillary excess and vertical macrogenia. There is overall anterior facial divergence with rather flat sagittal and parasagittal planes. The labiomental fold is shallow, which further accentuates the impression of vertical facial excess. In addition, there is lower face asymmetry with the chin point deviated to the right. There is a Class III anterior open-bite malocclusion with the mandibular midline shifted to the right.

The surgical correction included a two-piece Le Fort I osteotomy to advance the maxilla 7 mm; a 5-mm and 3-mm posterior and anterior maxillary intrusion, respectively; and a 5-mm transverse maxillary expansion. Sagittal split ramus osteotomy will set the mandible back approximately 4 mm and rotate it to the left to restore midline symmetry. However, there is no loss of mandibular sagittal volume due to mandibular autorotation. Osseous genioplasty will shorten the chin 4 mm. This patient underwent secondary Weir excisions of the alar base because of excessive alar flaring secondary to the maxillary surgery.

Postoperatively, there is a dramatic reduction in the height of the lower face with complete elimination of the lip strain. However, there is still adequate exposure of the incisors in repose and a full clinical crown exposed with smiling. There is still obvious anterior divergence to the profile but now with good convexity in both the midsagittal and parasagittal regions, as well as good relationships between the lips and adequate projection of the nose. The aesthetic results could have been further improved with augmentation of the nasal spine to open the nasolabial angle.

PREOPERATIVE **POSTOPERATIVE**

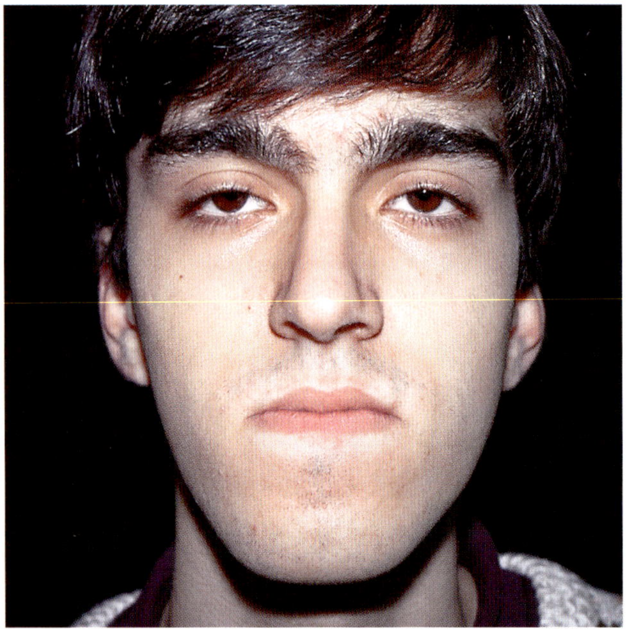

A

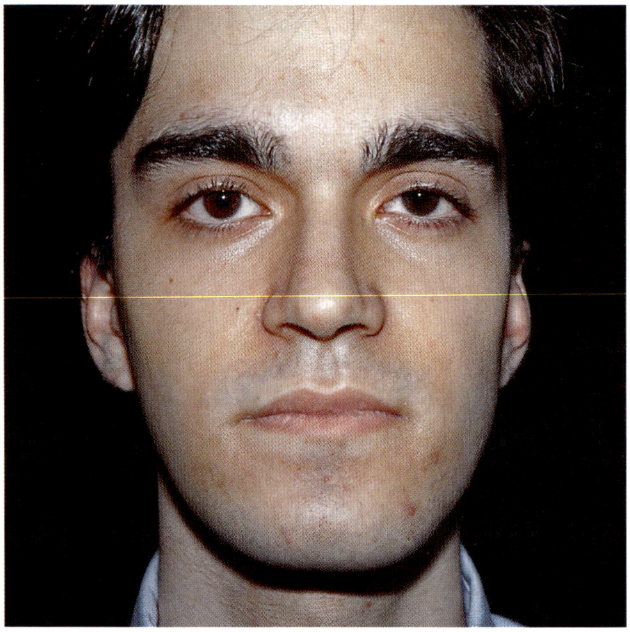

C

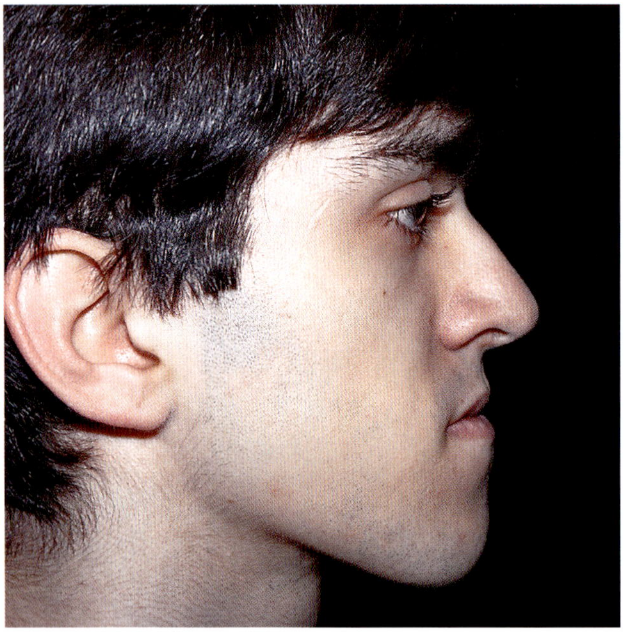

B

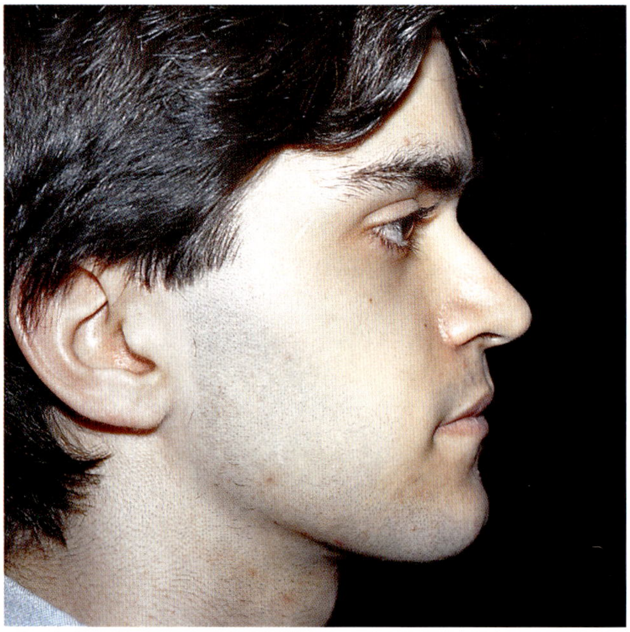

D

CASE STUDY 7-9 A & B: A long-face, prognathic patient with a Class III anterior open bite treated with double jaw surgery and genioplasty. C & D: Note reduction in facial height, overall facial convexity, and anterior divergent profile.

PREOPERATIVE POSTOPERATIVE

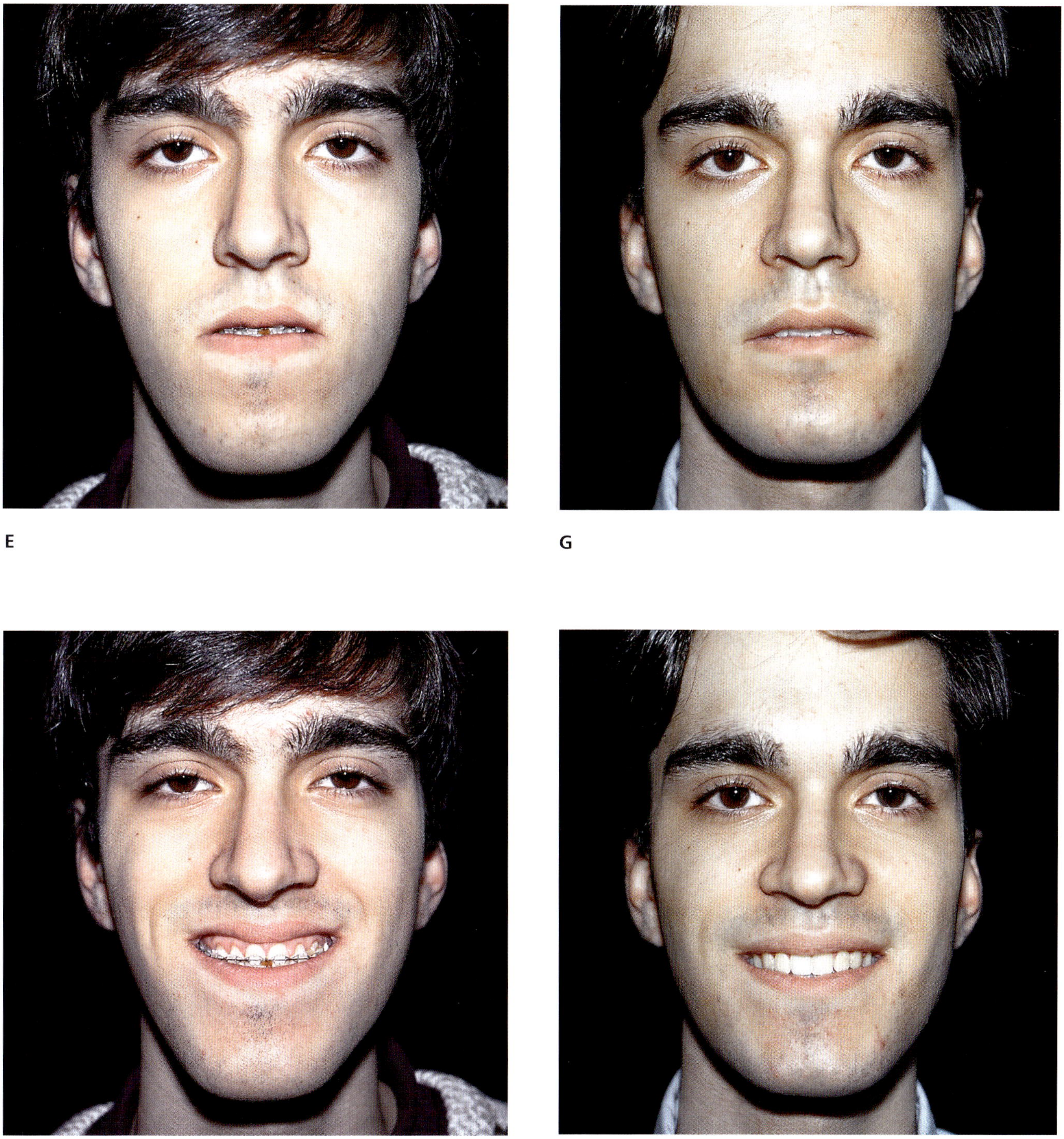

E G

F H

FIG. 7-9 (*Continued*) G & H: Note squaring off of the lower jaw due to vertical reduction of chin. In this instance, it is excessive.

PREOPERATIVE **POSTOPERATIVE**

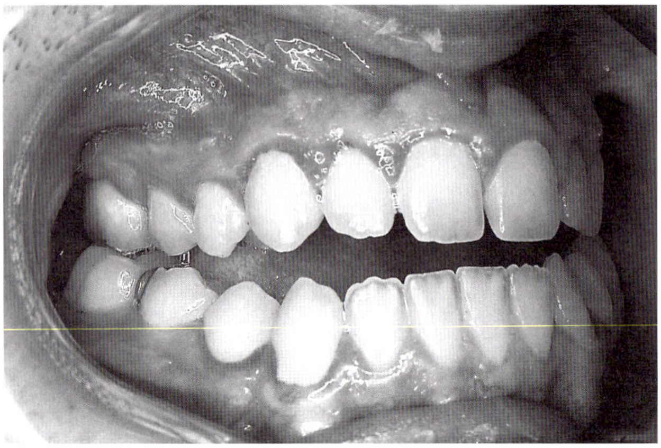

I

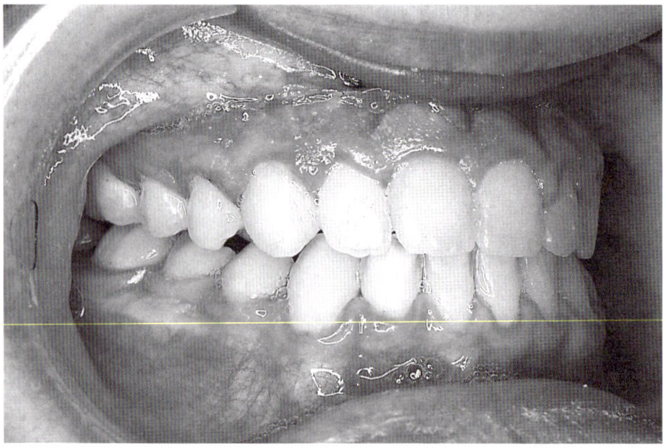

K

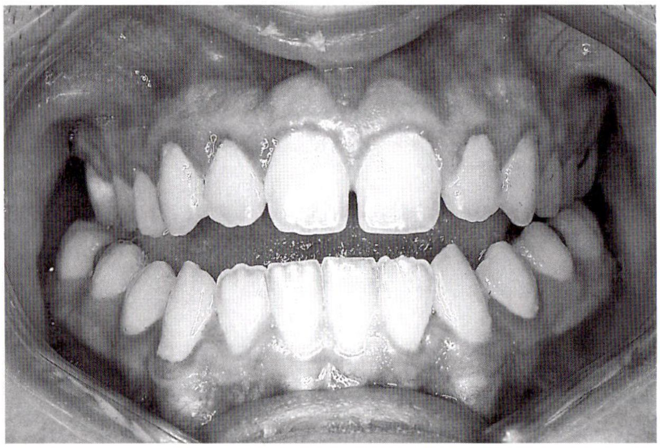

J

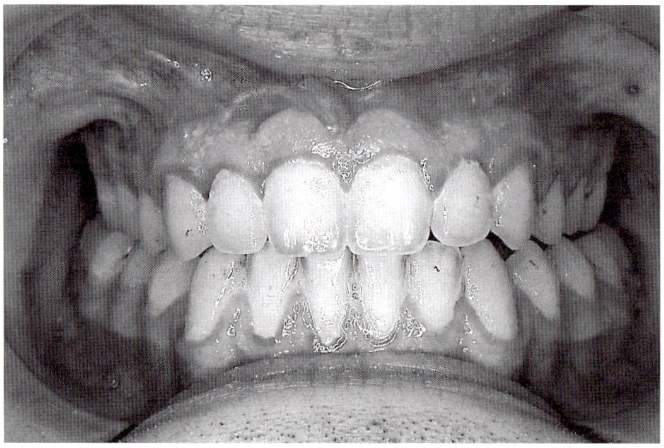

L

FIG. 7-9 (*Continued*) I & J: Class III anterior open bite with midline discrepancy and transverse deficiency. K & L: There has been relapse in the transverse plane. Note left-sided cross-bite. Occlusal result at five years. Orthodontist: Dr. James Ackerman

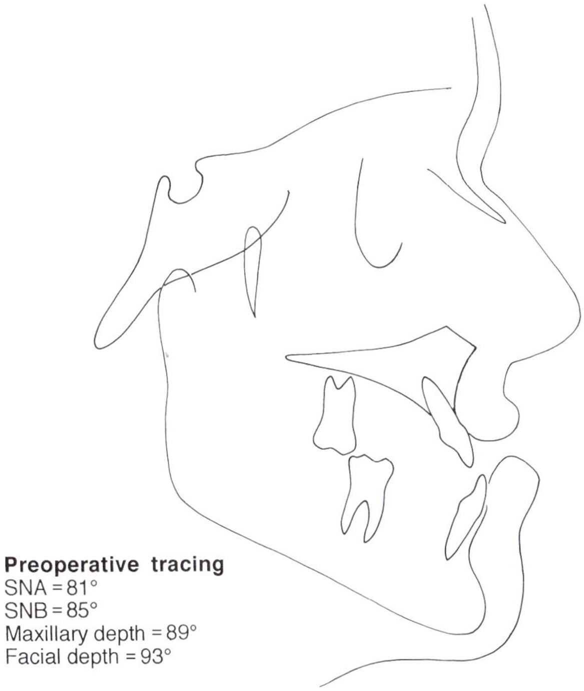

M

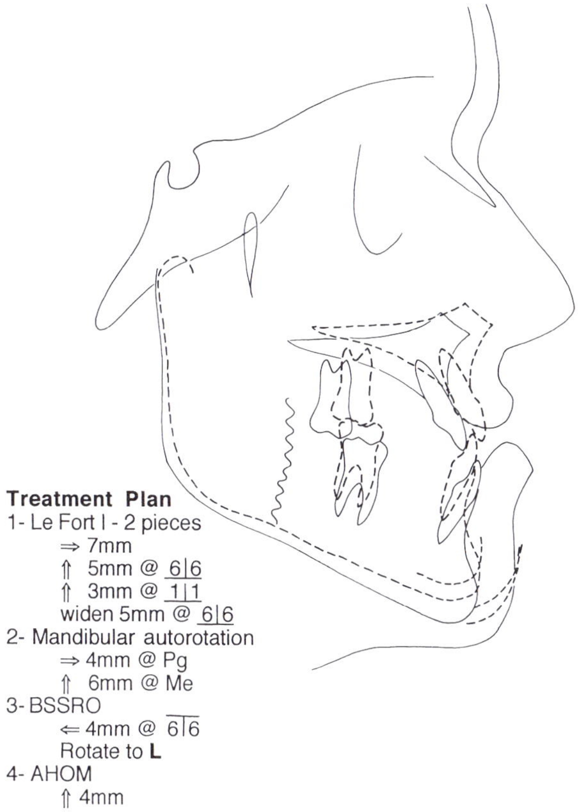

N

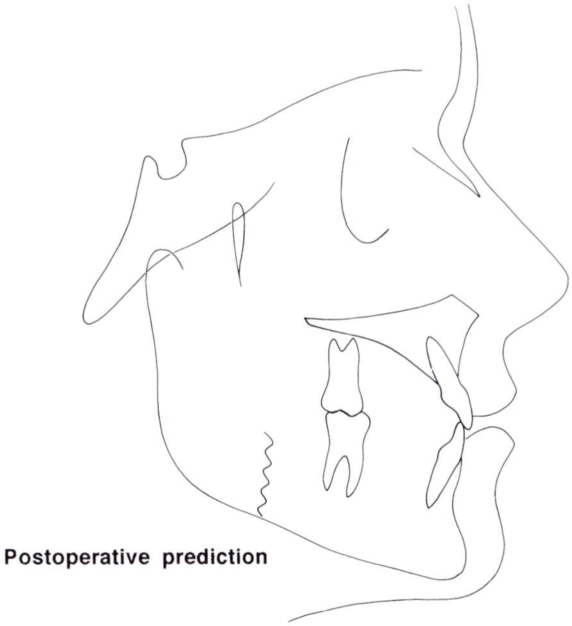

O

FIG. 7-9 (*Continued*) N: There is no planned reduction in the sagittal position of pogonion due to mandibular autorotation. O: The profile is anteriorly divergent and convex.

Case Study 7-10

The last and least common indication for double-jaw surgery in the individual with a Class III deformity is when a large maxillary advancement is aesthetically indicated, but only minimal to moderate advancement is possible because of the occlusal relationships. In this situation, mandibular advancement, in addition to maxillary advancement, may be undertaken so that adequate projection of the mid- and lower face can be achieved and, as a result, can provide adequate support of the soft tissues. This patient illustrates this treatment planning principle.

In the frontal view, there is normal vertical proportionality and good facial symmetry. There is inadequate facial mask support with an absence of angularity and definition to the soft tissues. This is manifested in a lack of zygomatic highlighting and the presence of jowl formation. The lateral view demonstrates a profile characterized by neutral divergence (pogonion falls on a vertical perpendicular dropped from glabella), modest midsagittal convexity, and parasagittal flattening. There is submental fullness. Following presurgical orthodontics, there is a Class III occlusal abnormality that will permit only 4 mm of maxillary advancement to achieve a Class I occlusion. This is confirmed in the preoperative tracing.

A 4-mm maxillary advancement would provide insufficient expansion of skeletal volume to achieve adequate support of the soft tissues. Accordingly, double-jaw surgery is indicated to advance both the maxilla and the mandible despite pogonion being in normal position. An 8-mm maxillary advancement and a 4-mm mandibular advancement are utilized. In addition, a 5-mm advancement genioplasty is planned. Despite normal preexisting vertical facial proportions, the maxilla will be lengthened 3 mm as it is advanced, and the chin will be lengthened 5 mm as it is advanced.

Postoperatively, there is quantitative excess in the height of the lower face, but it does not detract from the appearance of the patient because it is unaccompanied by lip strain or attenuation of the labiomental fold. The profile is now anteriorly divergent with overall facial convexity. The extent to which the midface and lower face project is excessive compared to normative standards. There is significant enhancement of the soft tissue support throughout the mid- and lower face; that is, there is zygomatic definition with a submalar hollow and elimination of the jowls and submental fullness. Isolated maxillary advancement surgery with or without genioplasty could not have achieved this aesthetic result as it relates to the support of the soft tissues. Soft tissue considerations have dictated skeletal proportions.

PREOPERATIVE | **POSTOPERATIVE**

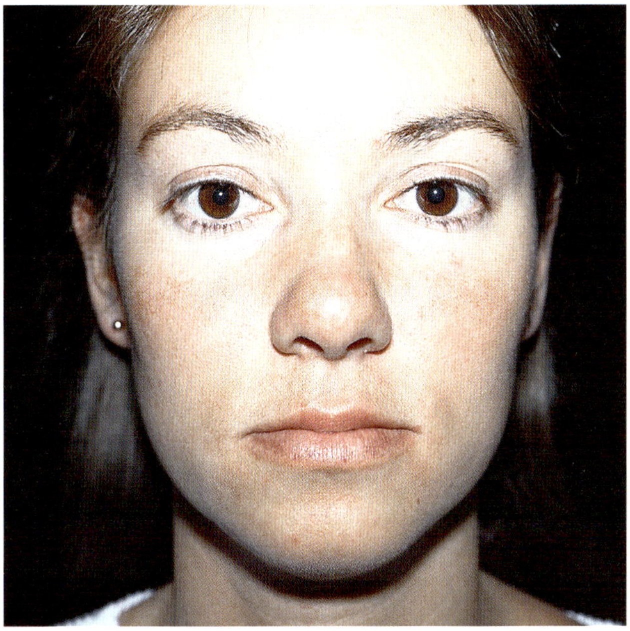

A

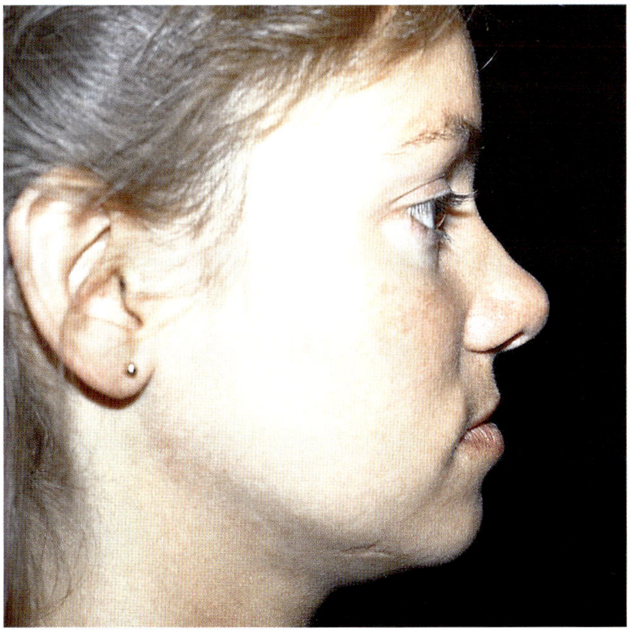

B

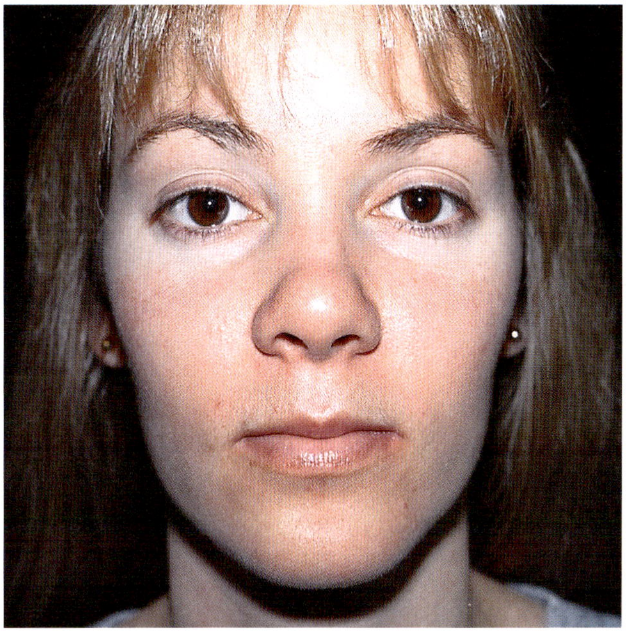

C

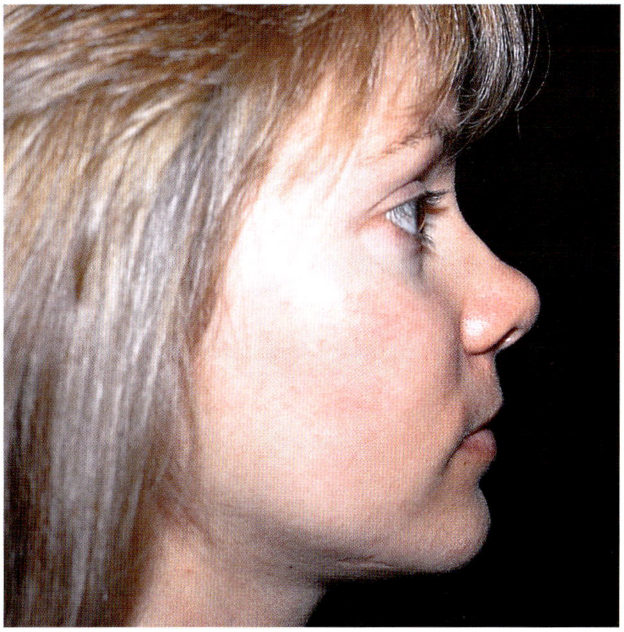

D

CASE STUDY 7-10 A & B: A Class III patient with "normal" vertical and sagittal facial proportions. The soft tissues are inadequately supported. The patient is treated with advancement of the mandible, maxilla, and chin, and vertical elongation of the maxilla and chin. C & D: Compared to normative standards, the face is now too long and projects excessively. Soft tissue support has been dramatically improved. Soft tissue goals have dictated skeletal displacements.

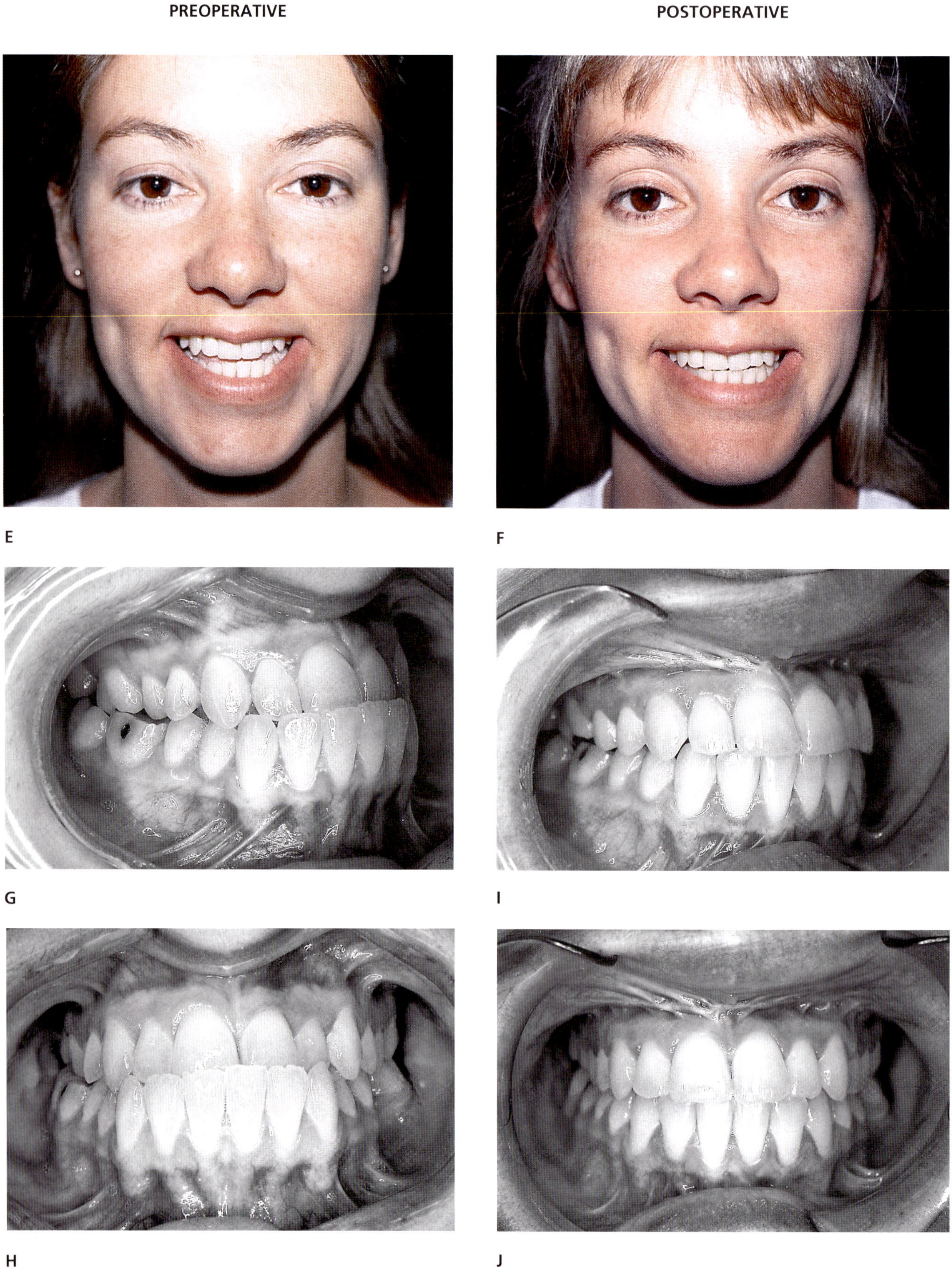

FIG. 7-10 (*Continued*) From: Rosen, H.: Facial skeletal expansion: Treatment strategies and rationale. *Plast Reconstr Surg* 89:798, 1992. Orthodontist: Dr. Julian Freeman

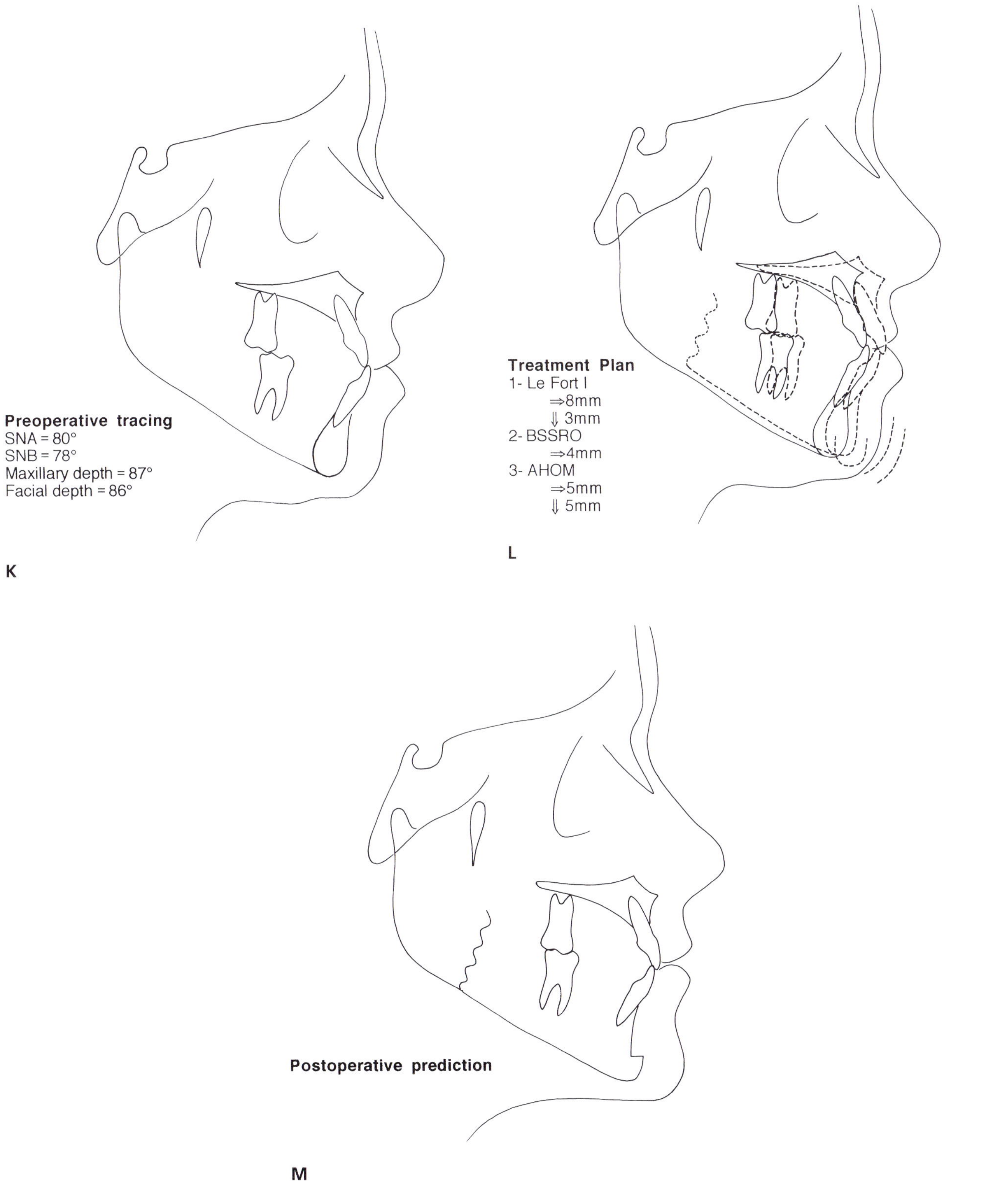

FIG. 7-10 (*Continued*) K, L, & M: The treatment plan is designed to achieve soft tissue support through bidimensional skeletal expansion.

Conclusion

As in all the dentofacial deformities, surgical correction of the Class III deformity must preserve skeletal volume as it relates to the overall support of the facial soft tissues. In the past, excessive reliance had been placed on mandibular setback surgery in treating the majority of these individuals. It is now maxillary advancement surgery that has become the workhorse for treating this malocclusion regardless of its skeletal etiology. This is based on the aesthetic premise that excessive anterior facial divergence can be an attractive morphologic trait provided it is seen in concert with an overall convex profile both sagittally and parasagittally, good labial relations, and an adequately projecting nasal dorsum and nasal tip. The aesthetic rationale of this approach is that it corrects the sagittal discrepancy without sacrificing skeletal volume. This, in turn, preserves and/or enhances the skeletal support of the facial mask.

Suggested Reading

1. Farkas, L.G., Sohm, P., Kolar, J.C., et al.: Inclinations of the facial profile: Art versus reality. *Plast Reconstr Surg* 75:509, 1985.
2. Gonzalez-Ulloa, M.: Quantitative principles in cosmetic surgery of the face (profileplasty). *Plast Reconstr Surg* 29:186, 1962.
3. Gonzalez-Ulloa, M.: A quantum method for the appreciation of the morphology of the face. *Plast Reconstr Surg* 34:241, 1964.
4. Legan, H.L. and Burstone, C.J. Soft tissue cephalometric analysis for orthognathic surgery. *J Oral Surg* 38:744, 1980.
5. Rosen, H.: Maxillary advancement for mandibular prognathism: Indications and rationale. *Plast Reconstr Surg* 87:823, 1991.
6. Rosen, H.: Facial skeletal expansion: Treatment strategies and rationale. *Plast Reconstr Surg* 89:798, 1992.
7. Rosen, H.: Bimaxillary Surgery. In M. Habal (Ed.): *Advances in Plastic and Reconstructive Surgery*. St. Louis: Mosby, Year Book, Chicago, 10:93, 1993.
8. Rosen, H.: Aesthetics in facial skeletal surgery. *Perspectives in Plastic Surgery* 6:1, 1993.
9. Salyer, K.E., Vasconez, H.C., Genecov, E., and Kolar, J.: Analysis of facial proportions in twenty beautiful faces. *Plast Surg Forum* 10:129, 1987.
10. Whitaker, L.A.: Aesthetic contouring of the facial support system. *Clin Plast Surg* 16:815, 1989.

Facial Convexity and Posterior Divergence: The Class II Deformity

CHAPTER 8

There is little question that the Class II skeletal deformity is the most common developmental deformity that confronts surgeons. It is estimated by the United States Public Health Service that approximately 5% of North American Caucasians have some degree of mandibular deficiency. Not infrequently, underdevelopment of the lower jaw is seen in association with vertical dysplasia of the maxilla, the most common condition being vertical maxillary excess with or without anterior open bite.

This chapter will deal only with those patients who have isolated mandibular deficiency and who are best corrected with single-jaw surgery. Mandibular deficiency accompanied by vertical maxillary excess or vertical maxillary deficiency will be addressed in Chapters 9 and 10, which deal with the long face and the short face, respectively. Admittedly, this is an arbitrary characterization of mandibular deficiency, and it is made only for organizational purposes. It should be appreciated that the spectrum of mandibular deficiency is incredibly varied and may be accompanied by quite subtle or severe maxillary problems. In fact, maxillary vertical dysplasia is responsible, in large part, for the tremendous variability in the facial morphology of the Class II patient. As the vertical dimension of the maxilla varies, so too do the overall height of the lower face, the extent of exposure of the maxillary incisor teeth, the mandibular plane angle, the depth of the labiomental fold, the competence of the oral commissure, and the overall support of the soft tissues.

However, regardless of whether mandibular deficiency is accompanied by vertical maxillary dysplasia, the common morphologic finding in almost all Class II patients is a profile characterized by excessive degrees of facial convexity accompanied by posterior facial divergence. The extent to which the face diverges posteriorly is usually excessive. (The most common exception to this morphology is the Class II, deep-bite patient with a prominent chin.) Because this is the common morphologic denominator among these patients, mandibular advancement must be the common surgical denominator for correction of this deformity. Double-jaw surgery (maxillary surgery in addition to mandibular advancement) is indicated in those Class II patients in whom a better aesthetic result would be obtained contrasted to that with mandibular surgery only. This obviously includes all those patients with maxillary deformities. In addition, it is indicated when adequate degrees of sagittal projection of the lower face can only be achieved with maxillary advancement and/or maxillary plane rotation in addition to mandibular advancement.

As stated before, this chapter will deal with those Class II patients who are adequately treated with mandibular surgery only. Although this approach tremendously simplifies the surgical correction, the aesthetic outcomes are to a large extent dictated by the occlusion because the maxillary dentition will dictate the position of the mandible in all three planes of space. In addition, the mandibular occlusal plane and, hence, inferior border of the mandible are also dictated by the unoperated position of the maxilla. As a result, osseous genioplasty becomes indicated in many of these patients to favorably influence the position of soft tissue pogonion in the sagittal, vertical, and transverse planes of space, as well as to favorably affect the aesthetics of the labiomental fold. Chin surgery represents the "icing on the cake" in many Class II patients treated with mandibular advancement surgery.

Limitations of Single-Jaw Correction

Let us first examine the aesthetic limitations of single-jaw surgery, because it is only in this way that one can fully appreciate appropriate surgical candidates for this approach. As already stated, the aesthetic results from a single-jaw correction are limited by the occlusal constraints placed on it; the extent to which one can advance the mandible is dictated by the available overjet.

The patient seen in Fig. 8-1 vividly illustrates these limitations. His facial morphology in the frontal view is characterized by good vertical proportionality and symmetry. The ears are prominent. The lower lip is procumbent causing the impression of a deepened labiomental fold. The profile is characterized by excessive facial convexity and posterior divergence with essentially no cervical mandibular angle. In addition, there is some paranasal flattening. There is a Class II, Division 1, malocclusion. The surgical correction involved an 8-mm mandibular advancement and an 8-mm advancement genioplasty. The occlusal result is excellent. However, the aesthetic result is poor due to deficient projection of the lower face. As a result, there remains a poorly defined cervical mandibular angle.

This patient should have been treated with double-jaw surgery involving bimaxillary advancement and occlusal plane counterclockwise rotation to achieve adequate degrees of projection of the lower face. Double-jaw

PREOPERATIVE **POSTOPERATIVE**

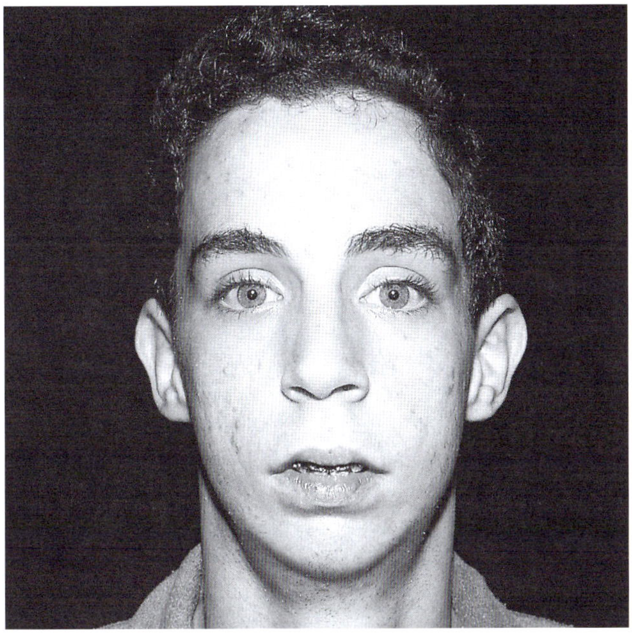

A

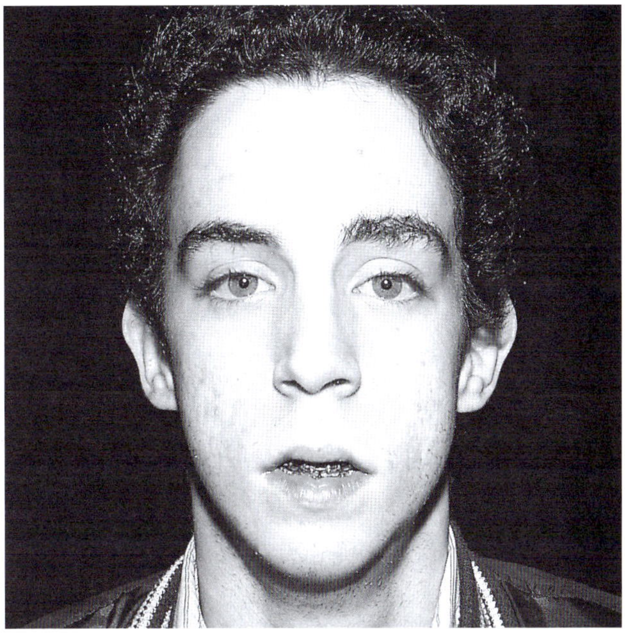

C

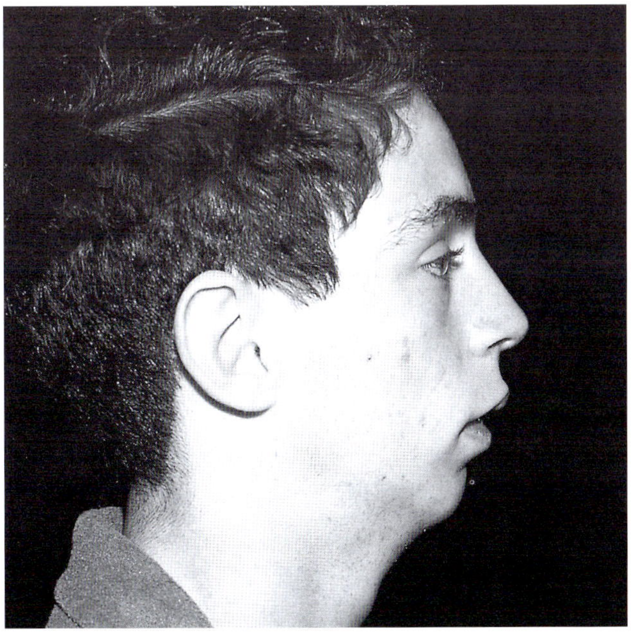

B

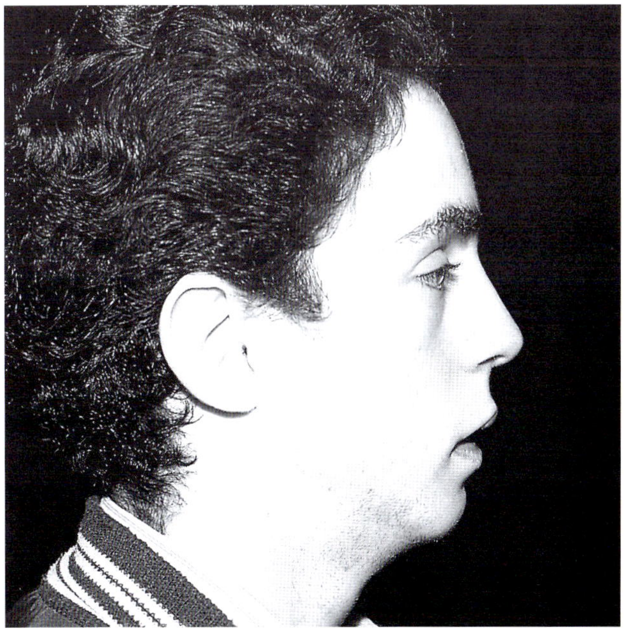

D

FIG. 8-1 A patient with mandibular deficiency demonstrating the aesthetic limitations of mandibular advancement and genioplasty. This patient should have undergone double-jaw surgery with maxillary advancement and counterclockwise rotation in addition to mandibular advancement and genioplasty.

PREOPERATIVE

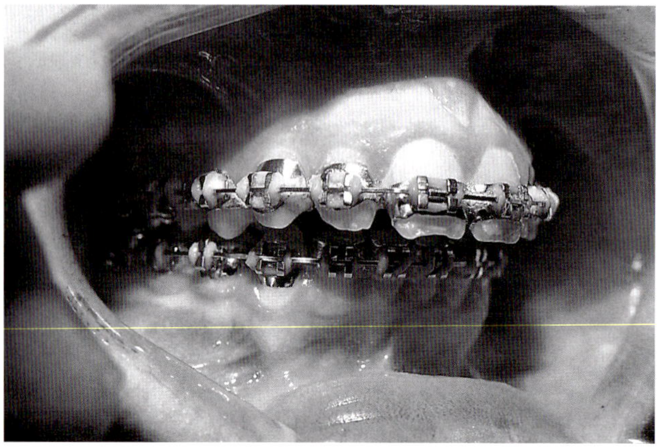

E

POSTOPERATIVE

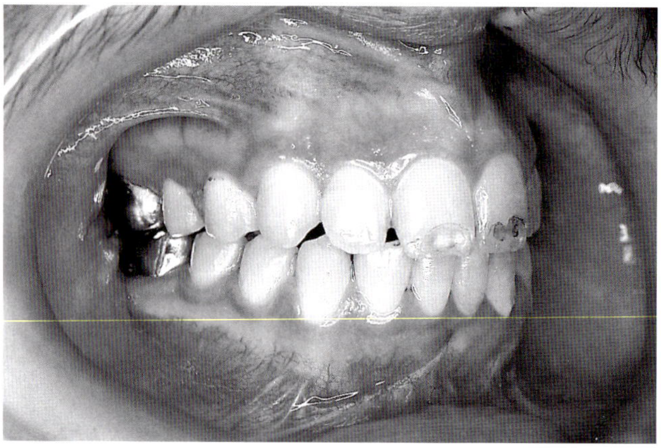

F

FIG. 8-1 (*Continued*) Orthodontist: Dr. Fred Thompson

surgery is frequently indicated in the mandibular deficient patient as it eliminates the restrictions that the occlusion places on the aesthetic results, thereby giving the surgeon control over both magnitude and direction of skeletal displacements. Occlusal constraints are entirely eliminated.

The patient seen in Fig. 8-2 provides another example of a suboptimal aesthetic result of a single-jaw correction of a Class II malocclusion. The surgical error committed in this patient was to perform maxillary setback surgery by way of four-piece Le Fort I osteotomy. In the frontal view, the patient displays good vertical proportionality and symmetry. There is a procumbent lower lip with a deepened labiomental fold. The profile is characterized by good projection of the midface and upper lip and a recessed and procumbent lower lip with a chin that is only modestly weak relative to the midface. Overall, there is facial convexity and posterior divergence. There is a Class II malocclusion. The patient has undergone orthodontic segmental preparation with premolar extraction spaces present.

As demonstrated by the pre- and postoperative radiographs, there has been surgical closure of the premolar extraction spaces by setback and intrusion of the anterior maxillary segments. There is a good occlusal result. However, the aesthetic result is deficient; that is, both the upper and lower lips are excessively recessed due to retraction of the anterior maxillary segment. This, in turn, causes excessive obtuseness of the nasolabial angle and deepening of the nasolabial fold. In summary, the lower face is now characterized by an unattractive sinking in and thinning out of the perioral soft tissues due to loss of foundational support. This is particularly evident in the frontal postoperative view and creates a premature aged appearance.

This patient illustrates a violation of the concept of the common surgical denominator for correction of the Class II deformity; that is, with very few exceptions, always perform mandibular advancement surgery and, if aesthetically appropriate, maxillary surgery.

PREOPERATIVE **POSTOPERATIVE**

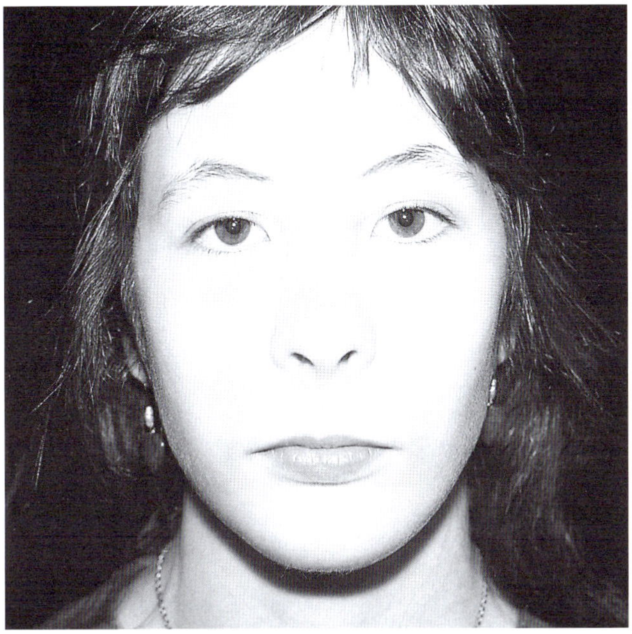

A

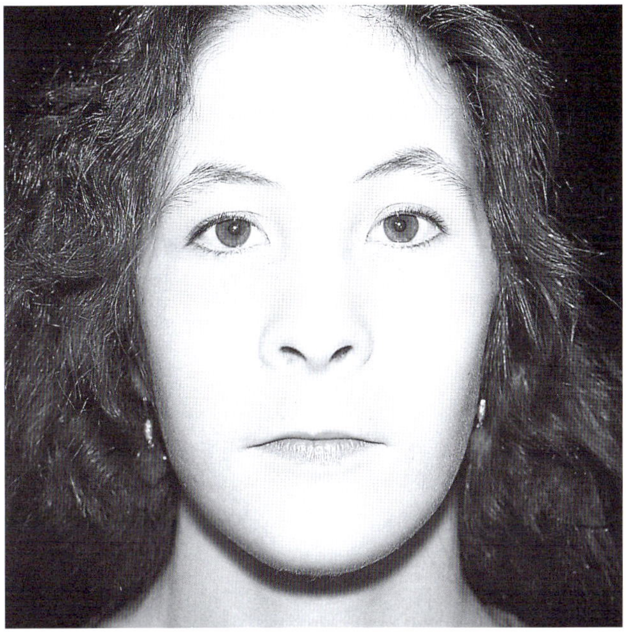

C

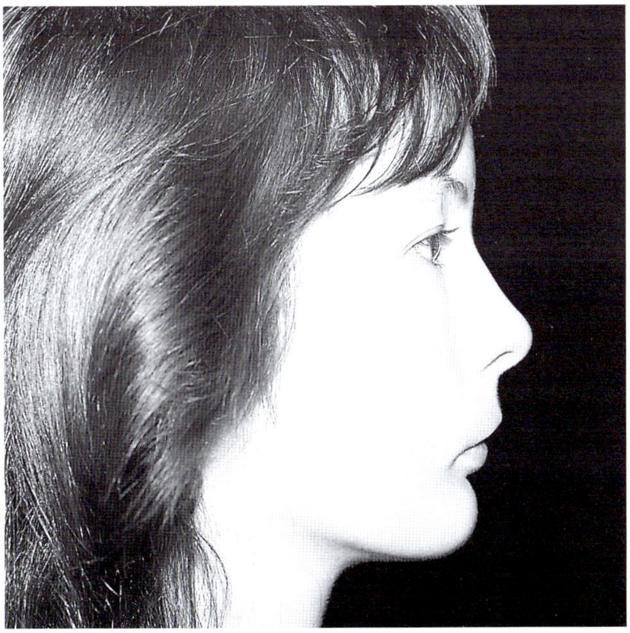

B

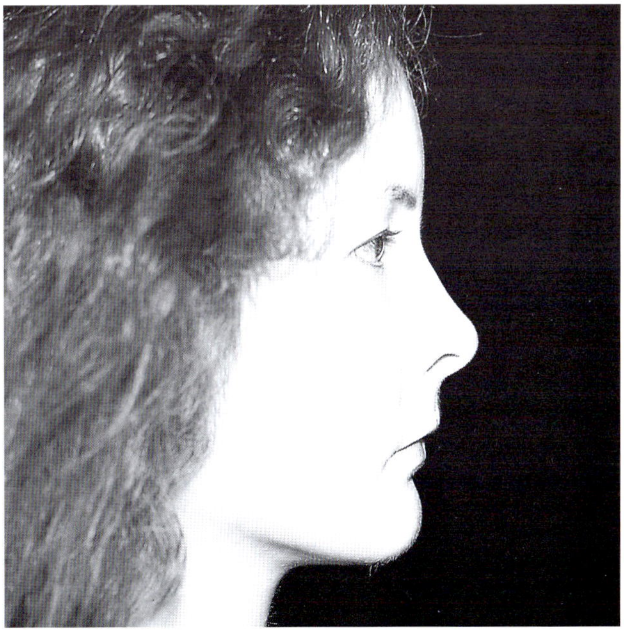

D

FIG. 8-2 A & B: A Class II patient demonstrating the aesthetic limitations of maxillary setback surgery. This is a clear example of the occlusal result taking priority over the aesthetic result. C & D: Postoperatively, the lips lack support, the nasolabial fold is deepened, and the patient appears to have aged. A better result would have been achieved with double jaw surgery. From: Rosen, H: Segmental osteotomies of the maxilla. *Clin Plast Surg* 16:785, 1989. Reprinted with permission.

PREOPERATIVE

POSTOPERATIVE

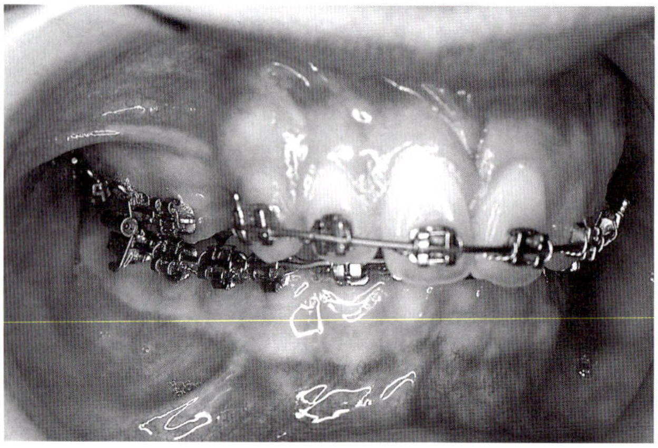

E

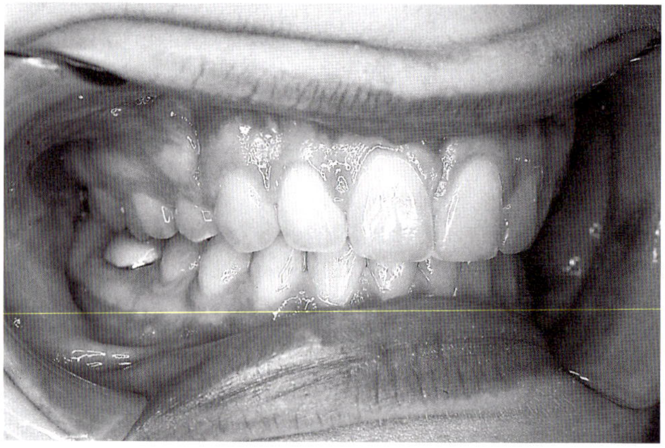

F

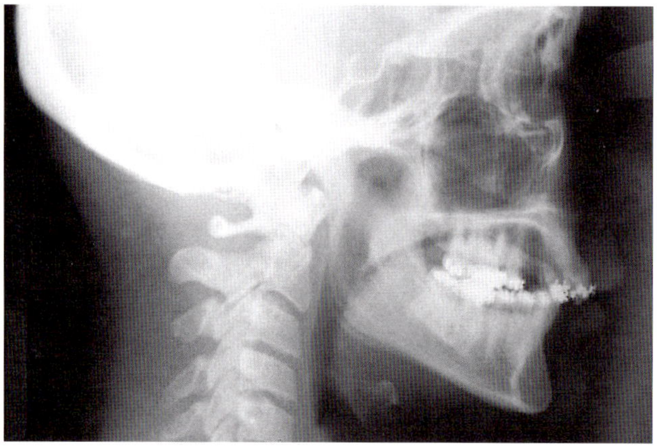

G

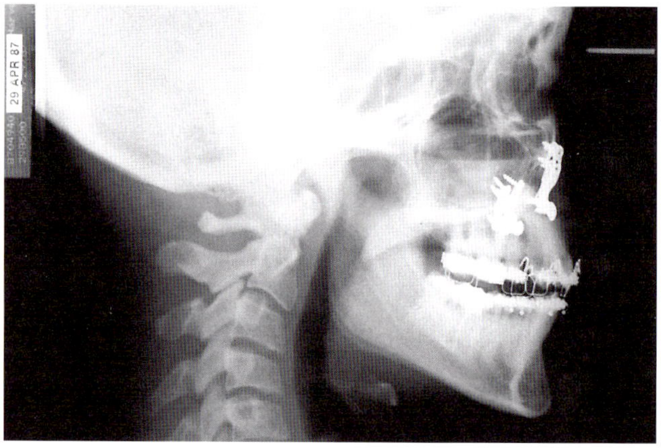

H

Fig. 8-2 (*Continued*) E & F: The occlusal result. G & H: Note surgical closure of extraction spaces by anterior maxillary setback. Orthodontist: Dr. James Ackerman.

Single-Jaw Correction of the Class II Deformity: Mandibular Advancement Surgery

Mandibular advancement surgery employing the sagittal split ramus osteotomy is the workhorse for the correction of the Class II deformity. Because the extent of overjet dictates the magnitude to which one can advance the mandible, adjunctive osseous genioplasty must be employed to achieve the best possible aesthetic result in many of these patients. When the advancement is accompanied by any rotational vector of movement, prediction tracings become imperative to predict the necessity as well as the vector and magnitude of any additional symphyseal displacement employing genioplasty techniques. These treatment planning principles are nicely illustrated in the following three patients.

Case Study 8-3

This patient has rather typical facial morphology of the Class II patient. In the frontal view, there is reasonable vertical proportionality and good facial symmetry. There is a modest component of mentalis hyperfunction. There is 5 to 6 mm of maxillary incisal show. The lateral view is dominated by excessive facial convexity and posterior facial divergence. The labiomental fold is deepened secondary to a procumbent lower lip. There is a modest Class II, Division 1 malocclusion allowing a mandibular advancement of only 4 mm. There is no rotational movement anticipated. This modest skeletal displacement will not achieve adequate projection of the lower face, and therefore, advancement genioplasty of 8 mm with no vertical shifting is also planned.

Postoperatively, there is a slight residual degree of mentalis hyperfunction secondary to modest vertical excess. The profile is now characterized by overall convexity with anterior facial divergence. There is a good soft tissue response. The aesthetic result achieved in this patient is a good one. Parenthetically, attempts to correct modest vertical maxillary excess in female patients, such as this one, who have 5 to 6 mm of incisal show should not be undertaken. Overcorrection is a great risk with subsequent lack of incisal show and a premature aged appearance.

PREOPERATIVE POSTOPERATIVE

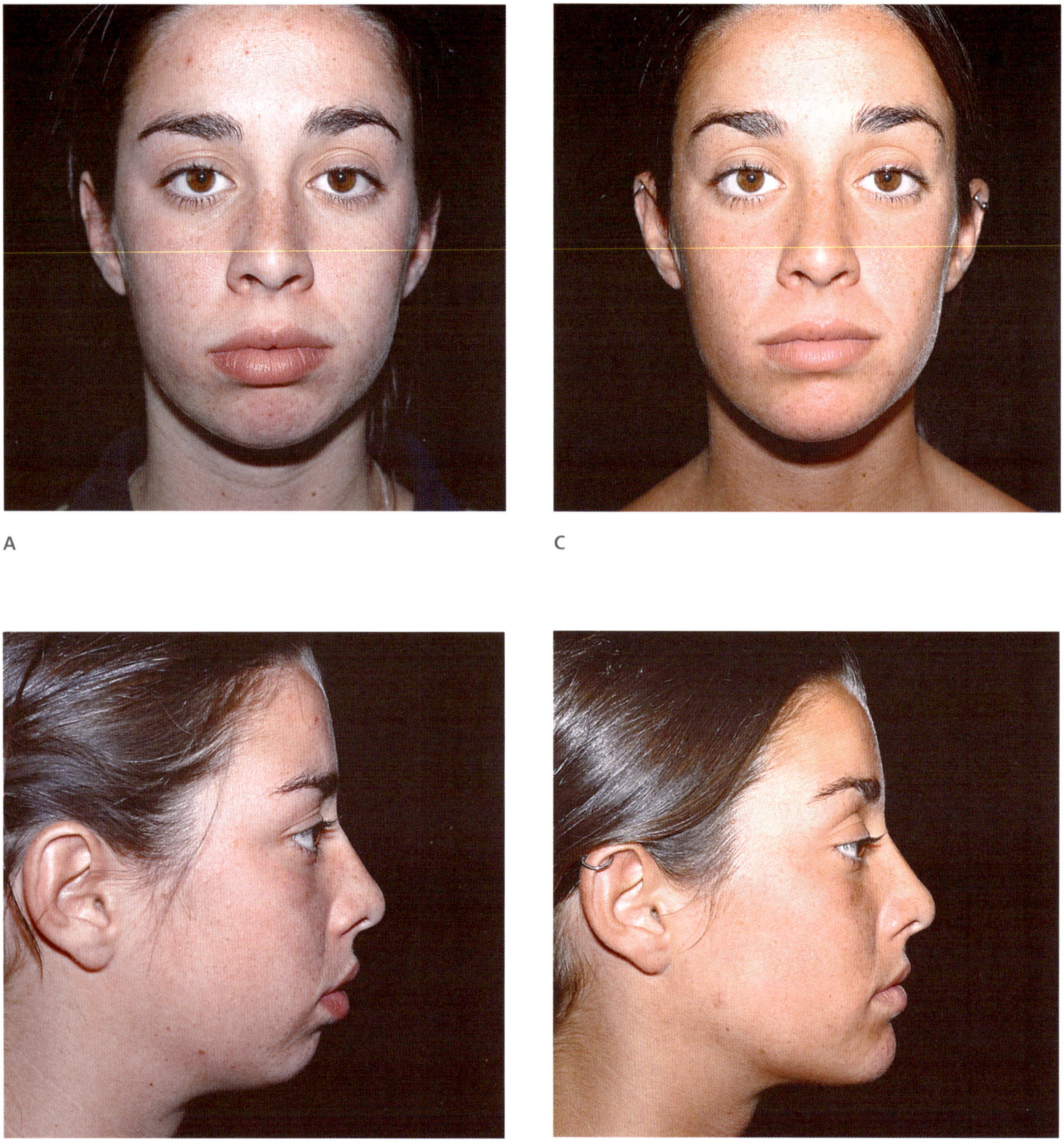

A C

B D

Case Study 8-3 A & B: A patient with mandibular deficiency treated with mandibular advancement with no rotation and advancement genioplasty. C & D: The postoperative result is characterized by facial convexity and anterior divergence.

PREOPERATIVE | **POSTOPERATIVE**

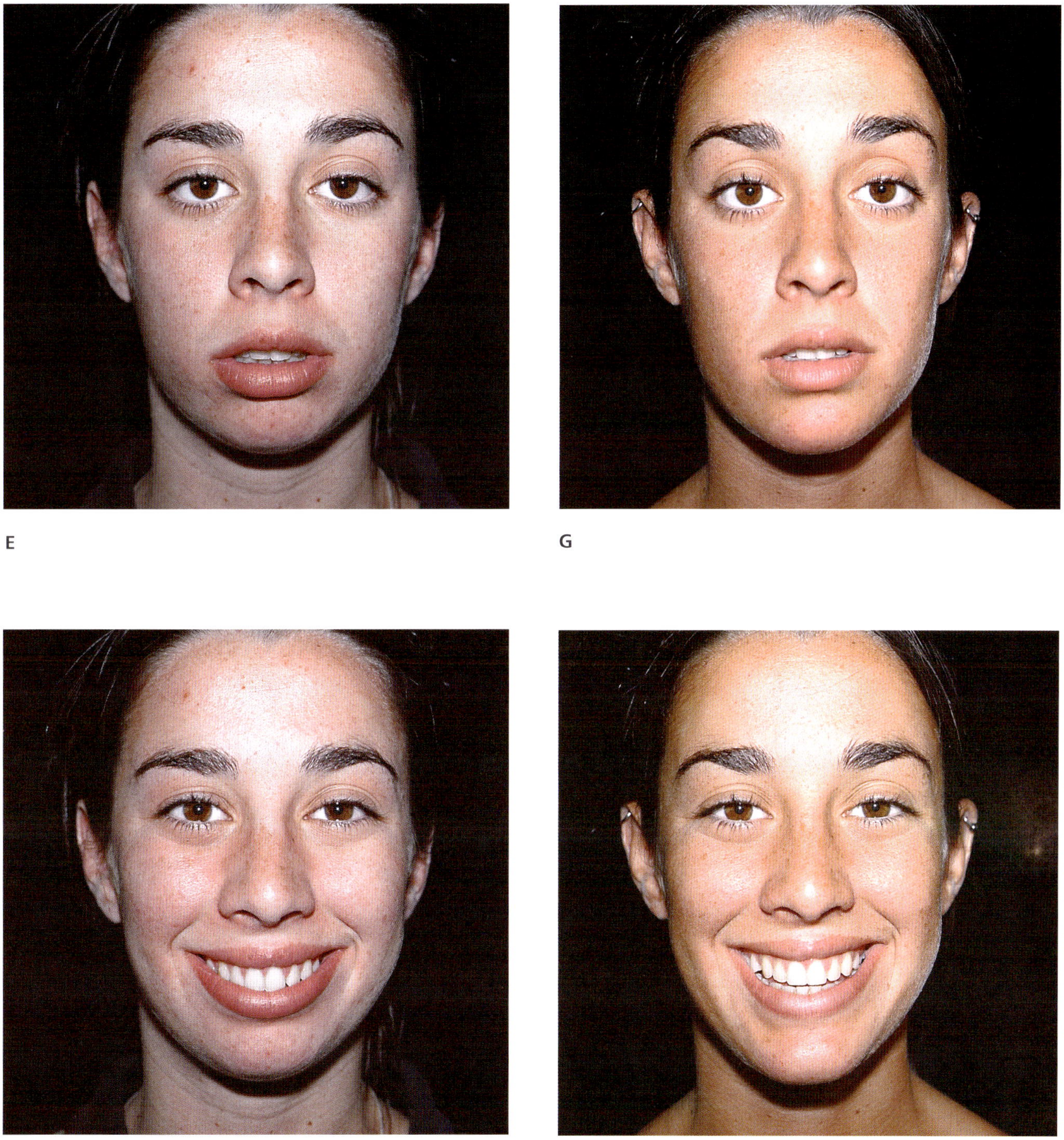

E G F H

FIG. 8-3 *(Continued)*

PREOPERATIVE

POSTOPERATIVE

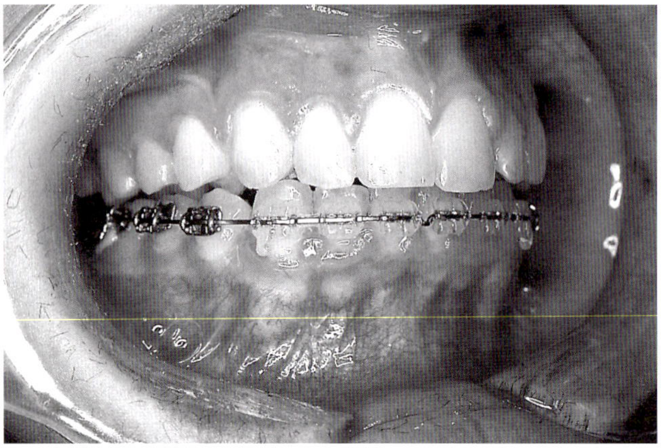

I

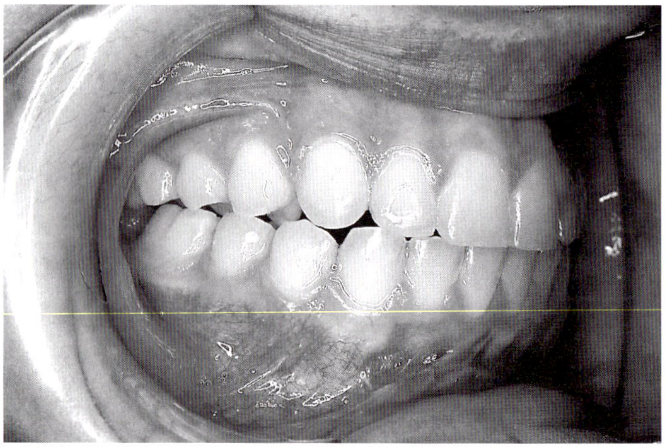

K

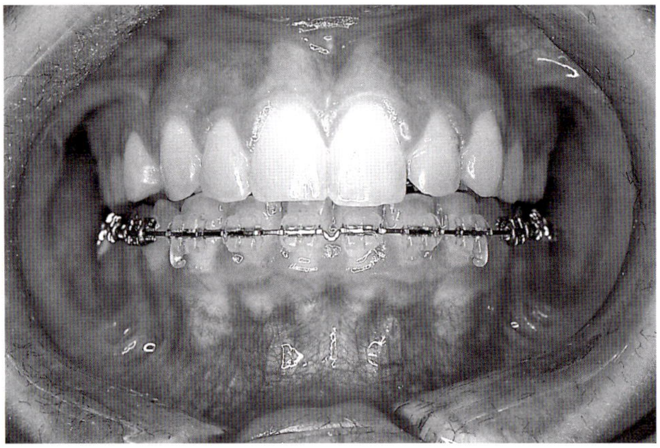

J

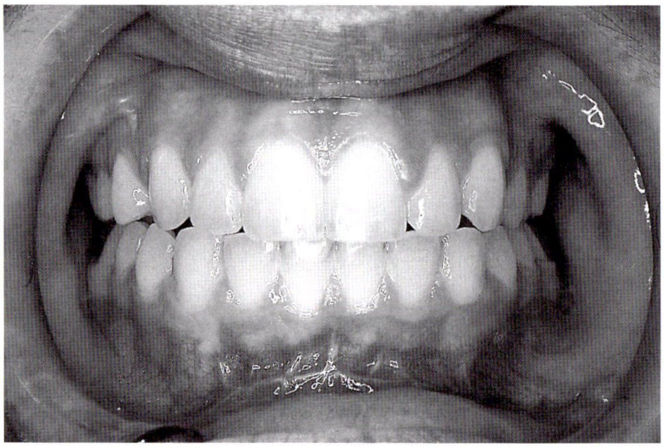

L

Fig. 8-3 (*Continued*) Orthodontist: Dr. Richard Drummond. Note: Orthodontic appliances removed early at request of patient.

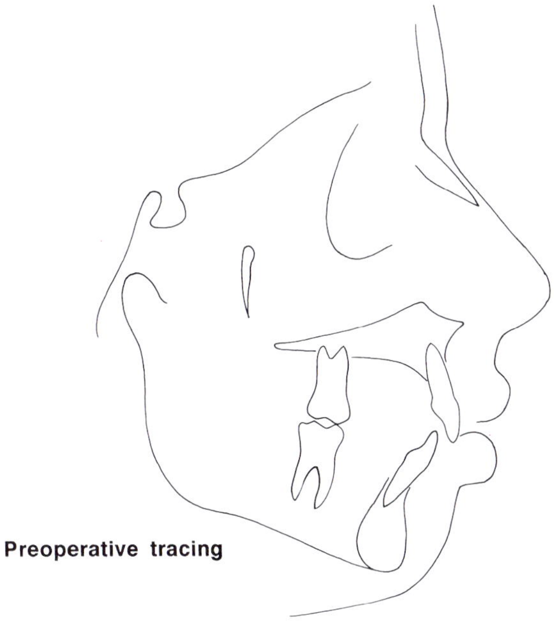

M

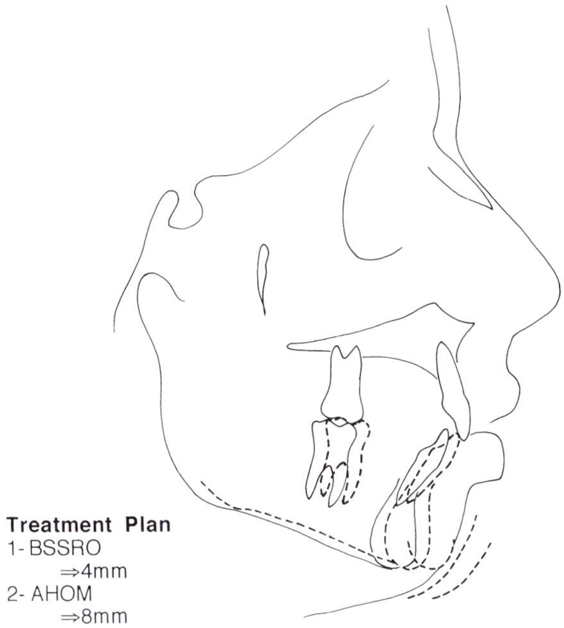

N

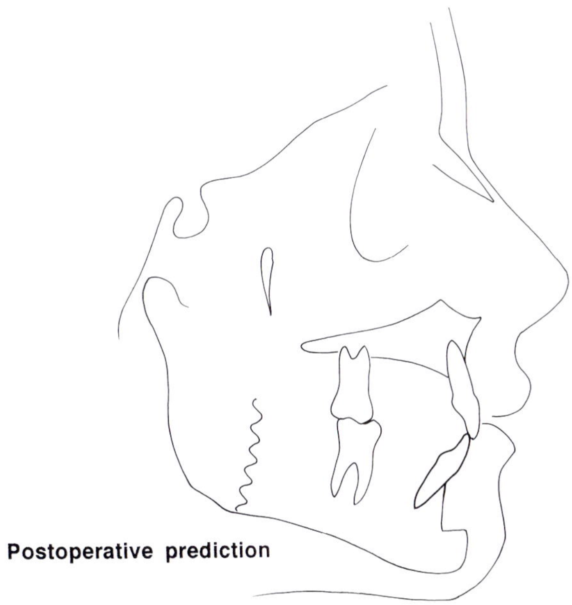

O

FIG. 8-3 (*Continued*) M: Note modest overjet that allows only 4 mm of mandibular advancement. N & O: Accordingly, the chin must be advanced to achieve adequate degrees of facial projection and anterior divergence.

Case Study 8-4

The patient in this figure has very different frontal morphology as compared to the previous patient. This patient has a Class II, Division 1 malocclusion with an anterior deep bite. Secondary to this, there is a lack of vertical dimension to the lower face, a procumbent lower lip, and a deepened labiomental fold. There is good lower face symmetry. The profile view demonstrates the common morphologic denominator of the Class II skeletal deformity, that is, excessive facial convexity with posterior facial divergence. Examination of his occlusion suggests that there will be a large, clockwise rotational component to his mandibular advancement. This rotational component is confirmed with a prediction tracing that demonstrates inferior displacement at pogonion with little sagittal advancement. As a result, there will be an attractive increase in vertical dimension of the lower face, but there will still be a lack of projection of the lower face. Accordingly, adjunctive osseous genioplasty to advance the chin an additional 8 mm, as well as liposuctioning of the submental and perimandibular regions, is planned.

Postoperatively, one can appreciate the significant increase in the vertical dimension of the lower face secondary to mandibular clockwise rotation. This is reflected in the postoperative occlusal result; that is, the deep bite has been eliminated and there is now a normal overbite relationship. Lateral facial morphology is now characterized by facial convexity and anterior divergence with normal posturing of the lower lip and a softening of the labiomental fold region. There has been significant enhancement of the soft tissue support due to a bidimensional expansion of the facial skeleton of the lower face.

PREOPERATIVE POSTOPERATIVE

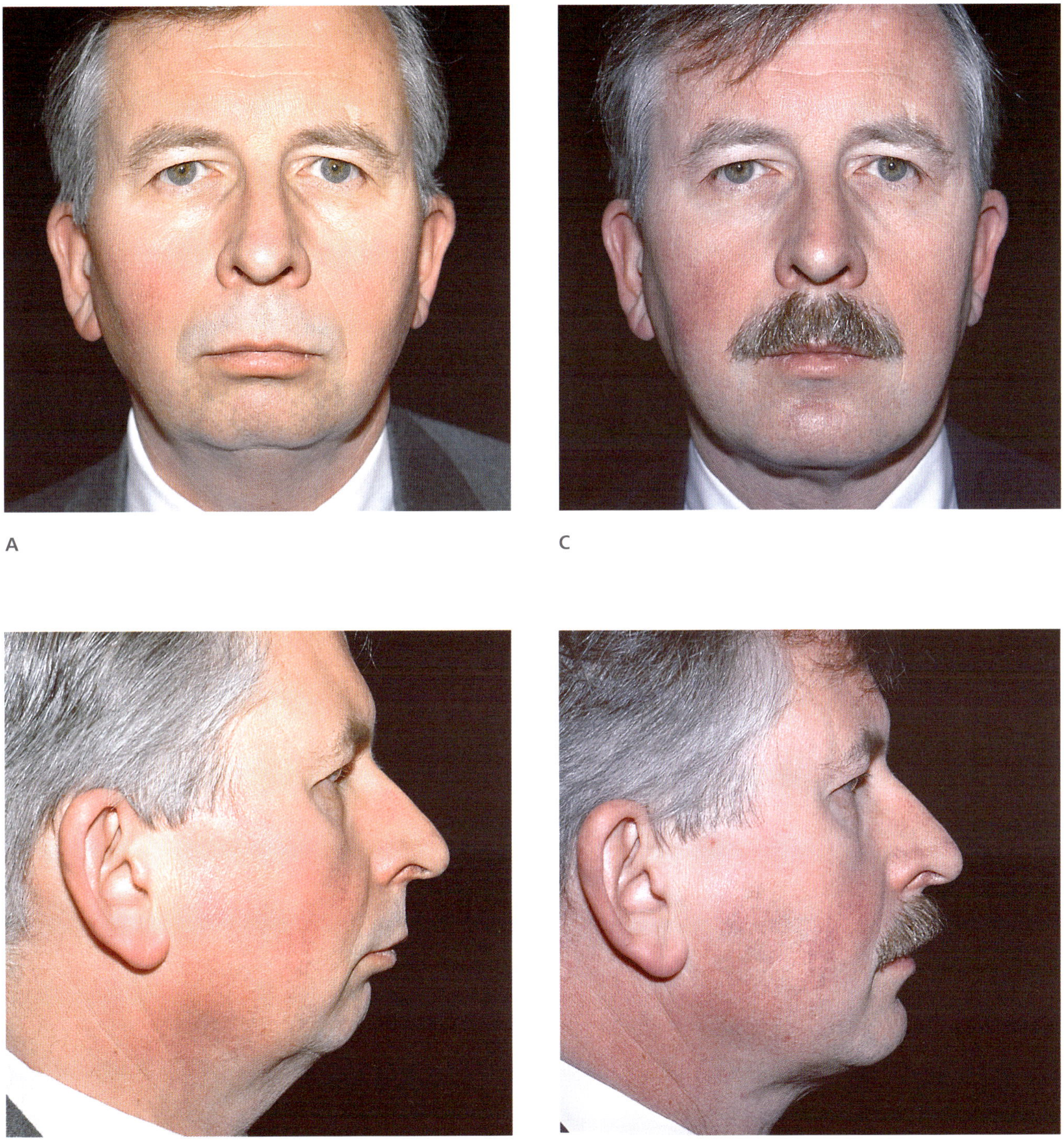

A C B D

CASE STUDY 8-4 A & B: A patient with mandibular deficiency and anterior deep bite. There is vertical deficiency of the lower face. C & D: Following mandibular advancement with clockwise rotation, advancement genioplasty, and liposuction of jowls and neck. Note increase in lower face height due to mandibular clockwise rotation.

PREOPERATIVE

POSTOPERATIVE

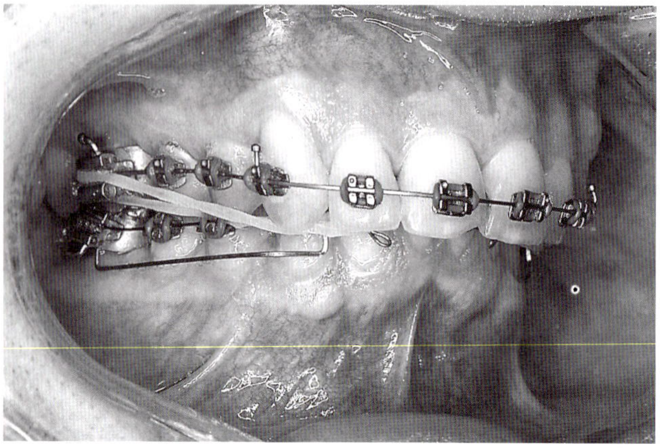

E

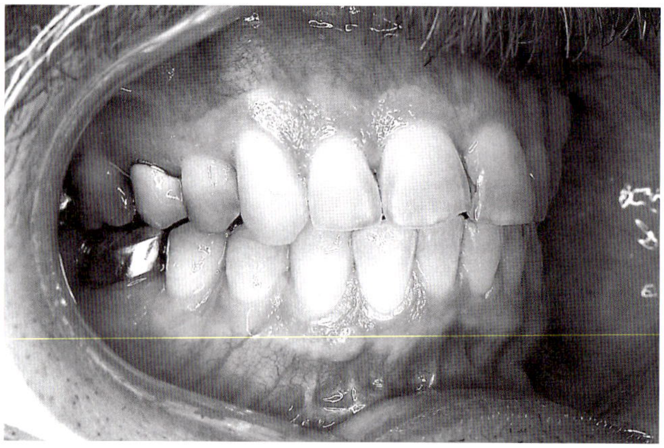

G

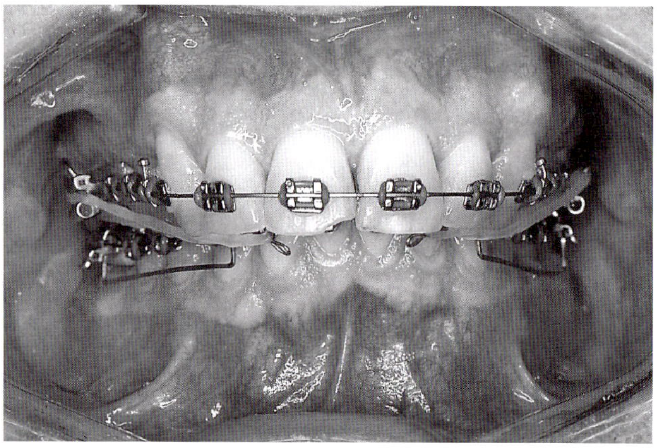

F

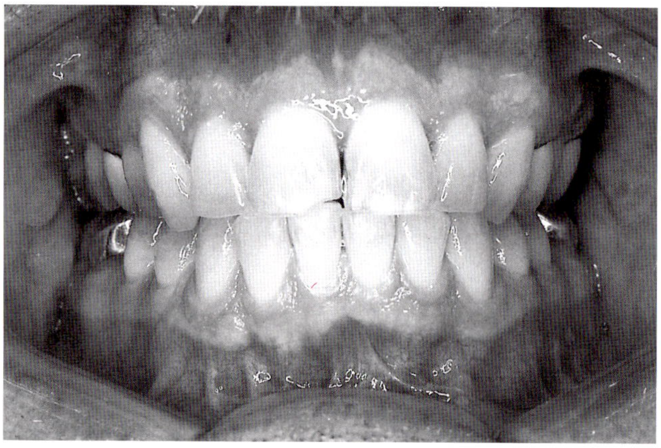

H

Fig. 8-4 (*Continued*) G & H: Note significant reduction in overbite, resulting in increase in vertical dimension of lower face as mandible is advanced. Orthodontist: Dr. Gerald Ginsberg.

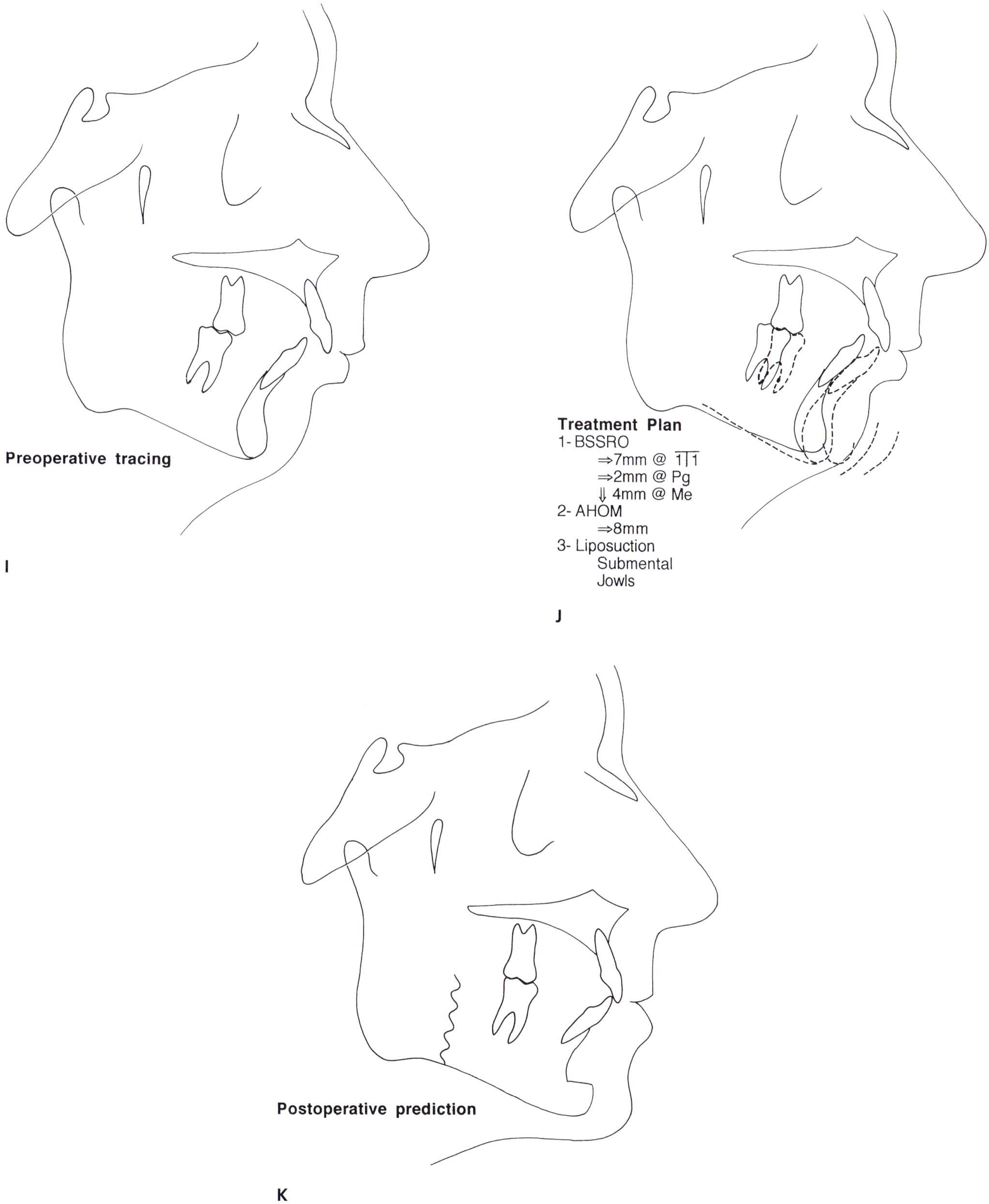

FIG. 8-4 (*Continued*) J: As the mandible is advanced, there is clockwise rotation due to incisal guidance. As a result, the lower face lengthens but there is only modest increase in projection. K: This necessitates an advancement genioplasty.

Case Study 8-5

This patient represents still another morphologic variation of the Class II skeletal patient. In the frontal view, there is a modest lack of lower face height and facial asymmetry with the soft tissue chin point deviated to the right of the facial midline. The labiomental fold is fairly deep. Common lateral morphology is again seen: excessive facial convexity and posterior facial divergence. There is reasonably good soft tissue support with only modest submental fullness. Occlusal examination demonstrates a Class II relationship with a modest anterior open-bite component.

The treatment plan must take into account that a counterclockwise rotational component will occur as the mandible is advanced. This is necessary to close the open bite and achieve a normal overbite relationship. As this occurs, soft tissue pogonion comes forward and upward; that is, anterior lower face height deficiency will be exaggerated. Accordingly, adjunctive genioplasty must not only advance the chin to achieve adequate projection of the lower face but also inferiorly displace it to achieve adequate vertical skeletal volume. The overall treatment plan involved a 6-mm mandibular advancement and osseous genioplasty to advance and inferiorly reposition the symphyseal segment 3 mm and 6 mm, respectively. In addition, a lateral shifting of the symphyseal segment to the left will help restore lower face symmetry.

Postoperatively, there are enhanced vertical relations (with an attractive decrease in the depth of the labiomental fold), and facial symmetry has been restored. The profile is now characterized by very modest overall facial convexity as well as anterior divergence. Ideally, the aesthetic result could have been improved with greater paranasal and upper lip support creating a more convex profile. The occlusal result is an excellent one. Note the deepening of the bite secondary to counterclockwise mandibular rotation.

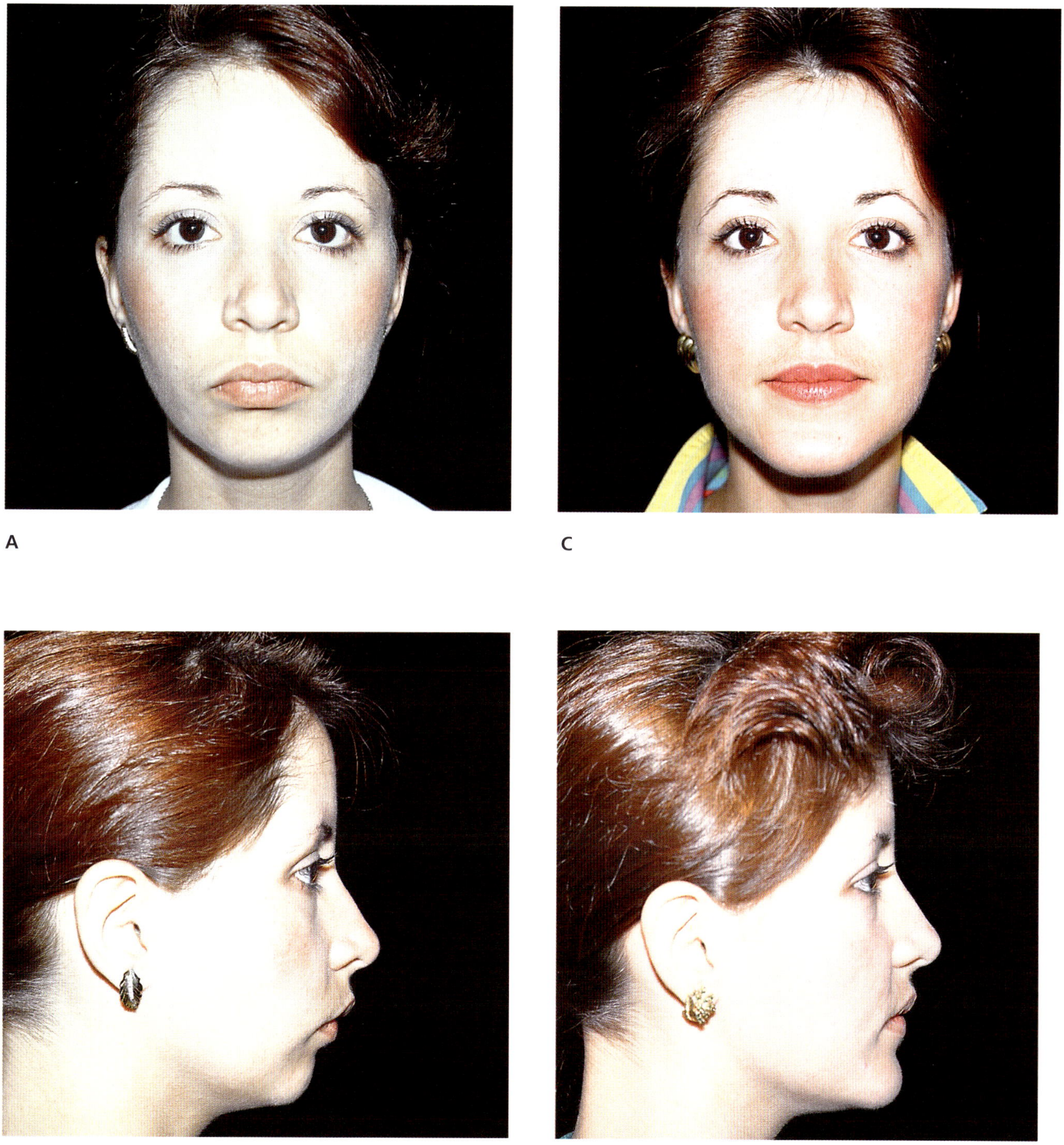

PREOPERATIVE

POSTOPERATIVE

A

C

B

D

CASE STUDY 8-5 A & B: A patient with mandibular deficiency and a Class II anterior open bite. Surgical correction involves mandibular advancement with counterclockwise rotation to close the bite. A lengthening genioplasty is indicated to provide for adequate height of the lower face. C & D: The postoperative appearance demonstrates excellent projection and height of the lower face. From: Rosen, H.: Surgical correction of the vertically deficient chin. *Plast Reconstr Surg* 82:247, 1988. Reprinted with permission.

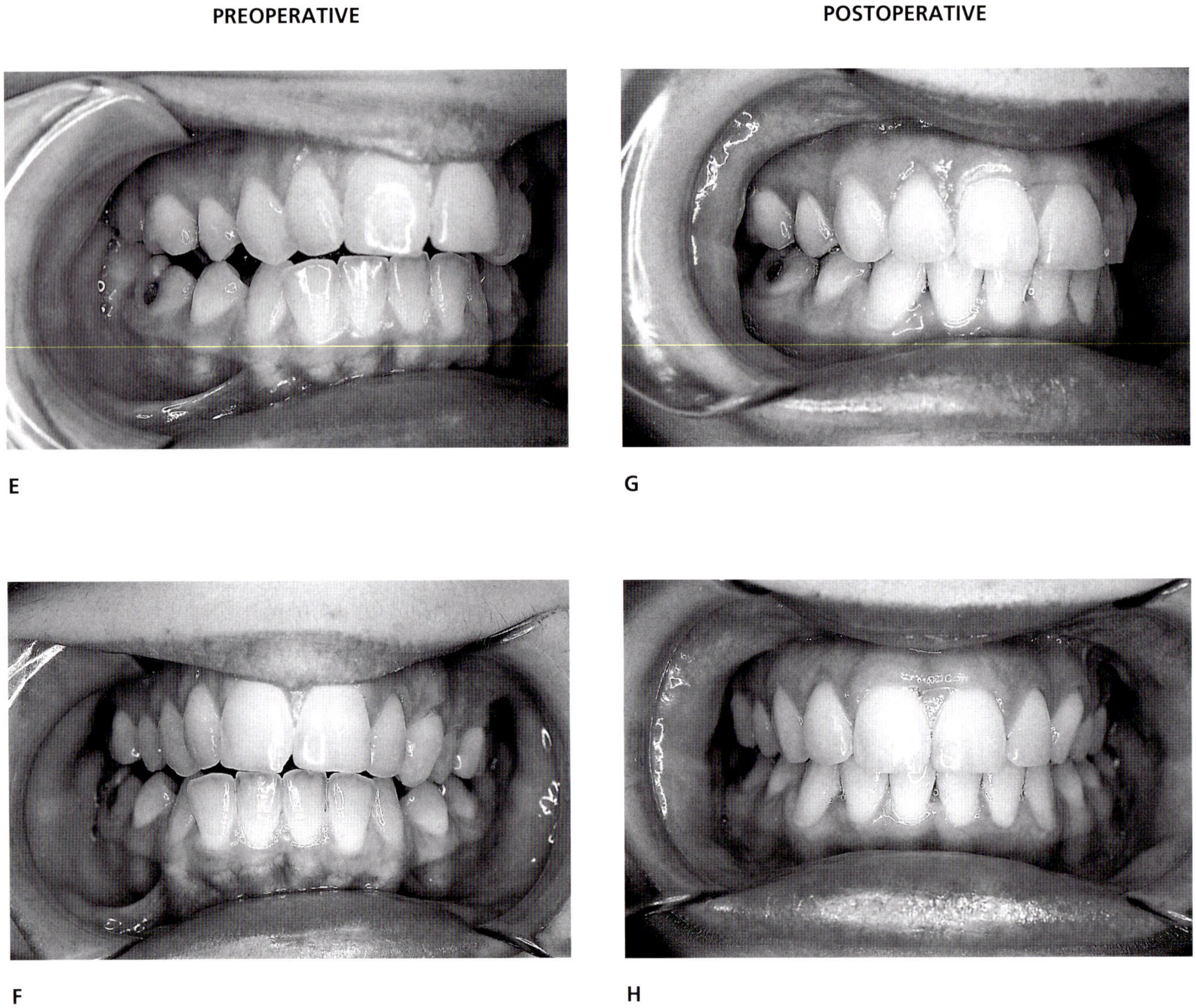

Fig. 8-5 *(Continued)* G & H: Note adequate overbite has been restored. A three-year occlusal follow-up. Orthodontist: Dr. William Roberts.

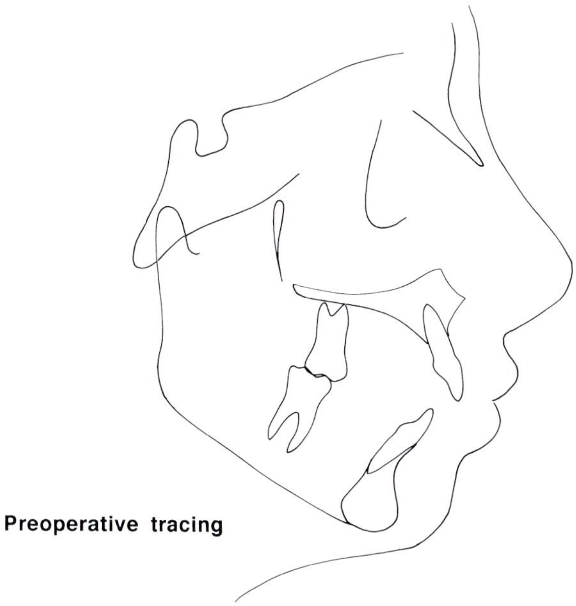

I

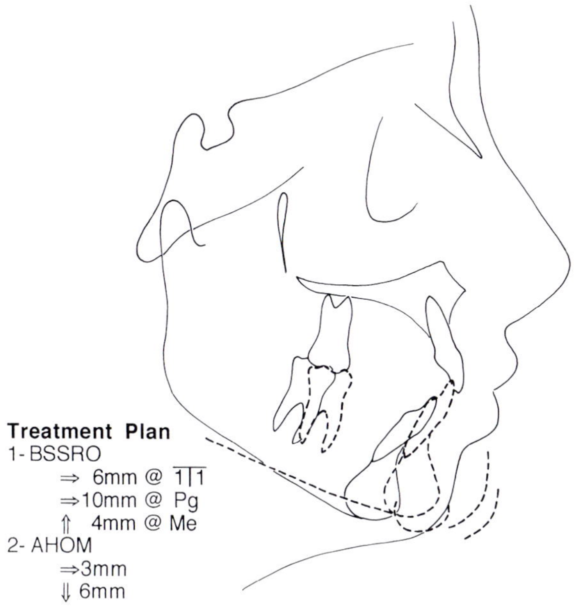

J

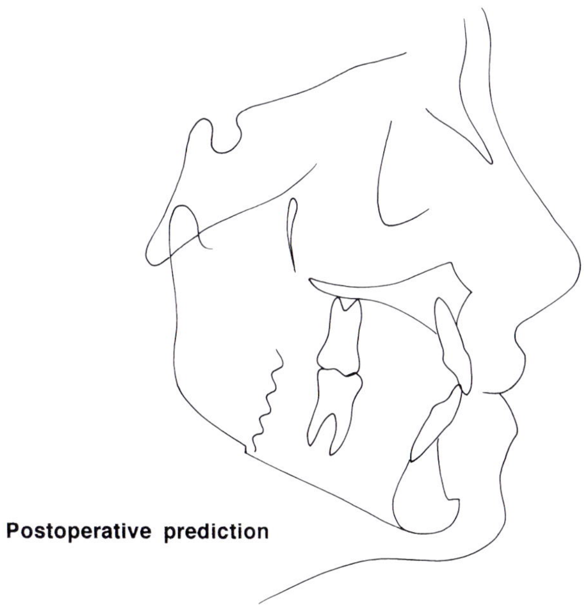

K

FIG. 8-5 (*Continued*) J: Counterclockwise rotation of distal mandible, as it is advanced, decreases facial height. A genioplasty involving 6-mm lengthening is indicated to achieve adequate facial height.

Conclusion

With very few exceptions, mandibular advancement surgery must remain the common surgical denominator of almost all Class II skeletal corrections. Maxillary setback surgery to correct excessive degrees of overjet is an operation that is outdated. It fails to address the excessive posterior divergence of the profile, it does not correct the inadequate soft tissue support in the lower face, and it is detrimental to the preexisting soft tissue support in the paranasal and upper lip regions.

However, mandibular advancement surgery does suffer from two basic limitations: (1) it fails to address any maxillary deformity, particularly in the vertical dimension, that frequently coexists with the mandibular deficiency; and (2) the extent to which one can advance the mandible is dictated by the desired occlusal relationship. As a result, osseous genioplasty is a key element to the surgical correction of many of these individuals as it "fine tunes" the position of the chin in all planes of space. Nevertheless, a large percentage of patients suffering from mandibular deficiency are still better served with double-jaw surgery.

Author's Note: Is posterior repositioning of the maxilla or anterior maxillary setback ever justified in the Class II individual? It is rarely indicated to perform such a skeletal displacement because the aesthetic consequences of doing so are unfavorable in terms of the soft tissue response. These include a lack of skeletal support to the upper lip that results in a thinning of the upper lip, a decrease in vermillion fullness, a decrease in incisal exposure, alar bases sunken into the midface, deepened nasolabial folds, and excessive flattening of the profile. In summary, an aged appearance results. In addition, anterior maxillary setback with surgical closure of extraction spaces is associated with a high incidence of periodontal and endodontic complications.

Nevertheless, there are certain instances in which total posterior maxillary repositioning is indicated. This should be reserved for the rare Class II patient who has grossly excessive skeletal support of the upper lip in whom orthodontic retraction of the anterior teeth is contraindicated. These contraindications may include the risk of excessive incisal uprighting, excessive root resorption in a patient who has undergone prior orthodontic tooth movement, or prolonged treatment time required to accomplish such tooth movement in the noncompliant, adult patient. It must be emphasized, however, that this combination of clinical circumstances is extremely unusual.

Ethnic differences must also be considered when posterior maxillary repositioning or anterior maxillary setback is contemplated. These procedures may be more aesthetically indicated in Class II and/or Class I bimaxillary protrusive patients in whom the perioral and perinasal tissues are thicker, i.e., patients of African and Asian descent. This is particularly true of Asian patients in whom a flatter midface is commonly seen.

Suggested Reading

1. Bell, W.H., Proffit, W.R., Chase, D.L., et al.: Mandibular Deficiency. In W.H. Bell, W.R. Proffit, and R.P. White (Eds.): *Surgical Correction of Dentofacial Deformities*. Philadelphia: Saunders, 1980, p. 684.
2. Epker, B.N. and Fish, L.C.: Essential of Treatment Planning. In *Dentofacial Deformities: Integrated Orthodontic and Surgical Correction.* St. Louis: Mosby, 1986, p. 97.
3. Rosen, H.: Occlusal plane rotation: Aesthetic enhancement in mandibular micrognathia. *Plast Reconstr Surg* 91:1231, 1993.
4. Wolfe, S.A., Hu, L., and Berkowitz, S.: In search of the harmonious face: Apollo revisited with an examination of the indications for retrograde maxillary displacement. *Plast Reconstr Surg* 99:1261, 1997.

Vertical Facial Excess

Chapter 9

The etiology responsible for the morphologic designation of the long face is multifactorial and variable. The most common skeletal cause of vertical facial excess is vertical maxillary excess. This may or may not be seen in association with an anterior open bite. Less common causes of the long face are vertical ramus deficiency in the mandibular micrognathic individual and anterior vertical mandibular excess, which is frequently seen in association with mandibular prognathism. These etiologies may occur alone or in some combination. Because the etiologic spectrum is so variable, the overall facial morphology seen in the long-face patient is also quite variable. However, the common morphologic denominator found in all of these patients, regardless of etiology, is a quantitative increase in the vertical dimension of the lower face relative to the midface. This is quantitatively defined as the vertical dimension from subnasale to menton being in excess of the vertical dimension measured from subnasale to glabella, as demonstrated in Fig. 9-1.

Surgical correction of the long-face patient, regardless of the malocclusion associated with it, presents the surgeon with a paradoxical therapeutic dilemma. On the one hand, the surgeon feels obligated to reduce the vertical dimension of the face. On the other hand, we know that vertical contraction of the facial skeleton can be associated with undesirable soft tissue aesthetic consequences as a result of creating relative soft tissue excess. These undesirable aesthetic consequences include the potential for deepening of the nasolabial fold; creation of jowls; deepening of the labiomental fold; thinning out of the upper lip; and, in general, creating an overall lack of skeletal support of the overlying facial mask. What then can we as surgeons do to mitigate these undesirable soft tissue consequences of skeletal contraction in the vertical dimension? The following treatment planning principles are suggested.

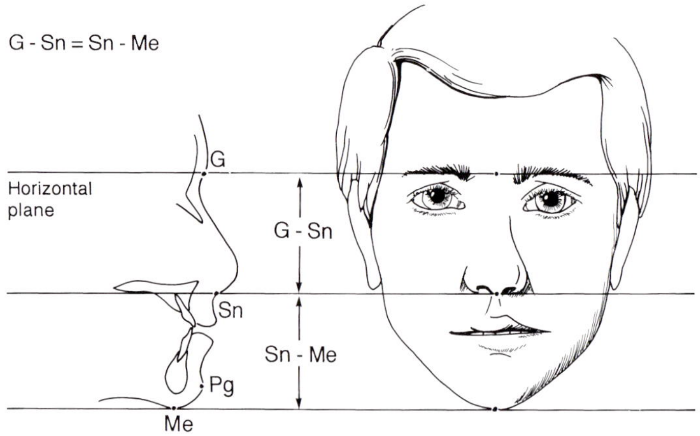

FIG. 9-1 Normal soft tissue vertical facial proportions. The height of the midface from glabella (G) to subnasale (Sn) should approximate the height of the lower face from subnasale to menton (Me). In the long-face patient, the height of the lower face exceeds that of the midface.

The first principle is to undercorrect the patient in the vertical dimension. This is particularly relevant as it relates to vertical maxillary impaction surgery, the most common operation performed to correct the long-face patient. Excessive maxillary impaction can be an aesthetic disaster resulting in insufficient or absent maxillary incisal show. It can create a premature aged appearance for three reasons: lack of incisal show makes the patient look edentulous, an excessive decrease of lower face height is associated with an older appearance because the aging process is accompanied by reduction in the height of the lower face, and excessive vertical reduction will increase the risks of inadequate skeletal support and relative soft tissue excess.

This phenomenon is clearly observed in the patient in Fig. 9-2, who has undergone maxillary impaction surgery for correction of an anterior open bite. Preoperatively, she is displaying approximately 5 mm of incisal show. Postoperatively, there is an absence of tooth show, a thinned upper lip, further deterioration of soft tissue support, and a prematurely aged appearance. Compare her postoperative result to the ideal circumoral aesthetics, seen in Fig. 9-3. There is 6 mm of incisal show, and the lips are full. These examples underlie two realizations. The first is that the often quoted "normal" degree of incisal show of 2 to 3 mm is utter folly. Secondly, there should be a healthy fear of maxillary impaction surgery in patients with 5 to 6 mm of incisal show. In fact, attempts to reduce such degrees of incisal show with maxillary impaction surgery are, in my opinion, ill advised.

It is important to remember that the response of the upper lip to vertical maxillary impaction surgery is not predictable. It certainly does not consistently shorten by 20% of the extent to which one superiorly repositions the maxilla, as some textbooks would have us believe. In some patients there is no shortening of the upper lip, and in others there may even be an increase in length of the upper lip. The best treatment planning ap-

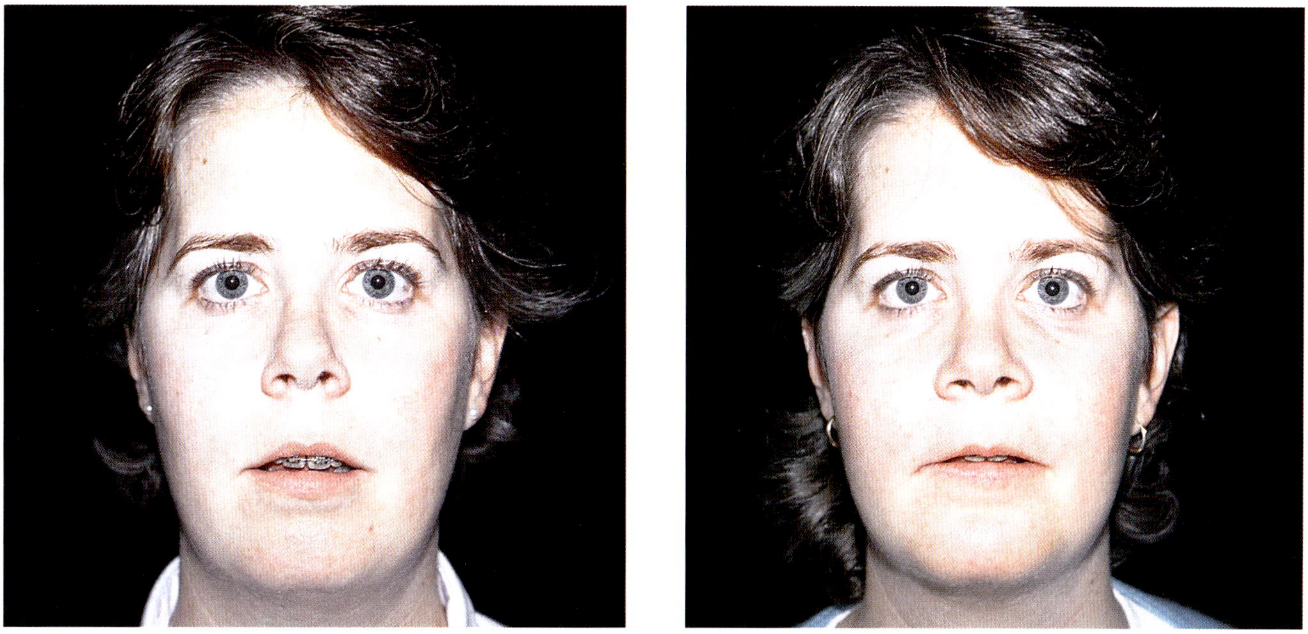

FIG. 9-2 A long-face patient overcorrected with maxillary impaction surgery. There is now an absence of incisal exposure, thinned upper vermillion, and a deterioration in skeletal support of the soft tissues. Also note unattractive alar flaring despite alar cinching. From: Rosen, H.: Lip-nasal aesthetics following Le Fort I osteotomy. *Plast Reconstr Surg* 81:171, 1988. Reprinted with permission.

proach is to assume that there will be no shortening of the upper lip accompanying maxillary impaction surgery. It is far better to leave the patient with modest "excessive" tooth show and perhaps some residual vertical facial excess than it is to risk overshortening the patient. This is particularly true as it relates to the response of the soft tissues.

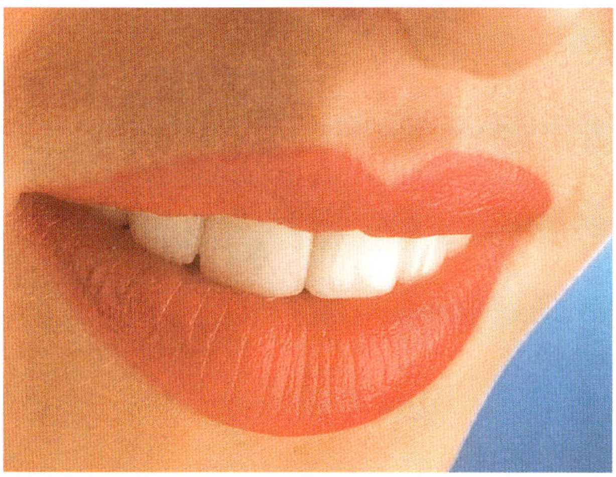

FIG. 9-3 Ideal circumoral aesthetics for a female patient. There is a generous degree of tooth exposure and full lips.

On occasion, it may even be appropriate to consider lengthening the long-face patient in an effort to insure a desirable response of the soft tissues. This is most frequently performed by vertical lengthening of the symphysis and, less commonly, by inferior repositioning of the maxilla as is commonly done in the short-face patient. Although this may seem contrary to our aesthetic goals, it is worthwhile to remember that quantitative vertical facial excess does not necessarily correlate with a face that visibly appears to be too long. As discussed in Chapter 2, quantitative vertical facial excess can be an extraordinarily attractive morphologic facial characteristic. Therefore, the surgeon is obligated to skeletally contract a quantitatively long lower face only when it is accompanied by its associated aesthetic stigmata that make the face look unattractively too long. These include an excessive degree of mentalis and orbicularis oris hyperfunction to achieve lip seal, excessive degrees of gingival show with full smiling, and an excessive attenuation or complete effacement of the labiomental fold. In the absence of these findings, one must not necessarily feel obligated to reduce vertical facial dimension and, in fact, may consider facial elongation given the appropriate set of aesthetic circumstances.

The second underlying principle is to expand the facial skeleton in the sagittal dimension to compensate for what one loses in the vertical dimension. In this way, one can maximize the chances of achieving or maintaining adequate support of the soft tissue. Not infrequently, one may even create an excessive degree of anterior facial divergence in an effort to compensate for the loss in the vertical dimension. Again, it is well worth remembering that excessive anterior facial divergence may, in fact, be extraordinarily attractive. However, it must be accompanied by overall facial convexity, normal maxillomandibular relations as evidenced by the positioning of the upper and lower lips, and an adequately projecting nose.

To summarize, when treating the long-face patient, undercorrect in the vertical dimension and overcorrect in the sagittal dimension.

The Long-Face, Class I Surgical Correction

For purposes of discussion, this patient group can be further divided into three categories. It includes the long-face, Class I, anterior open-bite patient, the long-face, Class I patient without an anterior open bite, and the long-face, Class I patient who has undergone prior orthodontic correction of a Class II malocclusion.

Long-Face, Class I, Anterior Open-Bite Correction

In general, these patients can be adequately corrected with single-jaw surgery involving maxillary impaction. In addition, many require osseous genioplasty of some type. However, double-jaw surgery in this patient group is usually unnecessary. This is because posterior maxillary impaction is necessary to close the anterior open bite. This, in turn, allows for significant mandibular autorotation that results in the desired decrease in height of the lower face as well as significant anterior displacement of pogonion. To accommodate the autorotating mandible, anterior displacement of the maxilla also must occur. This anterior movement of both the maxilla and the mandible compensates for the loss of skeletal volume in the vertical dimension and usually creates adequate projection of the lower face. Osseous genioplasty can then be performed to further enhance the final position of pogonion.

Case Study 9-4

An example of such a patient is illustrated. The facial morphology is characterized by a visibly long lower face and vertical asymmetry with the right side of the chin longer than the left side. With the lips in repose, there is approximately 3 to 4 mm of maxillary incisal show. In the lateral view, there is an anteriorly divergent facial profile with midsagittal convexity and modest parasagittal flattening. The soft tissues appear thick and fleshy with a lack of skeletal support throughout the face. There is an absence of facial angularity and highlighting. There is a Class I, anterior open-bite malocclusion. There is clearly an aesthetic rationale for vertical facial reduction in this patient. This includes the presence of mentalis and orbicularis oris hyperfunction to achieve lip seal and attenuation of the labiomental fold, which further exaggerates the visual impression of vertical facial excess. However, the lip–tooth relations are quite good.

The overall treatment planning approach in this patient will be to undercorrect in the vertical dimension and overcorrect in the sagittal dimension. A prediction tracing demonstrates that posterior maxillary impaction surgery will allow for an 8-mm advancement of pogonion secondary to mandibular autorotation. This must be accompanied by an 8-mm advancement of the maxilla at the nasal spine to maintain a Class I occlusion. This degree of maxillomandibular advancement will overcorrect the patient in the sagittal dimension as the patient has a pre-existing anteriorly divergent profile. No attempt will be made to change the height of the anterior maxilla in an effort to preserve adequate incisal show. Asymmetric vertical reduction genioplasty will be undertaken to restore lower face symmetry, but no attempt will be made to achieve one-to-one vertical proportionality between the mid- and the lower face; that is, the patient will remain undercorrected in the vertical dimension.

Postoperatively, there has been an obvious reduction in the height of the lower face, with subsequent elimination of lip strain and an attractive deepening of the labiomental fold. A deeper fold serves to deemphasize the residual quantitative vertical facial excess secondary to vertical undercorrection. With the lips in repose, there is adequate exposure of the upper incisors. There is a full clinical crown exposed with smiling. The postoperative profile is characterized by an excessive degree of anterior facial divergence compared to normative standards. However, there is good facial convexity, normal relationships between the upper and the lower lips, and an adequately projecting nose. Throughout the face there has been an enhancement of the soft tissue support with improved definition in the submalar, perimandibular, and cervical regions. The occlusal result is an excellent one. Weir nasal excisions would improve the result.

This patient serves to emphasize the treatment planning concept that soft tissue concerns dictate skeletal proportions; that is, the patient was intentionally undercorrected vertically and overcorrected sagitally because of the soft tissue problems with which she presented and the desire to achieve an angular, well-defined result.

PREOPERATIVE

POSTOPERATIVE

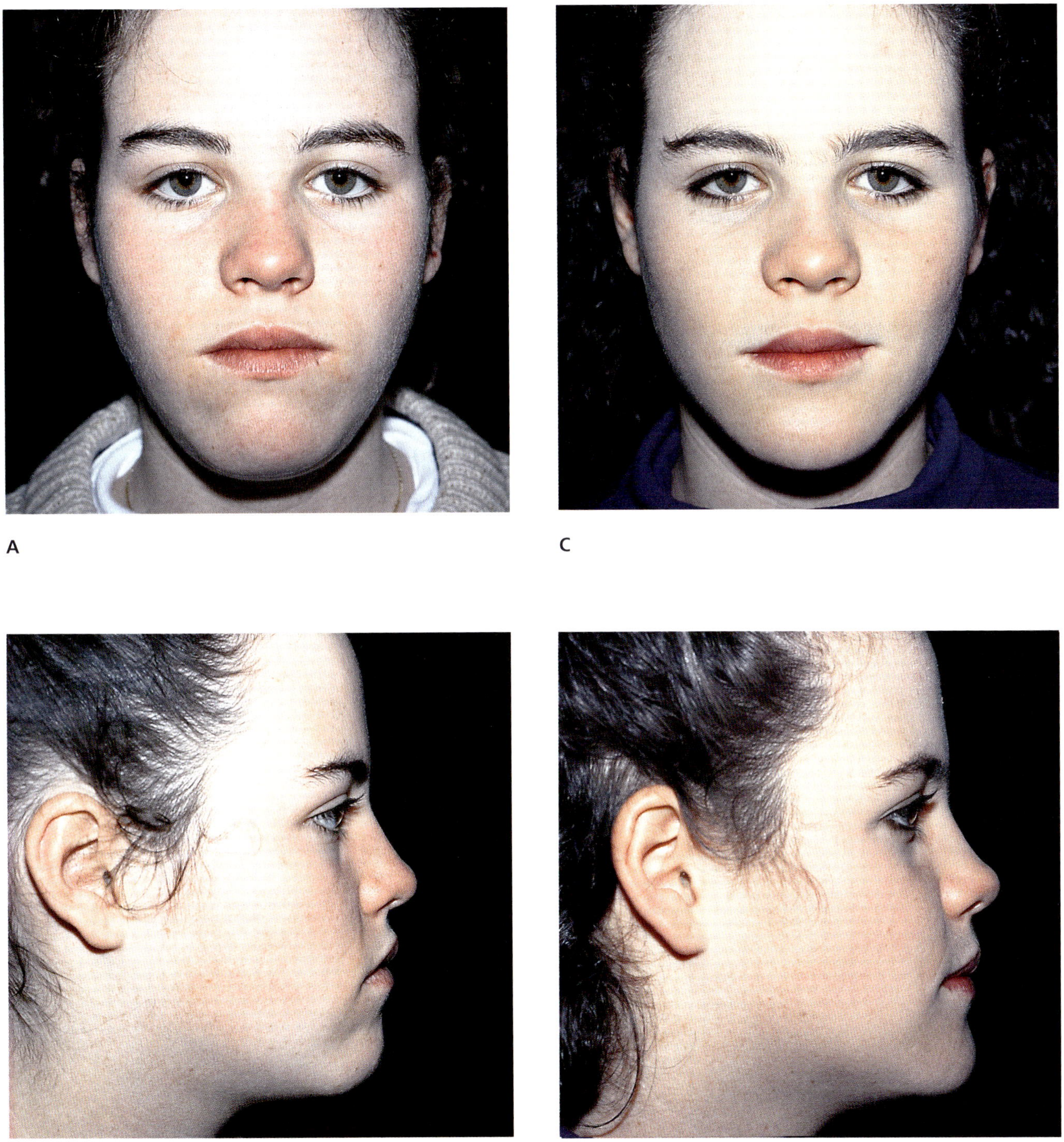

A C B D

CASE STUDY 9-4 A & B: A long face, Class I anterior open bite patient treated with maxillary intrusion, mandibular autorotation, and genioplasty. C & D: The postoperative appearance is characterized by residual vertical facial excess and anterior divergence.

PREOPERATIVE

POSTOPERATIVE

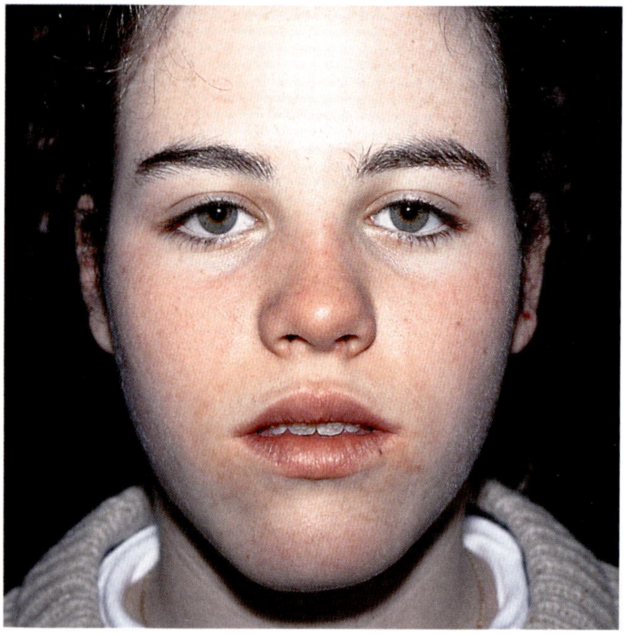

E

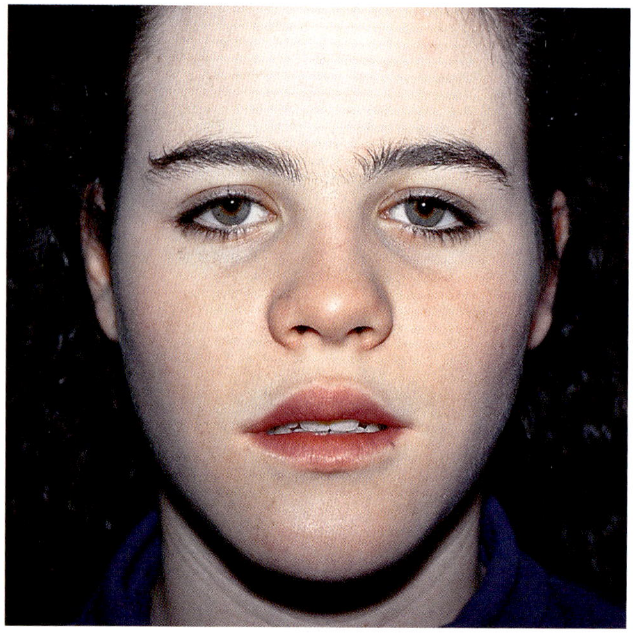

G

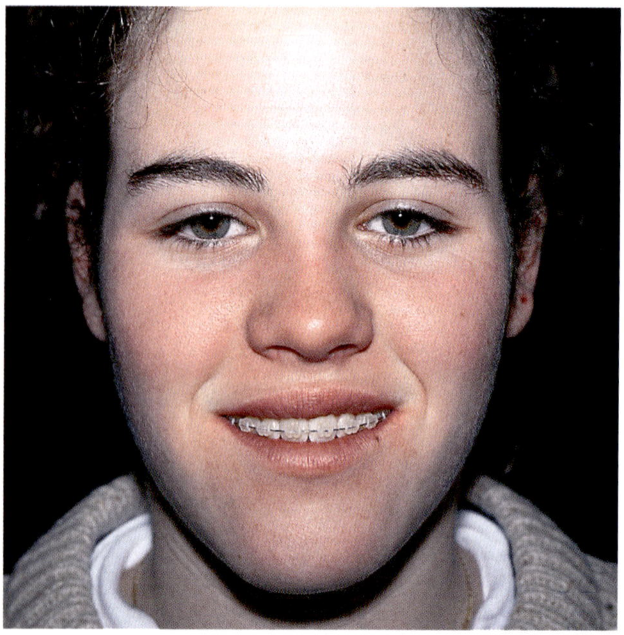

F

H

FIG. 9-4 (*Continued*) G & H: The degree of maxillary incisal show has been preserved. From: Rosen, H.: Aesthetics in facial skeletal surgery. *Perspectives in Plastic Surgery* 6:1, 1992. Reprinted with permission.

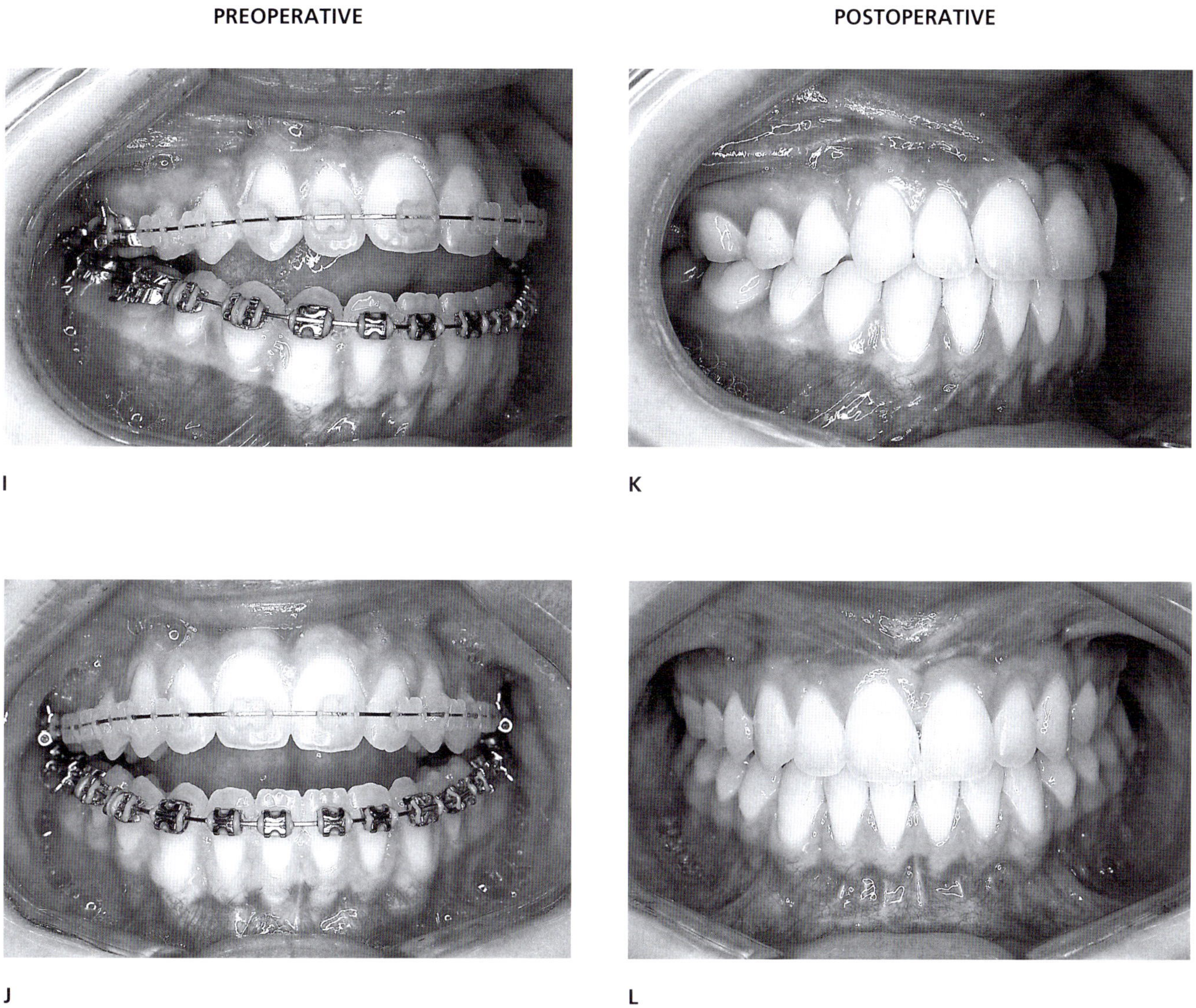

FIG. 9-4 (*Continued*) K & L: Four years following surgery, the anterior bite has remained closed. Orthodontist: Dr. Stephen McConnell

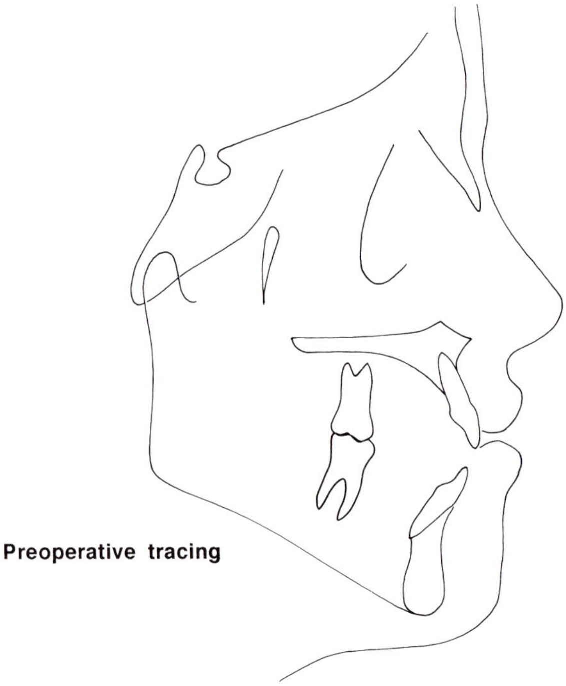

M

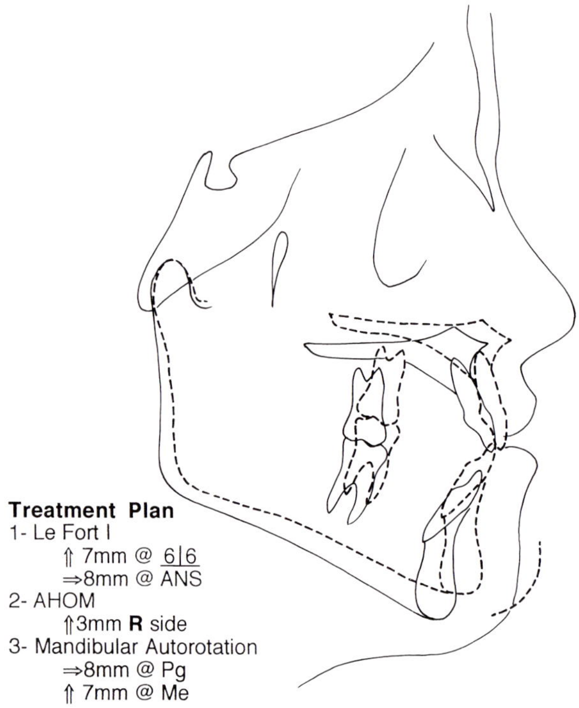

N

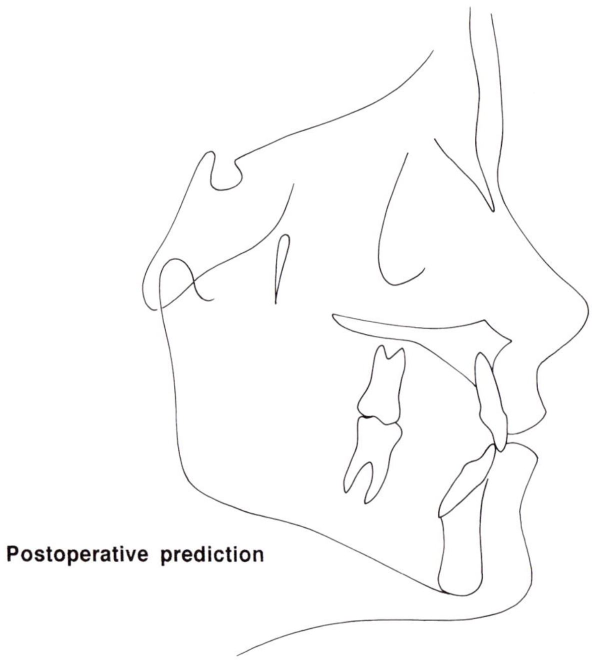

O

FIG. 9-4 (*Continued*) N: The autorotating mandible must be accompanied by an 8-mm advancement of the maxilla to maintain a Class I occlusion.

Case Study 9-5

This patient is similar to the previous patient in that she presents with a Class I, anterior open-bite occlusion. However, unlike the previous patient, the soft tissues are neither thick nor poorly supported; they are rather thin, delicate, and well defined. She is to undergo a very similar operation to the previous patient, that is, posterior maxillary impaction and maxillary advancement of 6 mm to allow autorotation of the mandible to close the anterior open-bite and a vertical reduction genioplasty. Overall, there will be an 8-mm reduction in lower face height.

Postoperatively, the frontal result is quite acceptable. There is preservation of adequate maxillary incisal show with the lips in repose. There is a marked enhancement of the patient's appearance with full smiling. However, viewed laterally, there is an apparent deepening of the nasolabial fold secondary to relative redundancy and ptosis of the soft tissues of the medial cheek. The occlusal result is an excellent one.

This patient serves to emphasize the ever-present danger of sacrificing soft tissue support secondary to vertical facial reduction. This potential exists even in patients with favorable preexisting soft tissue characteristics.

Long-Face, Class I Surgical Correction Without Open Bite

These patients fall into two basic groups. The first, and the less common, is the true Class I patient with vertical maxillary excess. They typically complain about a toothy and/or gummy appearance with smiling. Although maxillary impaction surgery with or without advancement genioplasty will usually provide an acceptable aesthetic result, one must again err on the side of undercorrecting the extent to which the maxilla is superiorly repositioned. This is prudent because it avoids less-than-desirable exposure of the maxillary incisors and the development of inadequate soft tissue support in the midface. In general, this operation should not be undertaken in individuals with less than 6 to 7 mm of incisal show or in individuals with gummy smiles but without excessive incisal show in repose.

The second, and more common, Class I, long-face patient is that individual who has undergone prior orthodontic correction of a Class II, Division 1 malocclusion due to mandibular deficiency. The long-face morphology in these patients is due to concomitant vertical maxillary excess. Orthodontic correction, by retraction of maxillary incisors with or without premolar extractions, achieves a Class I occlusion but leaves the patient with insufficient projection of the lower face. In addition, the upper lip may be inadequately supported, and the excessive incisal show may be aggravated as the incisors are retracted.

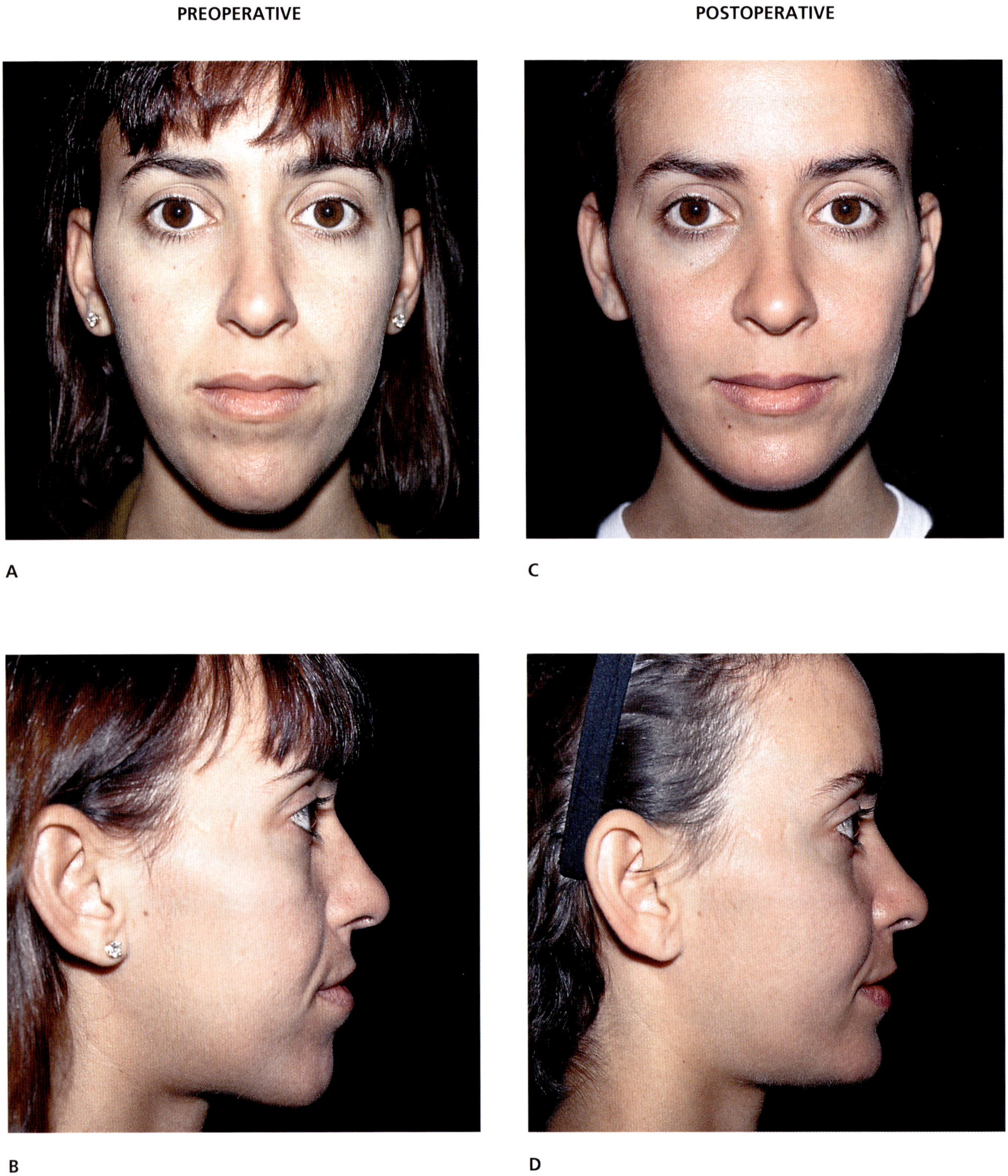

CASE STUDY 9-5 A & B: A long face, Class I anterior open bite patient treated with maxillary impaction, mandibular autorotation, and genioplasty. C & D: There has been an 8-mm reduction in facial height with secondary soft tissue excess in the region of the malar fat pad.

PREOPERATIVE

POSTOPERATIVE

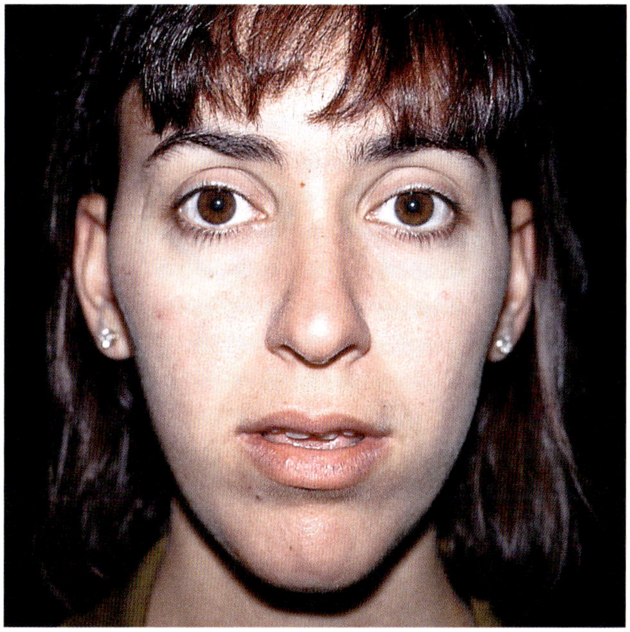

E

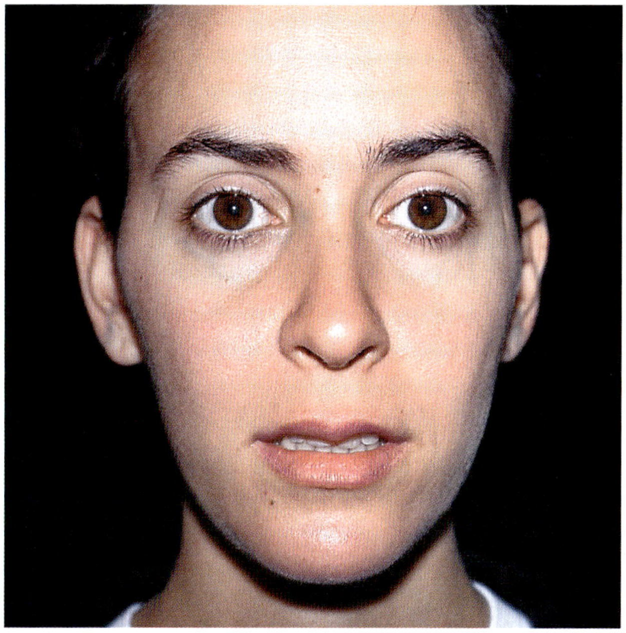

G

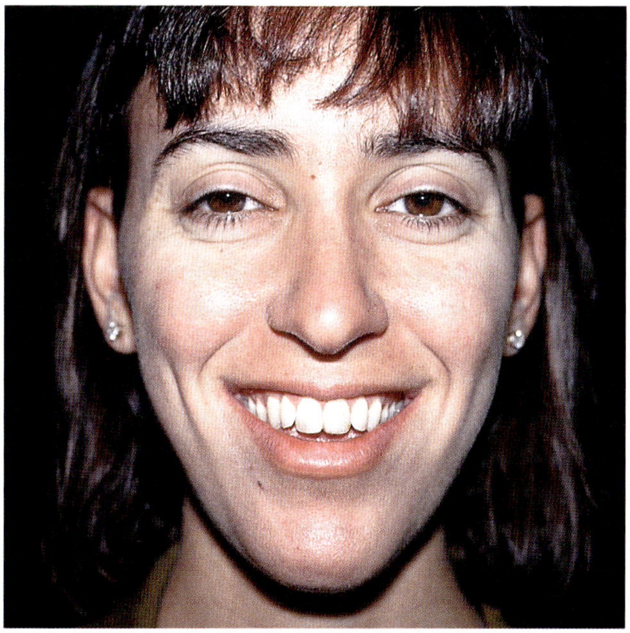

F

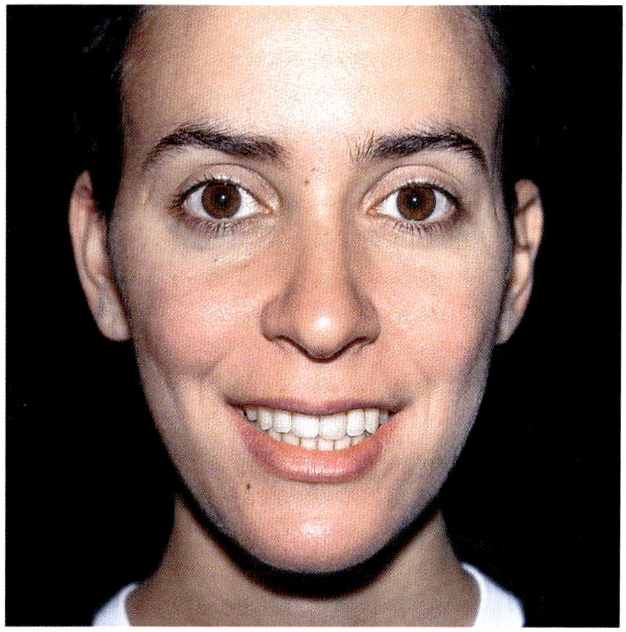

H

FIG. 9-5 (*Continued*) E & F: The extent of maxillary incisal show has been preserved.

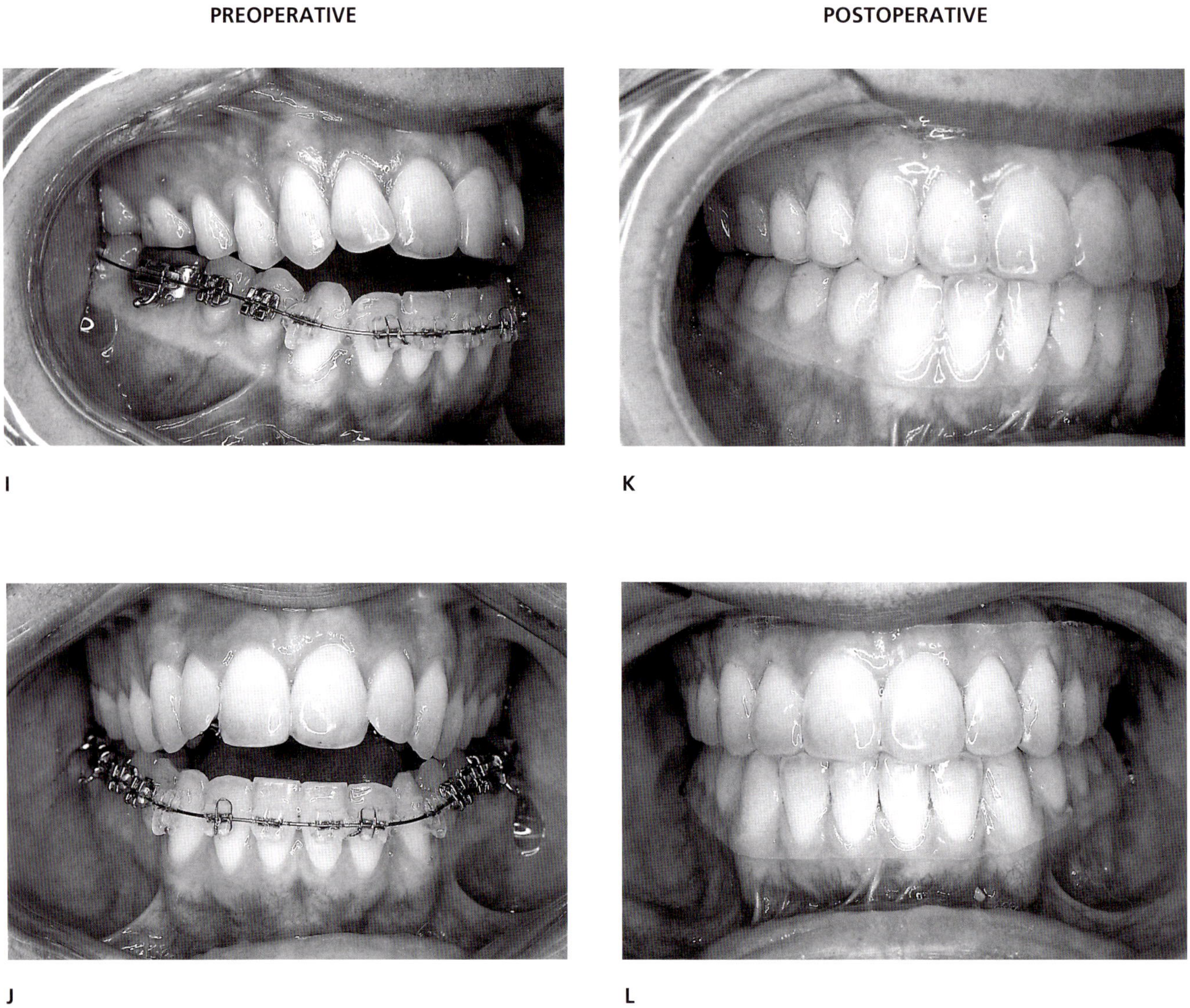

FIG. 9-5 (*Continued*) Orthodontist: Dr. Richard Drummond

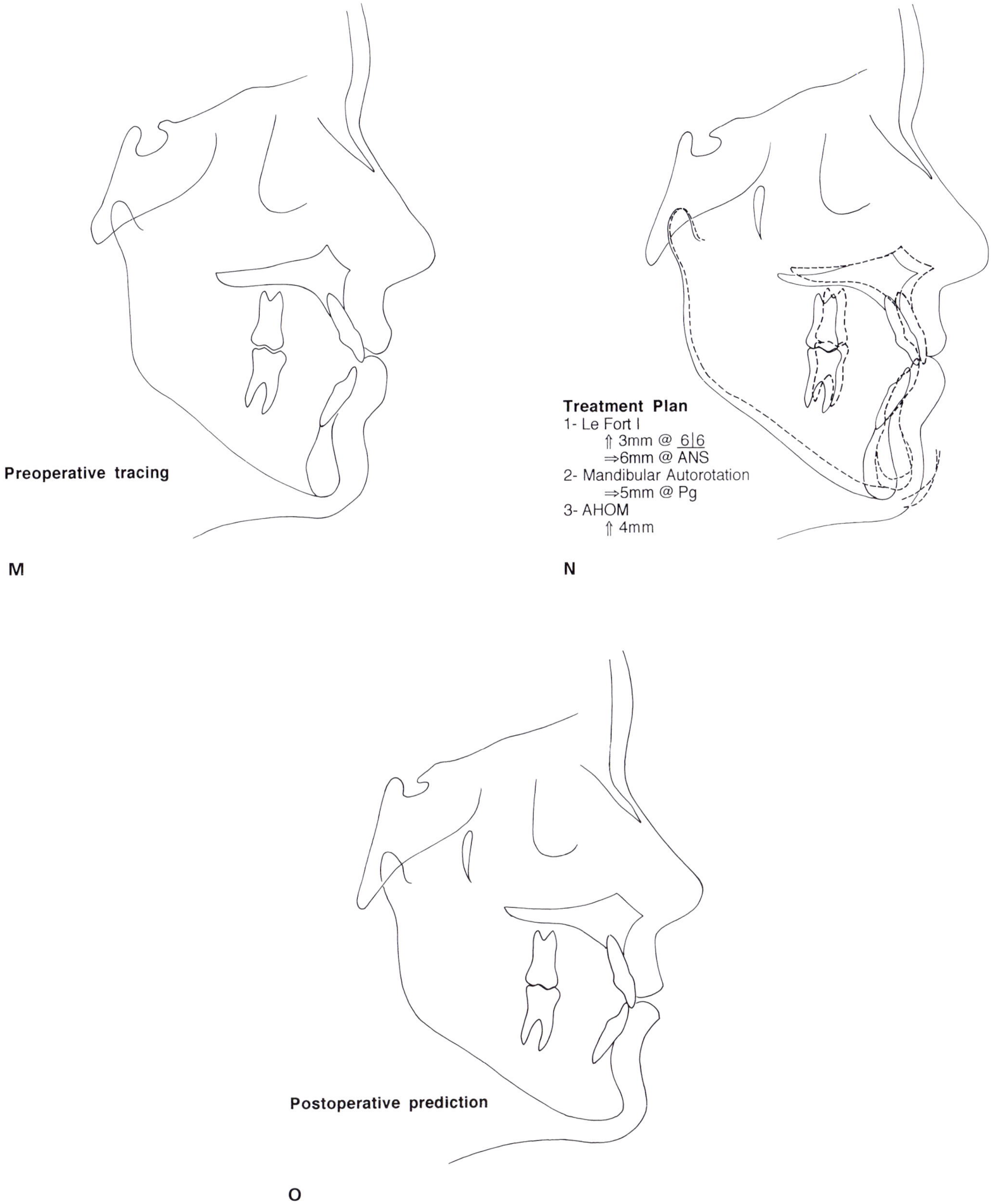

FIG. 9-5 *(Continued)*

Although the patient is now a Class I, long-face patient following orthodontic correction, isolated maxillary impaction surgery, with or without advancement genioplasty, is aesthetically deficient; that is, adequate projection of the lower face cannot be achieved with mandibular autorotation alone. As a result, there is a tendency to overimpact the maxilla to compensate with autorotation for what should be accomplished with mandibular advancement surgery. The basic aesthetic principles of long-face correction are, thus, violated; one has overcorrected vertically and undercorrected sagittally. Excessive dependence is also placed on advancement genioplasty to project the chin, causing it to appear too large relative to the lower lip.

Case Study 9-6

This patient has undergone prior orthodontic correction of a Class II, Division 1 malocclusion. He complains about a toothy appearance and a weak chin. In the frontal view, his lower face is long with lip strain and 7 mm of incisal exposure. The profile is characterized by excessive facial convexity and posterior divergence. There is a solid Class I occlusion. The surgical correction involved a maxillary impaction with 5 mm of anterior shortening and modest advancement to accommodate the autorotating mandible. A 7-mm advancement genioplasty was also performed in an effort to achieve adequate lower face projection. Finally, simultaneous tip rhinoplasty was undertaken.

Postoperatively, there is an unacceptable aesthetic result. The maxillary incisor teeth are now totally hidden beneath the upper lip. There is only 4 mm of exposure with smiling. The profile is characterized by a recessive lower lip, a chin that looks too large relative to the recessed lower lip, and a lack of adequate projection of the entire lower facial third. The postoperative occlusal result pictured is unchanged from his preoperative occlusion. In retrospect, the postoperative prediction tracing was unrealistic; the result depicted could not have been achieved with the surgery that was performed.

To avoid this frequently seen result, these patients should be treated with double-jaw surgery despite a Class I occlusion. This should involve a modest maxillary impaction and generous maxillary advancement, mandibular advancement, and possible modest advancement genioplasty. Occasional counterclockwise rotation of the occlusal plane may be necessary to achieve adequate lower face projection and proper axial inclination of the maxillary incisors.

PREOPERATIVE

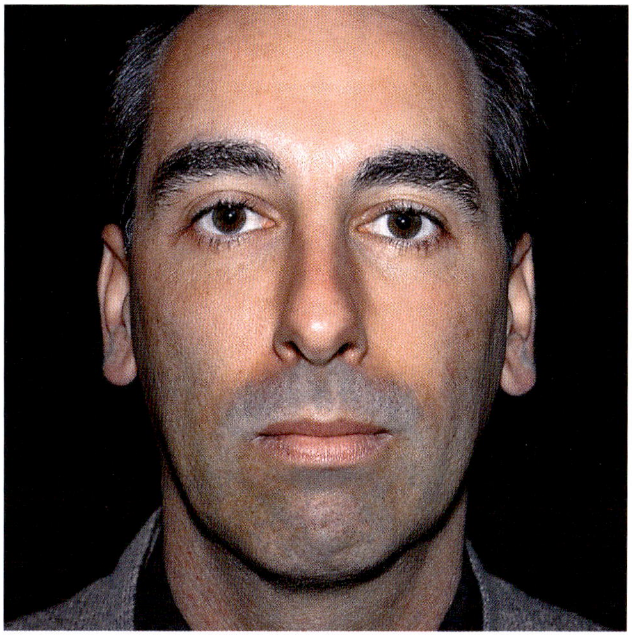

A

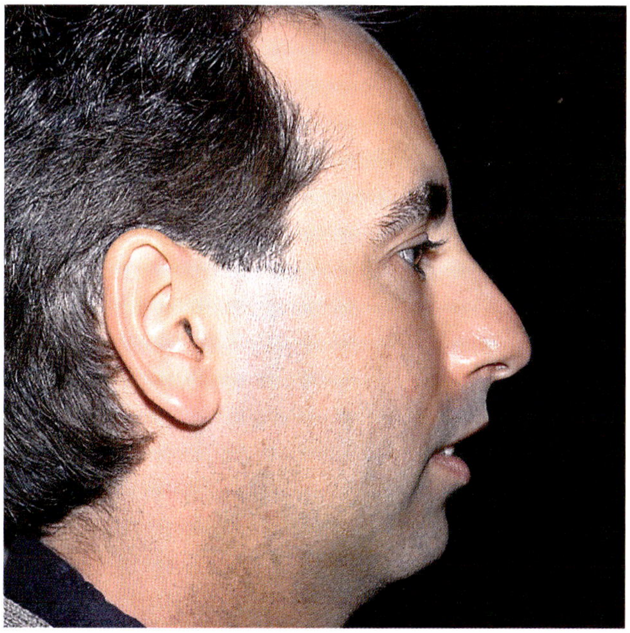

B

POSTOPERATIVE

C

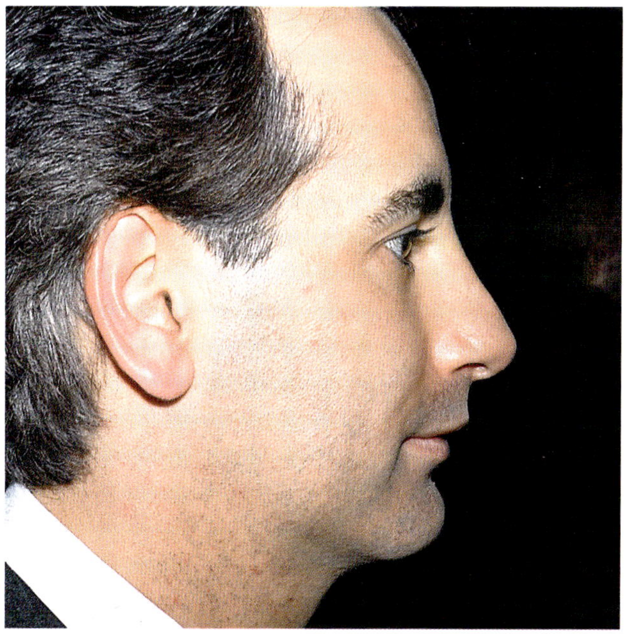

D

CASE STUDY 9-6 A & B: A long face, Class I patient who has undergone prior orthodontic correction of a Class II malocclusion. The patient was treated with maxillary impaction, mandibular autorotation, and advancement genioplasty. C & D: The patient still lacks adequate projection of the lower face. This patient should have been treated with double jaw surgery involving mandibular advancement.

PREOPERATIVE

POSTOPERATIVE

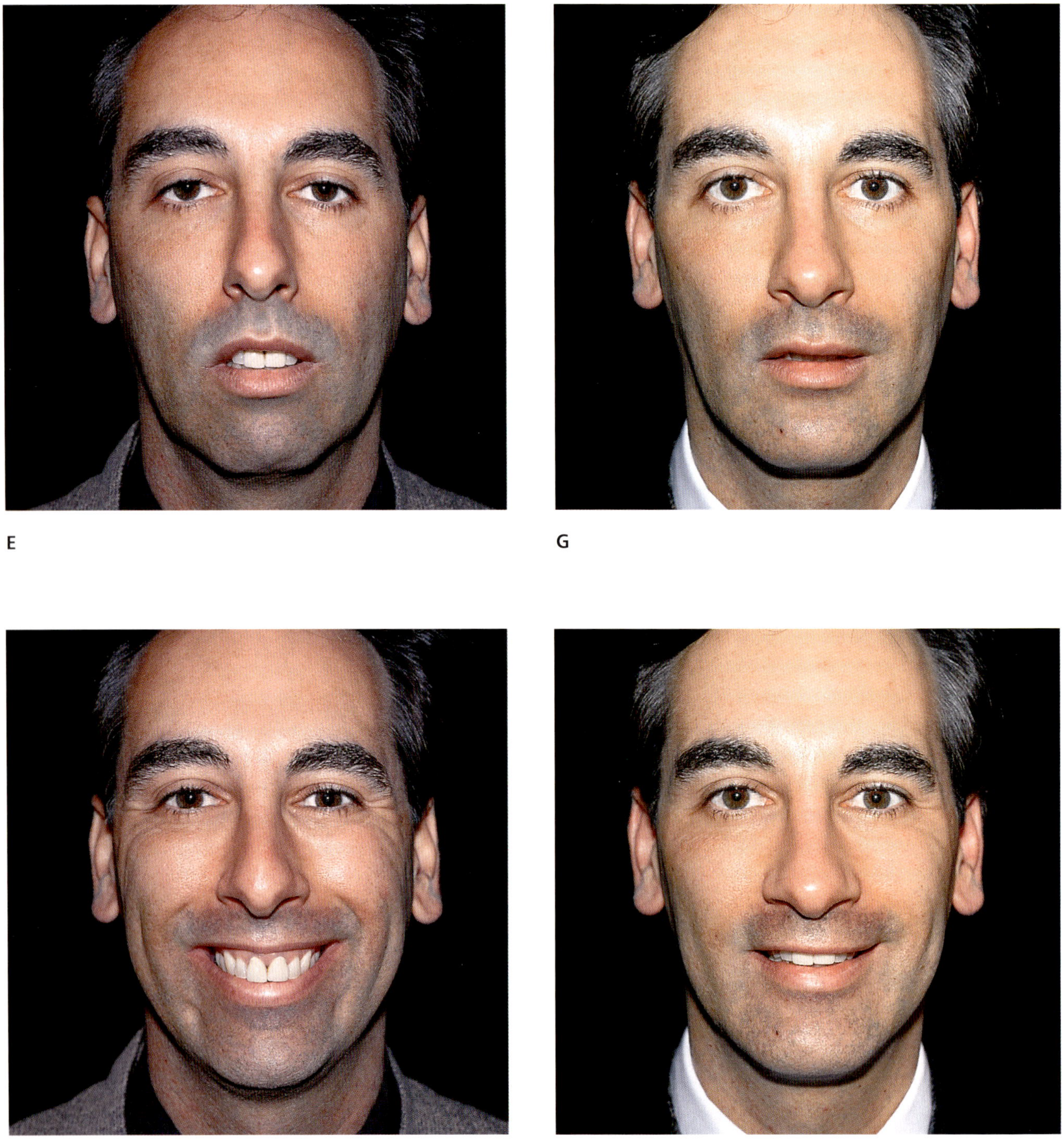

E

G

F

H

FIG. 9-6 (*Continued*) E through H: The patient has been overcorrected vertically. There is now insufficient incisal exposure.

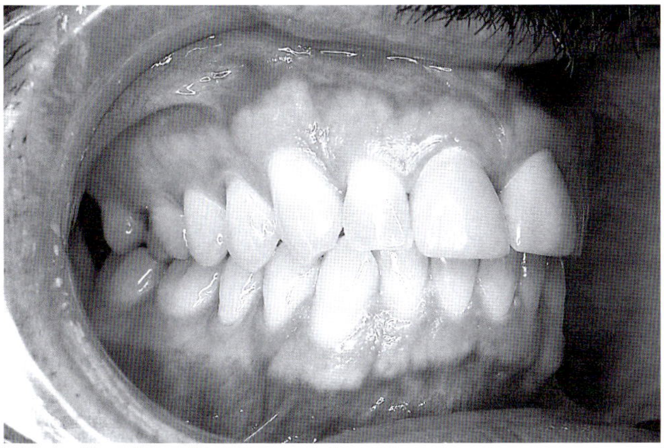

I

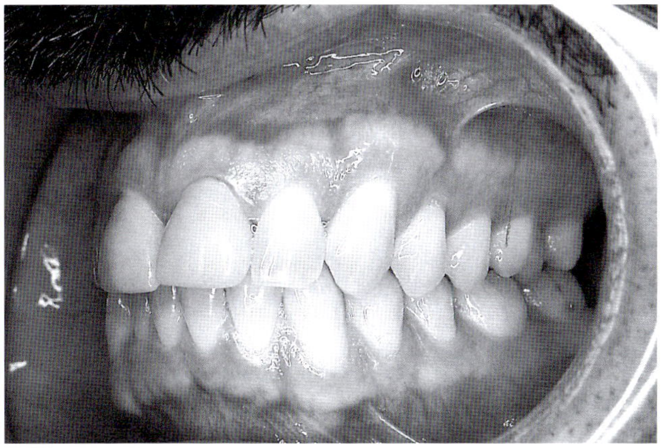

J

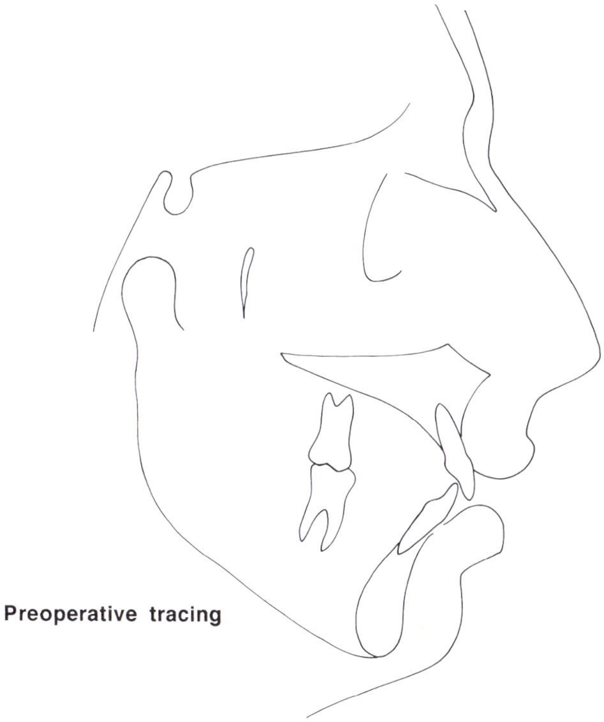

K

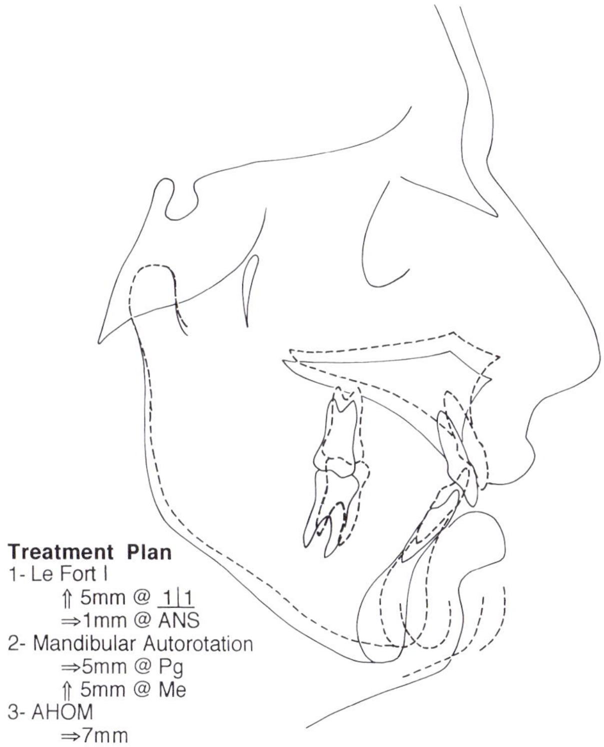

L

FIG. 9-6 (*Continued*) I & J: The postoperative occlusion is identical to the preoperative occlusion. Orthodontist: Dr. James Ackerman

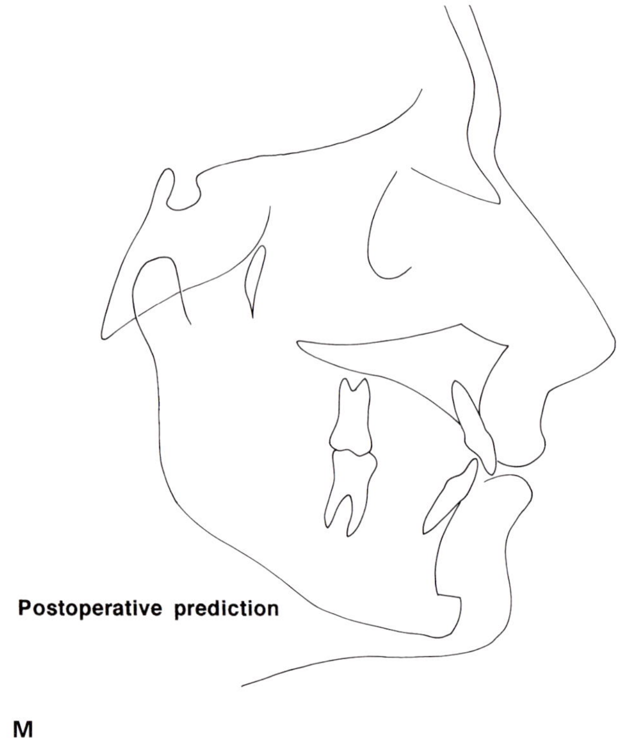

FIG. 9-6 *(Continued)*

Long-Face, Class II Surgical Correction

As a general rule, these patients should be treated with double-jaw surgery, because it is the one surgical option that will consistently provide for adequate projection of the lower face. This is unlikely to occur with maxillary impaction surgery and concomitant mandibular autorotation and genioplasty only. Utilizing this latter surgical option, it may be necessary in many patients to move the maxilla posteriorly to achieve a Class I occlusal relationship. This not only risks lack of projection of the lower face but inadequate support of the upper lip. It results in inadequate soft tissue support because minimal increase in sagittal skeletal volume has taken place to compensate for the loss in the vertical dimension. For all of these reasons, double-jaw surgery is definitely indicated in this patient group.

It is also recommended that a significant advancement of the maxilla, in addition to its impaction, be planned despite its position in the sagittal dimension relative to the mid and upper face. This is suggested for three reasons. The first is that maxillary impaction surgery is always technically easier when accompanied by maxillary advancement because it minimizes the amount of bone needed to be removed in the maxillary tuberosity region as the maxilla is advanced. Secondly, Class II patients not infrequently have a dual bite, that is, a discrepancy between where the patient habitually closes and the true anatomic, seated position of the condyles in the glenoid fossa. If maxillary advancement surgery in such an individual is not planned, posterior repositioning of the maxilla will actually take place during the time of surgery because this discrepancy will have been built into the intermediate splint. A planned anterior vector of maxillary movement will minimize or prevent this tendency. Finally, maxillary advancement will allow for a greater degree of mandibular advancement, which compensates for the loss of skeletal volume in the vertical dimension.

To reiterate, the overall treatment planning philosophy in these patients should be to undercorrect in the vertical dimension and overcorrect in the sagittal dimension. Soft tissue considerations must influence skeletal proportions.

Surgical Corrections Utilizing Mandibular Autorotation and Mandibular Advancement

Mandibular autorotation, defined as rotation of the entire mandible with the center of rotation at the condyle following superior repositioning of the maxilla, is an integral component of surgical corrections of the long-face, Class II patient. Mandibular autorotation decreases the sagittal discrepancy between the maxilla and mandible and reduces posterior face height at gonion. Accordingly, this technique is most appropriately utilized in patients who have increased posterior face height as well as increased anterior face height. Following mandibular autorotation, remaining excessive overjet is eliminated with sagittal split ramus osteotomy to advance the mandible.

Case Study 9-7

This patient has a long face secondary to vertical maxillary excess and a posteriorly divergent profile with excessive convexity due to a small mandible. The maxilla is well positioned in the sagittal plane. The soft tissues are poorly supported with a lack of facial definition. They appear thickened and redundant. There is a clear aesthetic rationale to vertically contract the facial skeleton; that is, there is 8 mm of incisal show with the lips at rest, excessive exposure of gingiva with full smiling, and severe lip strain in an effort to achieve lip seal. There is a Class II, Division 1 malocclusion with artificial opening of the bite due to the presence of maxillary lingual appliances.

The presence of poor soft tissue support in the context of the need to vertically contract the facial skeleton dictates that the principle of facial disproportion be utilized in an effort to achieve the soft tissue goals. The maxilla will be impacted only 4 mm anteriorly despite 8 mm of tooth show. This will leave the face quantitatively long. Despite being in good sagittal position, the maxilla will be advanced a full 6 mm to allow pogonion to come forward 12 mm due to 4 mm of autorotation and 8 mm of mandibular advancement. The maxilla will also be split to achieve 5 mm of posterior widening.

The patient's frontal, postoperative appearance is characterized by residual quantitative vertical facial excess. However, this does not detract from her appearance as the lip strain has been eliminated. In addition, the excessive degree of gingival show with smiling has been entirely eliminated. Surprisingly, there is now only 2 mm of tooth show despite a minimal 4-mm anterior maxillary impaction. The upper lip has clearly lengthened in this patient! The profile is now anteriorly divergent to an excessive degree but with good overall convexity, normal labial relationships, and an adequately projecting nasal dorsum and tip. Secondary to having maintained vertical facial excess and having created sagittal disproportion, there has been a marked enhancement of the overall support of the soft tissues with improvement in the facial highlighting and definition. However, there is still a minimal amount of excess soft tissue in the perimandibular region. Modest liposuctioning in the jowl region would have improved the result. The occlusal result is excellent.

PREOPERATIVE POSTOPERATIVE

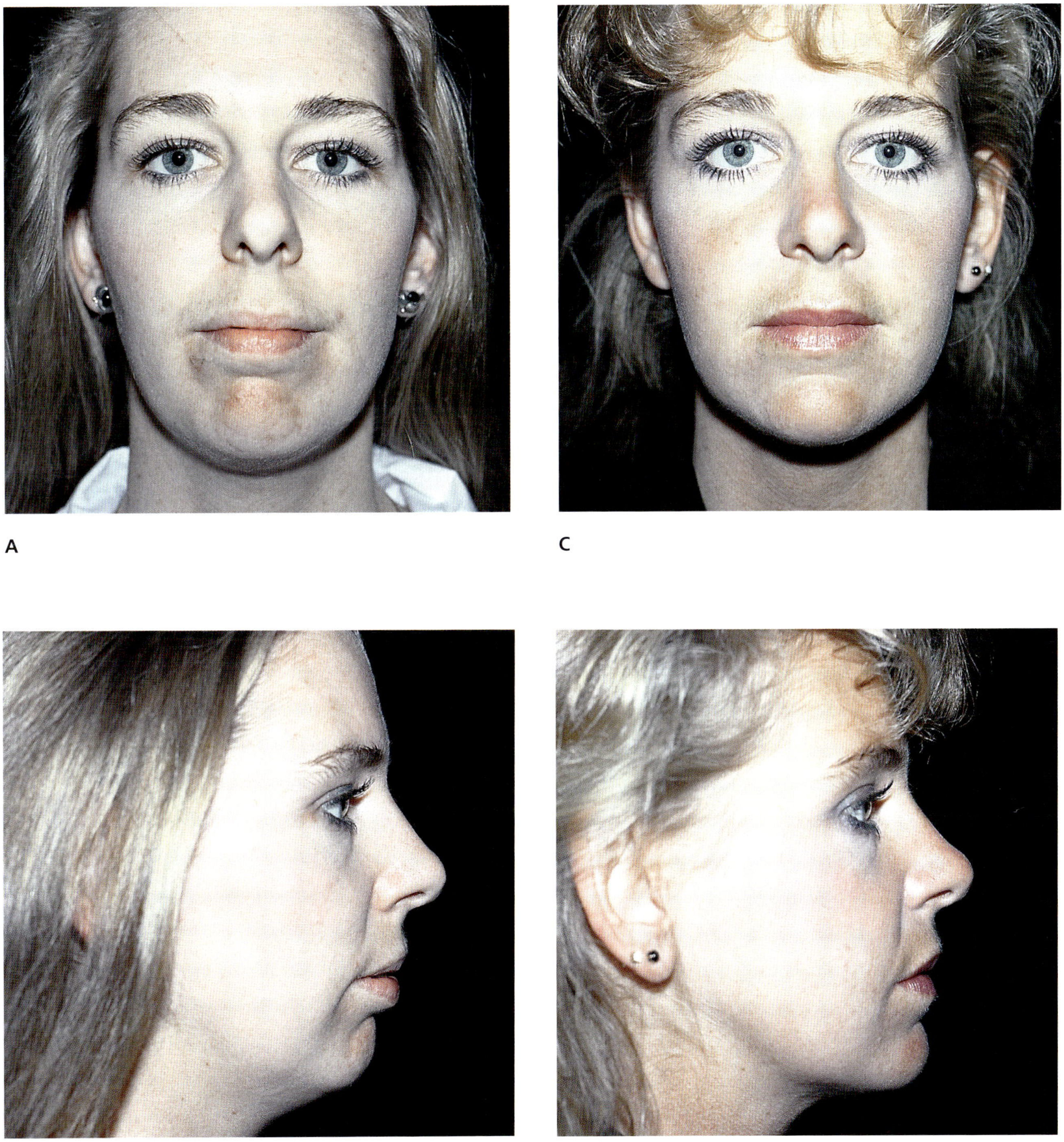

A C

B D

CASE STUDY 9-7 A & B: A mandibular deficient, vertical maxillary excess patient treated with double jaw surgery. C & D: The patient has intentionally been undercorrected vertically and overcorrected sagitally in an effort to enhance skeletal support of the soft tissues.

PREOPERATIVE **POSTOPERATIVE**

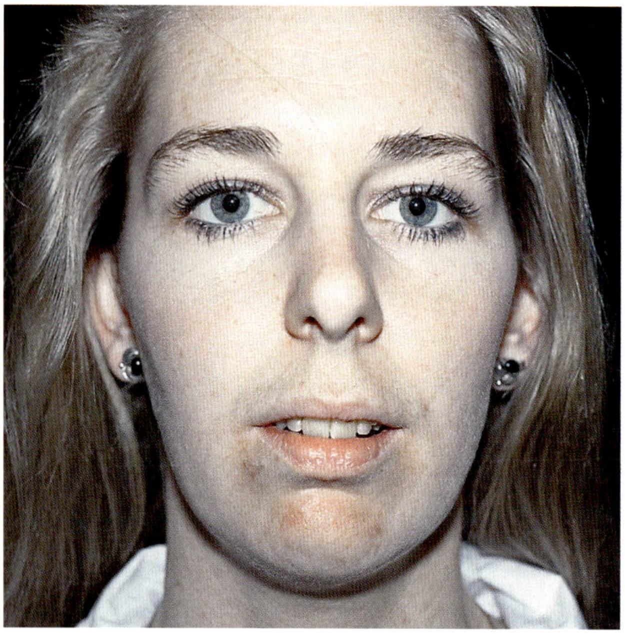

E

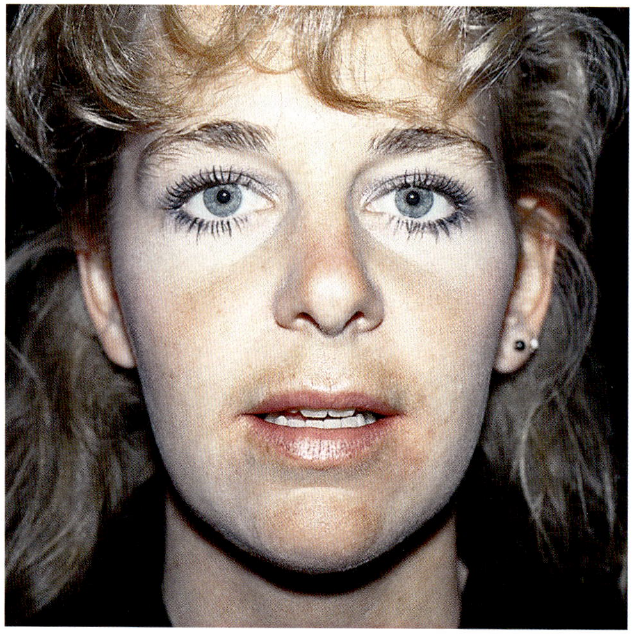

G

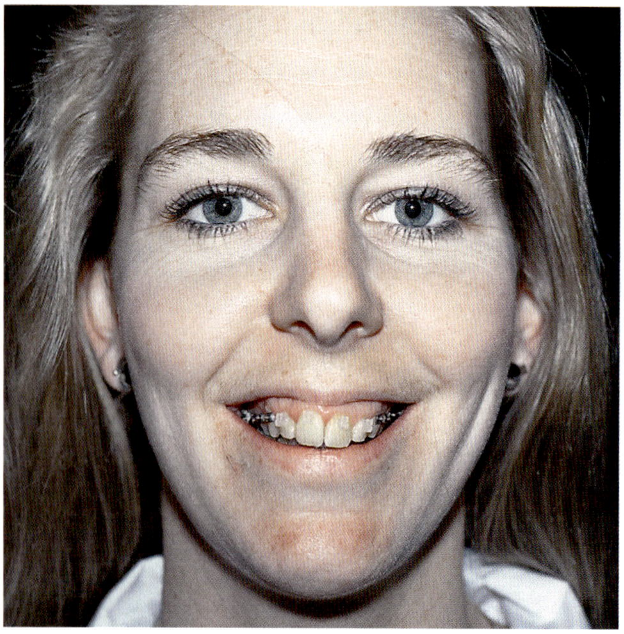

F

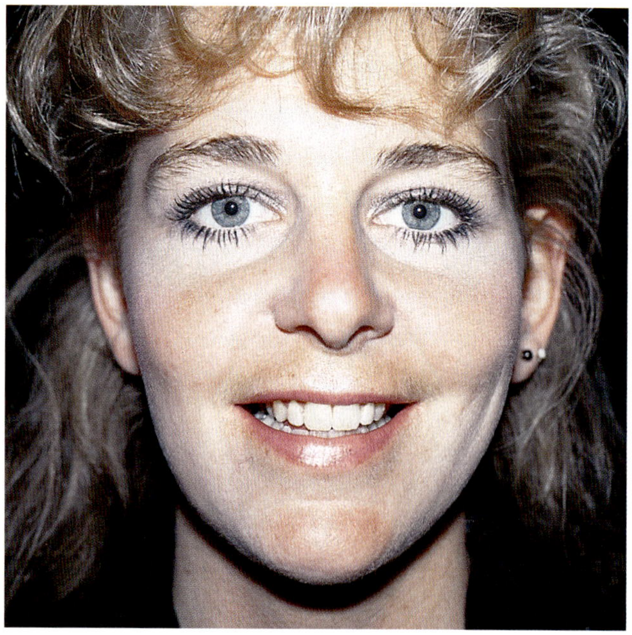

H

FIG. 9-7 *(Continued)* G & H: Despite a modest 4-mm anterior maxillary impaction, there is now only marginal incisal exposure. The smile has been dramatically improved. Note the correction of the occlusal cant. From: Rosen, H.: Facial skeletal expansion: Treatment strategies and rationale. *Plast Reconstr Surg* 89:798, 1992. Reprinted with permission.

PREOPERATIVE POSTOPERATIVE

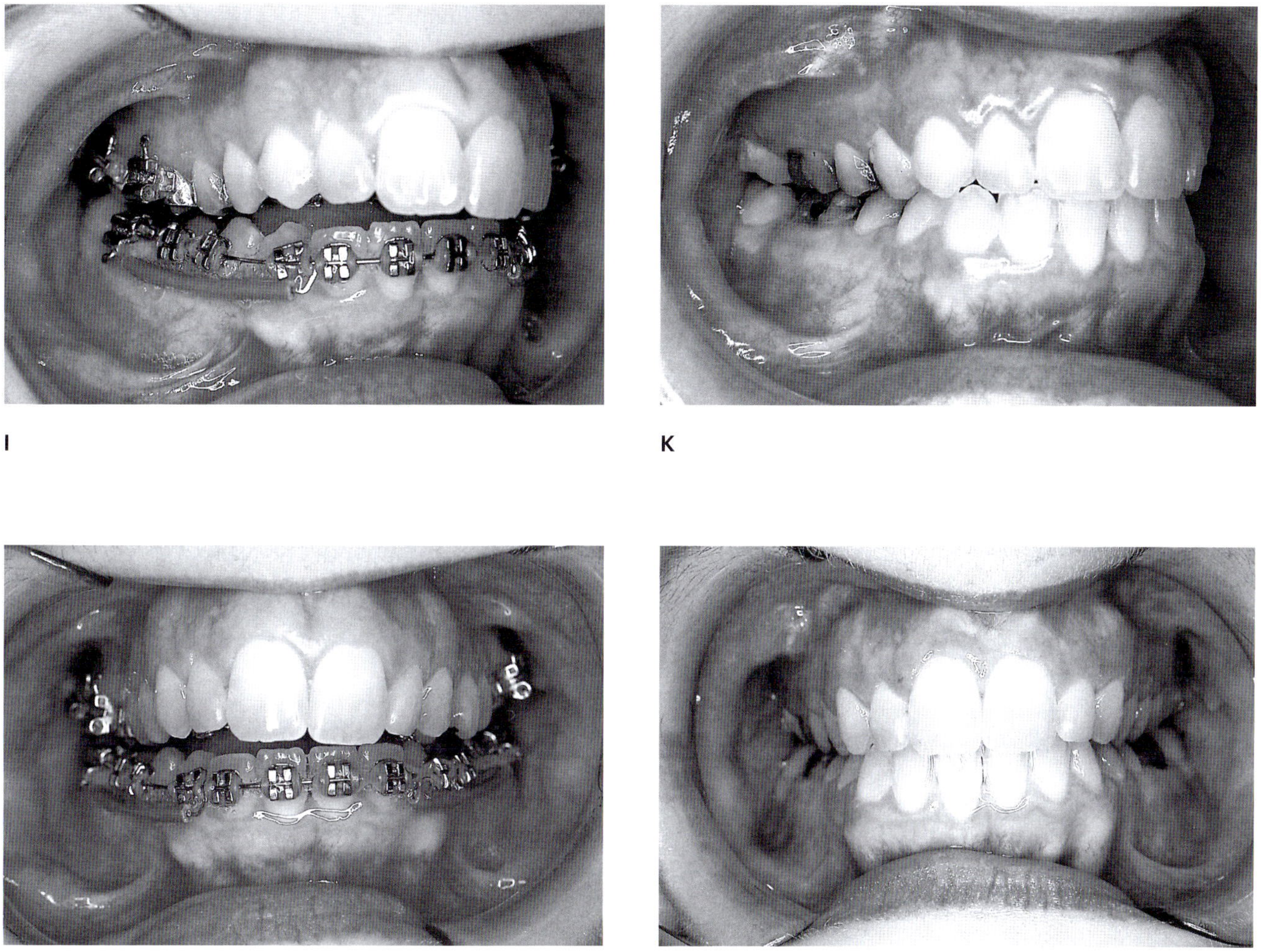

I K J L

FIG. 9-7 (*Continued*) I & J: The bite is artificially open due to lingual appliances. K & L: Postoperative occlusal result at twenty months. Orthodontist: Dr. Gerald Ginsberg

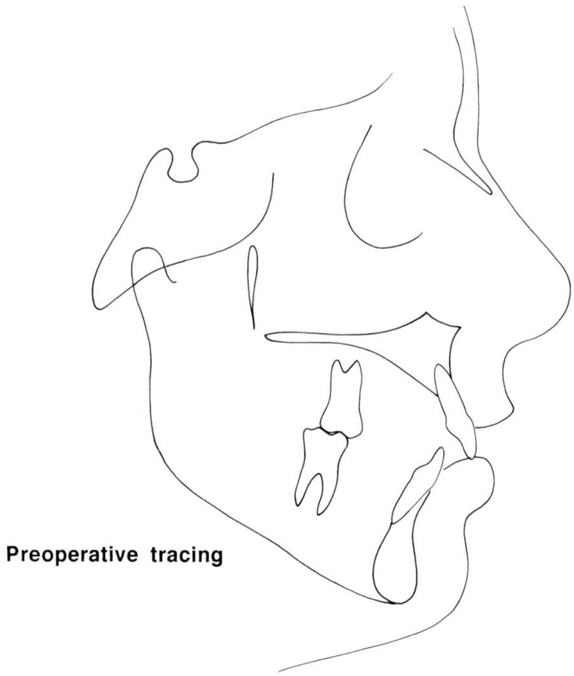

M

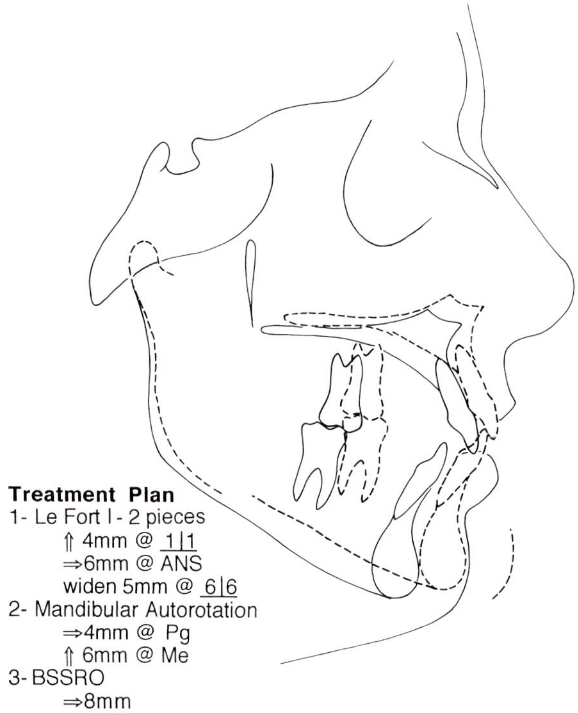

N

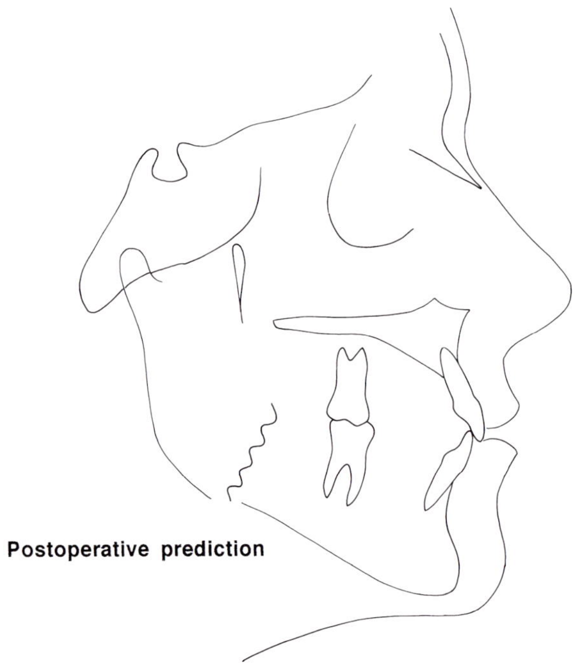

O

Fig. 9-7 *(Continued)*

Case Study 9-8

This patient presents with similar paradoxical treatment planning concerns. There is a clear aesthetic need to vertically contract the facial skeleton, seen in the presence of soft tissue excess, poor skeletal support, and a total lack of facial angularity and highlighting. There is approximately 10 mm of incisal show at rest and excessive gingival show with smiling. The maxilla is reasonably well positioned sagittally, although there is some paranasal flattening. An additional aesthetic concern is that the soft tissue chin is sagittally prominent relative to the lower lip. There is a Class II, Division 1 malocclusion with excessive uprighting of the anterior maxillary teeth due to previous orthodontic attempts to correct the malocclusion.

One would ideally like to deal with the relative chin prominence by clockwise rotation of the occlusal plane as the mandible is brought forward so that the lower lip advances more than pogonion. As demonstrated by the patient's cephalometric tracing, clockwise rotation of the occlusal plane is not feasible, because this would further upright the maxillary anterior teeth. The second alternative is to vertically elongate the chin (a seemingly counterproductive approach in a long-face patient) to soften the area of the labiomental fold and create the visual illusion of a less prominent chin relative to the lower lip.

The chosen surgical correction involves a two-piece Le Fort I osteotomy to advance the maxilla 6 mm and impact anteriorly 6 mm and posteriorly 6 mm and to widen 4 mm in the first molar region. The mandible will be advanced 8 mm. An osseous genioplasty to vertically elongate the symphysis by 4 mm is planned. Submental lipectomy will also be performed.

Postoperatively, the patient's lower face remains quantitatively long, and there is, perhaps, a very small amount of residual mentalis hyperfunction; the patient has been vertically undercorrected. There is approximately 5 mm of incisal show with the lips in repose and a marked enhancement of the aesthetics of the patient's smile. The profile is again characterized by an overcorrection in the sagittal dimension but with good overall facial convexity and adequate projection of the nose. The position of the upper lip relative to the lower lip is excellent. Notice the visual impression of decreased chin prominence relative to the lower lip. There is a good occlusal result. Overall, there has been marked improvement of the soft tissue support with increased definition and angularity.

PREOPERATIVE | POSTOPERATIVE

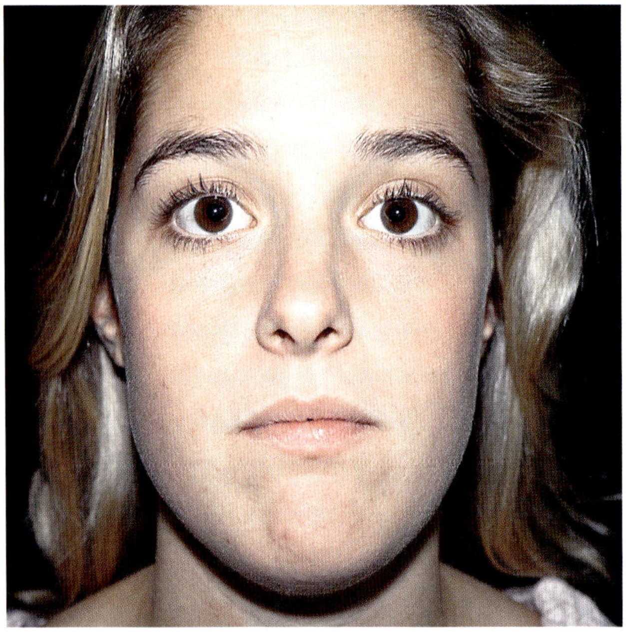

A

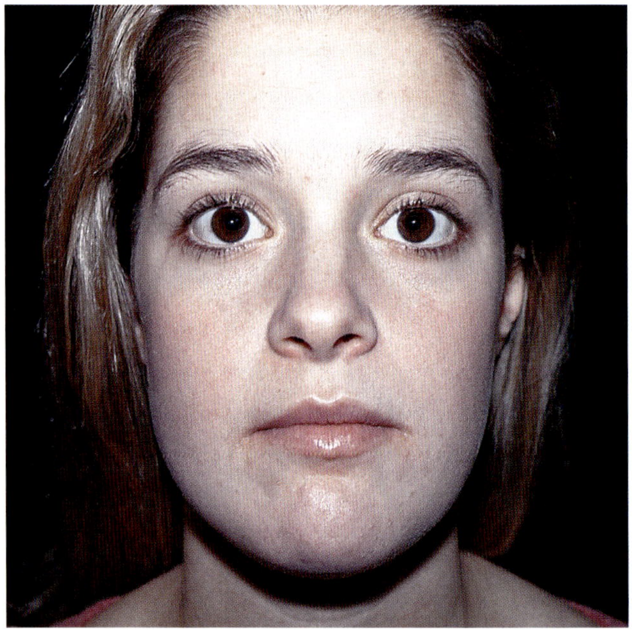

C

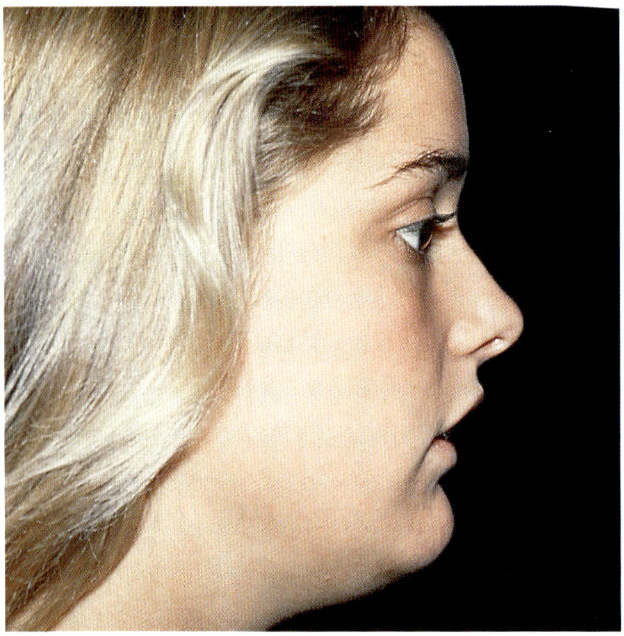

B

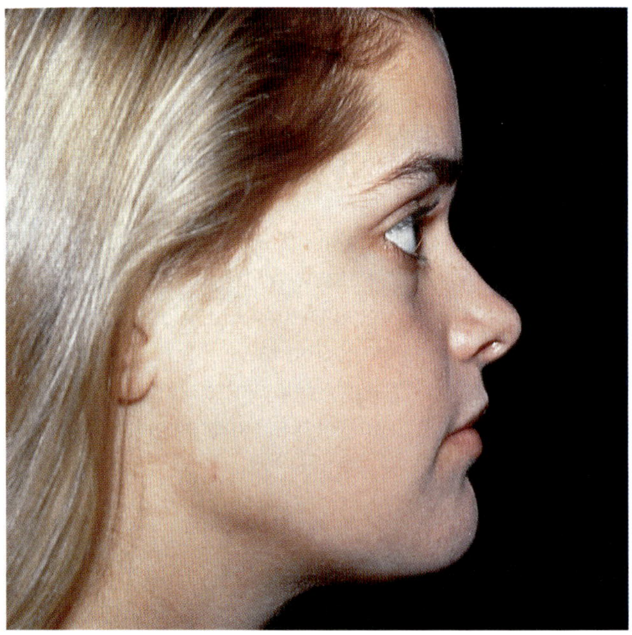

D

Case Study 9-8 A & B: A long face, Class II patient treated with double jaw surgery and vertical lengthening genioplasty. C & D: There is residual vertical facial excess (undercorrection) and an anteriorly divergent profile with overall convexity. The definition and angularity of the soft tissue has been markedly improved.

PREOPERATIVE

POSTOPERATIVE

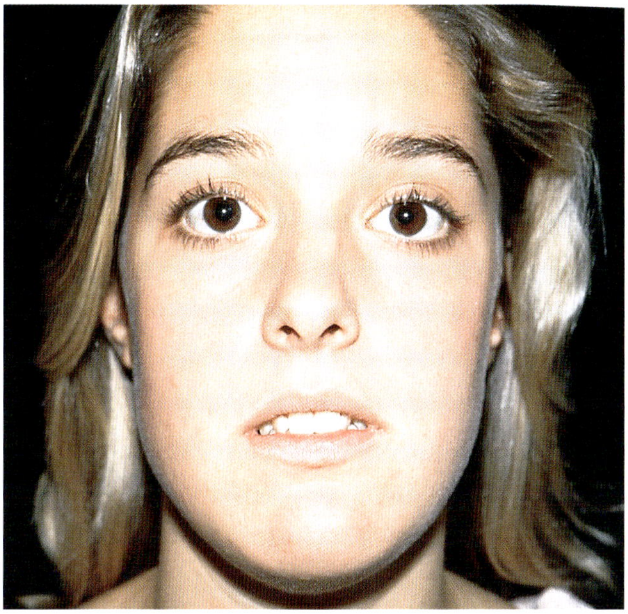

E

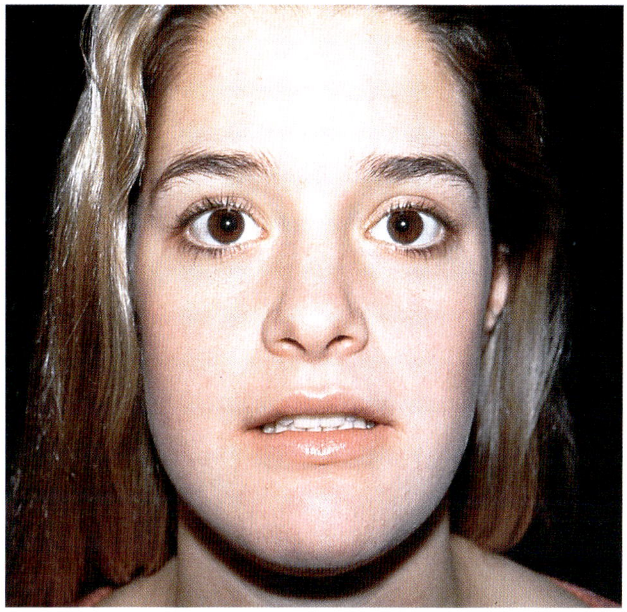

G

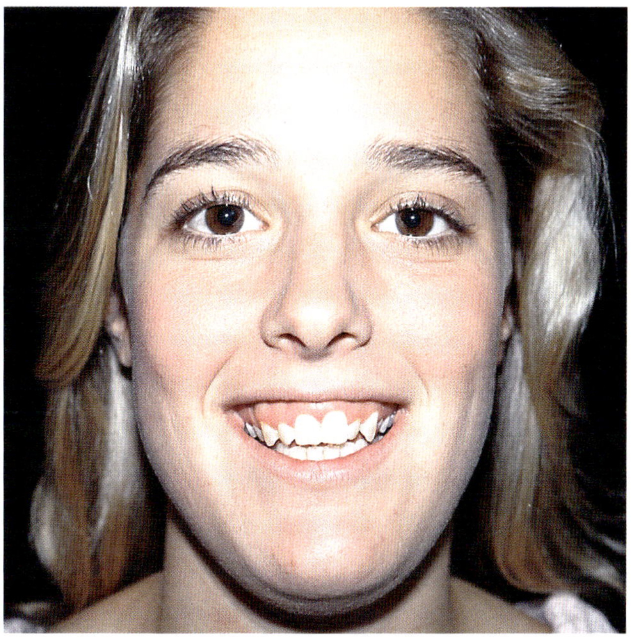

F

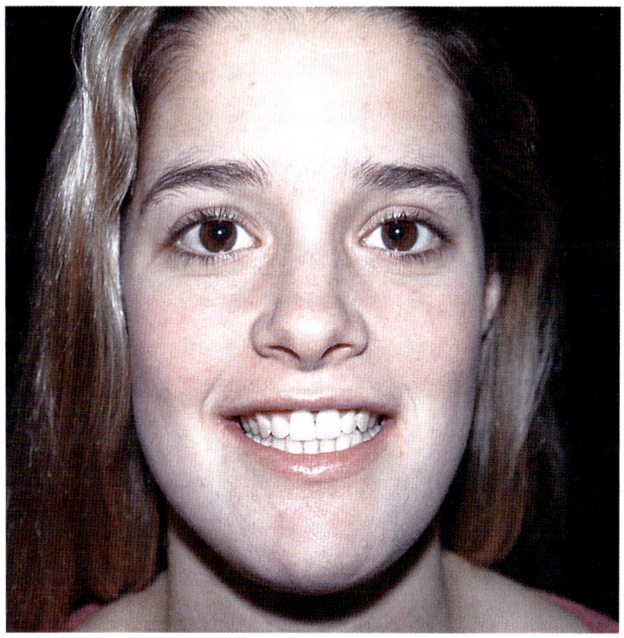

H

FIG. 9-8 *(Continued)* G & H: There is still a generous display of the maxillary teeth in both repose and full smiling. From: Rosen, H.: Aesthetics in facial skeletal surgery. *Perspectives in Plastic Surgery* 6:1, 1992. Reprinted with permission.

PREOPERATIVE | POSTOPERATIVE

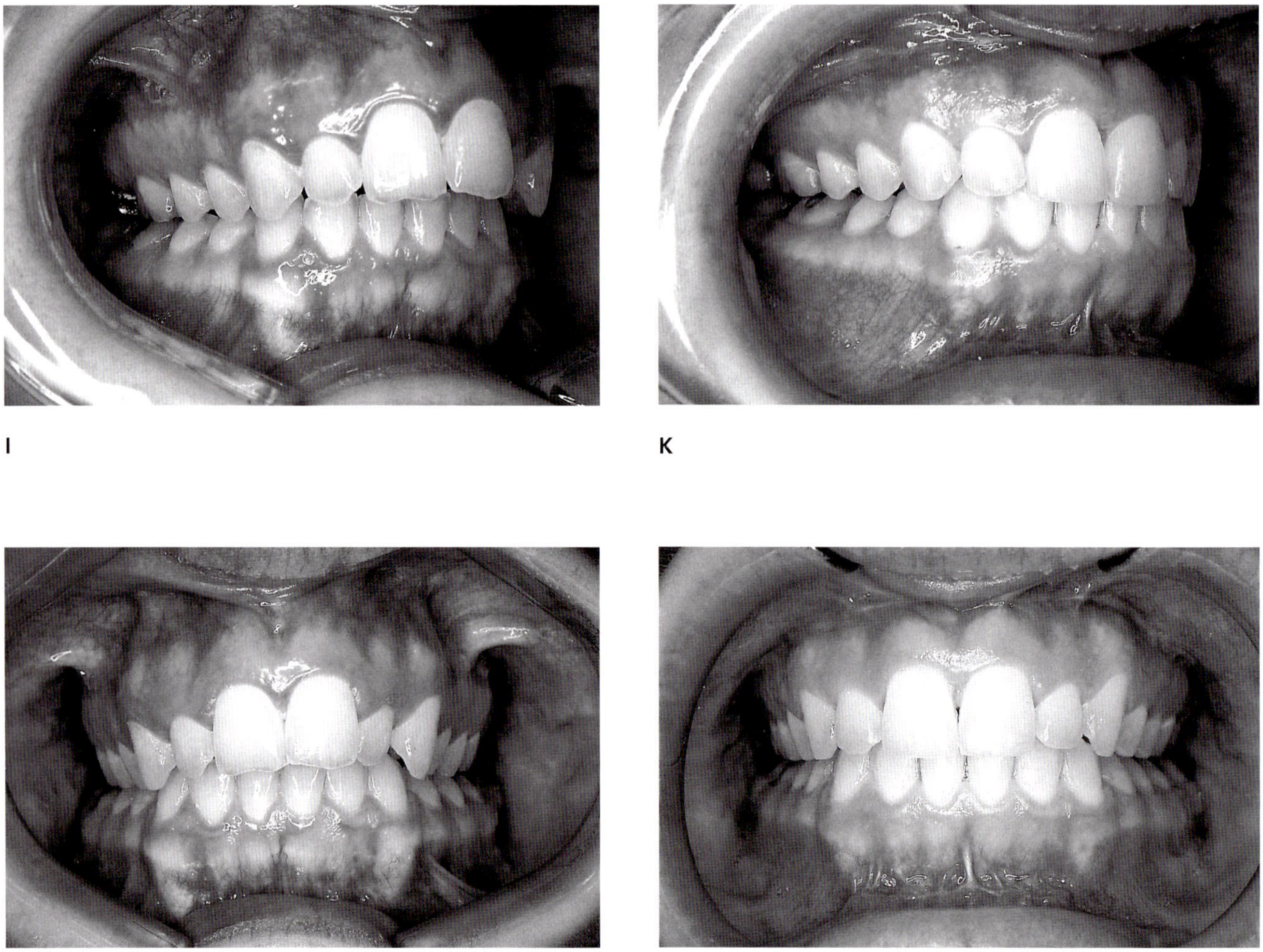

I, K, J, L

Fig. 9-8 (*Continued*) I & J: Note the uprighted maxillary incisors. K & L: The postoperative occlusion at twenty-four months. Orthodontist: Dr. James Ackerman

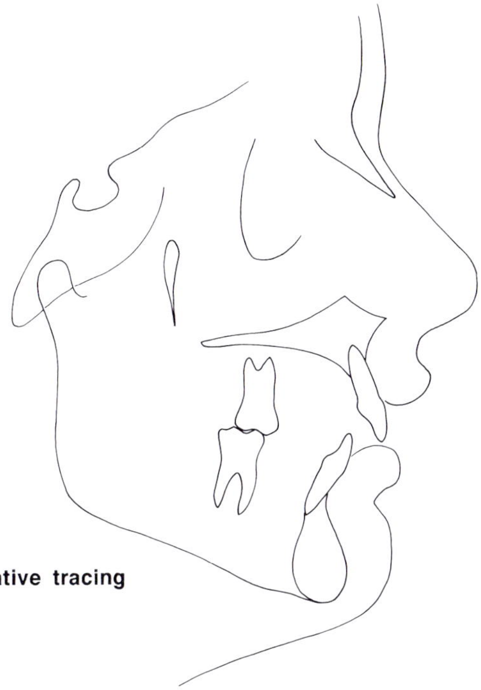

M

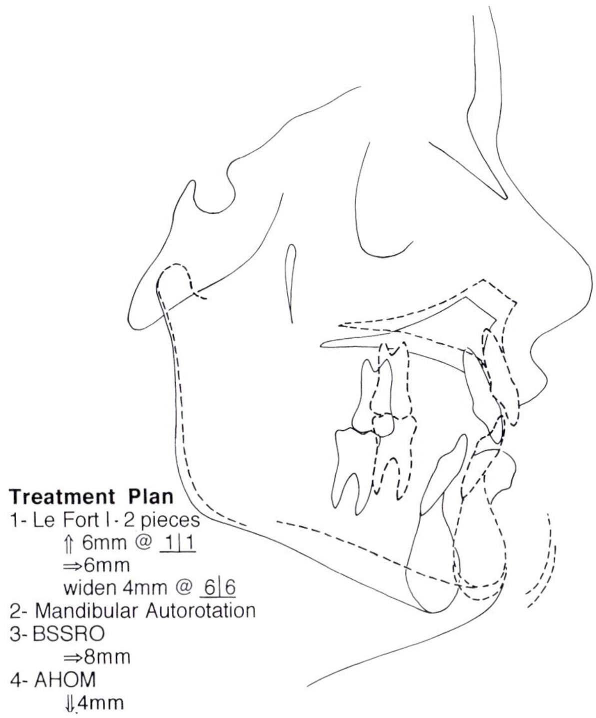

N

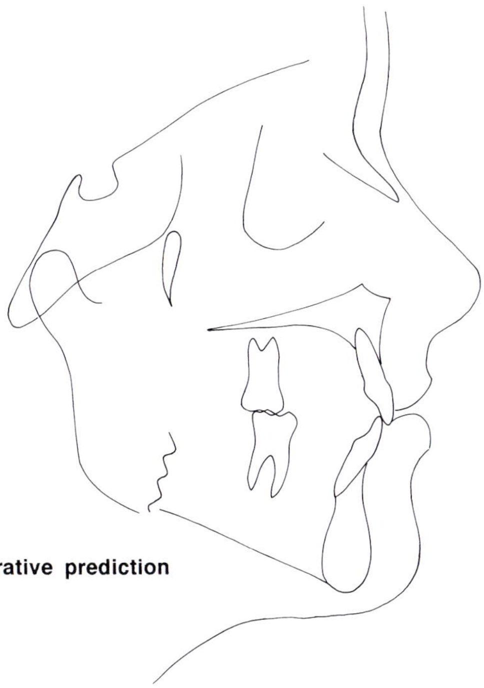

O

FIG. 9-8 (*Continued*)

Surgical Corrections Utilizing Counterclockwise Rotation of the Distal Mandible

Perhaps one of the most difficult problems that the orthognathic surgeon faces is the long-face patient seen in association with severe mandibular micrognathia due not only to horizontal deficiency of the mandibular corpus but vertical deficiency of the ramus as well. This deformity is in marked contrast to the long-face, Class II patients in Fig. 9-7 and 9-8, in whom severe ramus insufficiency is not present. The lateral cephalometric tracing in Fig. 9-9 is taken from an individual demonstrating vertical deficiency of the ramus. The short ramus results in deficient posterior facial height and an excessive vertical vector of mandibular corpus growth. This, in turn, causes excessively steep occlusal and mandibular planes and excessive anterior facial height. It is as if the entire maxillomandibular complex has been rotated in a clockwise fashion. If maxillary dentoalvealor compensation is complete, the anterior bite is closed with excessive tooth show. If it is incomplete, there is an anterior open bite.

The traditional surgical correction of this deformity is depicted in Fig. 9-10. It involves Le Fort I maxillary impaction to decrease maxillary incisal show and/or allow open-bite closure secondary to mandibular autorotation, and mandibular advancement surgery. This approach has been shown to be aesthetically deficient because it fails to address the clockwise

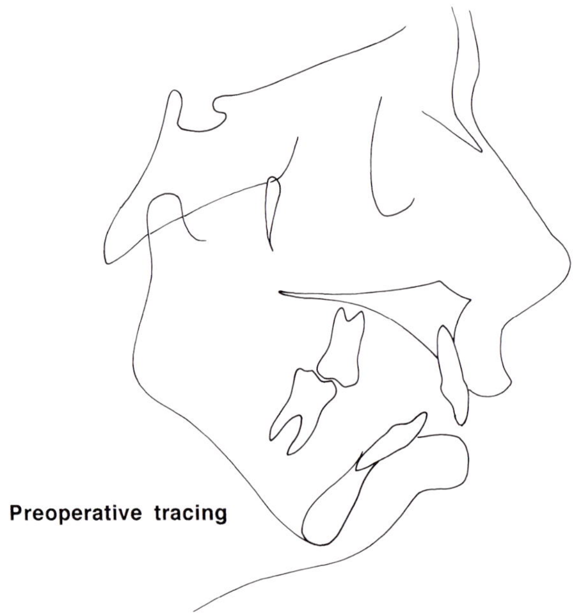

FIG. 9-9 Lateral cephalometric tracing from a patient with mandibular deficiency involving vertical ramus deficiency with secondary decrease in posterior face height, steepened occlusal plane and increased anterior face height. The entire maxillomandibular complex demonstrates a clockwise rotational deformity.

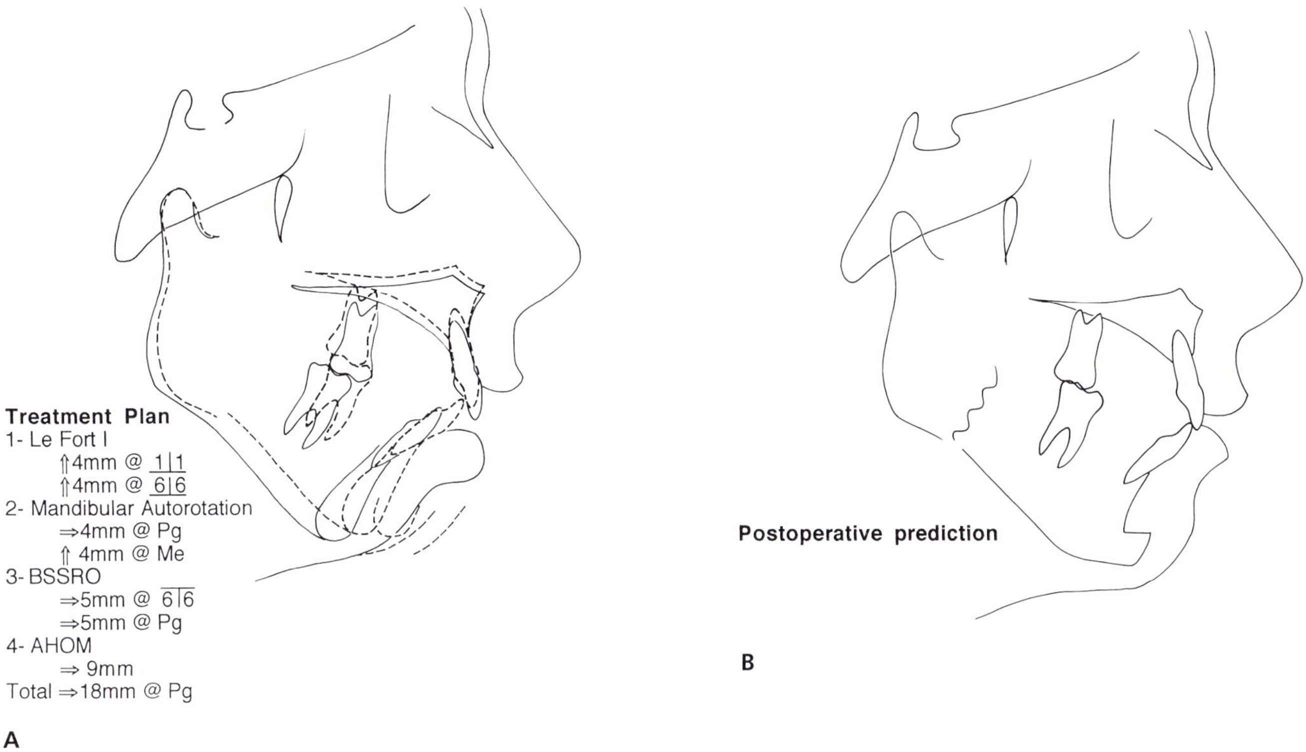

FIG. 9-10 A: Traditional surgical correction of Class II patient with vertical ramus deficiency. B: Adequate projection of the lower face cannot be achieved because the clockwise rotational deformity of the maxilla and mandible remains uncorrected.

rotational deformity. As a result, the mandible is advanced forward and downward along the uncorrected steep occlusal plane, and adequate projection of the lower face is never achieved. This is true even when a generous advancement genioplasty is added. In addition, the deficient height of the posterior face is further reduced because gonion moves superiorly as the mandible autorotates.

Adequate aesthetic correction of these patients must involve counterclockwise rotation of both maxilla and mandible. Maxillary rotation is accomplished with Le Fort I surgery to impact the anterior maxilla only. This normalizes lip–tooth relations, opens the anterior bite, and normalizes the steepened maxillary occlusal plane. It allows no mandibular autorotation at the condyle because the posterior dentition remains in contact. The Class II, anterior open bite must then be closed with sagittal split ramus osteotomies to advance and counterclockwise rotate the distal mandibular segment with the point of rotation at the ramus osteotomy. The rotational movement provides for a differential advancement of the mandible with greater movement anteriorly because it is further away from the point of rotation. Adequate projection of the lower face is, therefore, achieved, and posterior face height is preserved. This surgical correction is illustrated with model surgery in Fig. 9-11. The articulated models demonstrate a Class II

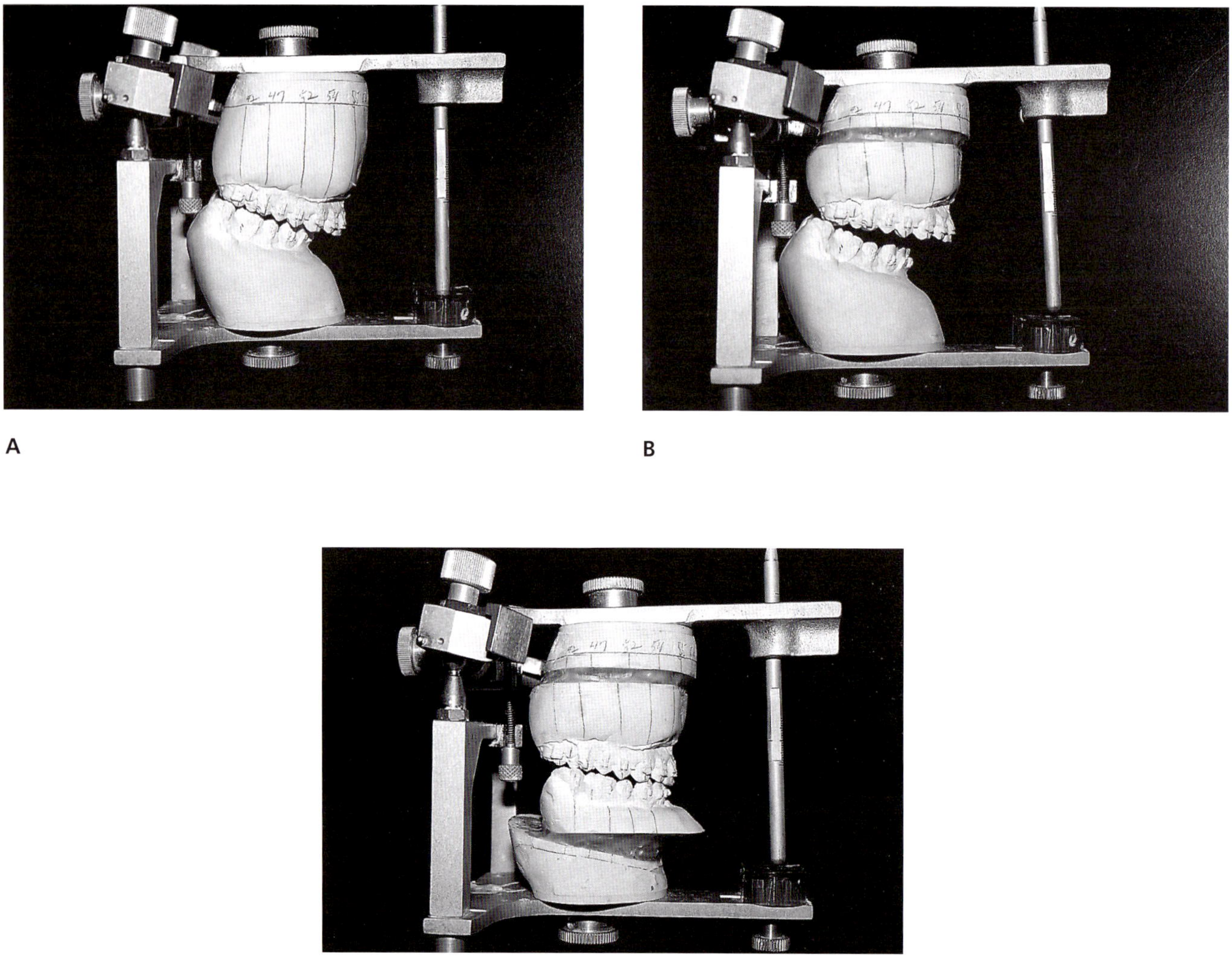

A

B

C

FIG. 9-11 Model surgery utilizing counterclockwise rotation of the maxilla and mandible. A: The occlusion is Class II with a 9-mm overjet. B: Le Fort I surgery designed to impact the anterior maxilla. This levels the maxillary occlusal plane and opens the anterior bite. C: Mandibular surgery must now close the open bite with counterclockwise rotation and advancement of the distal segment. This results in a large forward displacement of the anterior mandible.

malocclusion with a 9-mm overjet and closed anterior bite. Anterior maxillary impaction opens the anterior bite. Note the large anterior displacement of the mandibular cast due to the counterclockwise vector of rotation as it is advanced to correct the Class II, open-bite malocclusion.

Although this type of skeletal displacement was, heretofore, considered unstable, it has now been demonstrated to achieve predictably stable results. Technical modifications responsible for this enhanced stability include the following:

1. Rigid fixation of the osteotomy site. Although a variety of methods have been described, the author prefers 2.7-mm lag screw fixation

with compression of the proximal and distal cortices. These screws are placed along the anterior border of the proximal segment. This screw placement achieves high degrees of friction within the osteotomy site and yet minimizes the risk of nerve compression.

2. Extended split. By extending the osteotomy into the region of the first molar, adequate bone contact remains for screw placement despite large advancements.
3. Suprahyoid myotomy. Subperiosteal release of the suprahyoid muscles through the genioplasty osteotomy reduces the muscular forces thought to be responsible for open-bite recurrence following mandibular counterclockwise rotation.
4. Hunsuck modification of sagittal split ramus osteotomy. This technique places the vertical osteotomy of the medial ramus just posterior to the lingula. This allows the maximum amount of the pterygomasseteric muscle sling to remain on the proximal segment and, therefore, minimizes relapse due to stretching of this muscle group.
5. Class II elastics. Although intermaxillary fixation is not employed, Class II elastics are placed postoperatively. They are continued for approximately 2 weeks. This minimizes condylar loading and compression during the initial period of soft tissue adaptation.

Despite these technical modifications, one great variable remains in achieving stable results when utilizing counterclockwise rotational vectors of the distal mandibular segment. This is the technical ability of the surgeon. This factor must not be minimized when attempting to explain an unstable surgical result.

Case Study 9-12

This patient demonstrates surgical counterclockwise maxillary and mandibular rotation. It is the same patient whose tracings are depicted in Figs. 9-9 and 9-10. The facial morphology is characterized by a long lower face with severe lip strain, deficient posterior facial height, severe posterior facial divergence, and facial convexity. There is essentially no cervical mandibular angle because of the deficient and vertically inclined mandible. Occlusal exam reveals a Class II, Division 1 malocclusion. Thyroid work-up is negative. It is clear that there is not enough overjet to accommodate more than 8 mm of mandibular advancement. In addition, one cannot advance the maxilla to increase the available overjet as it is already somewhat prominent. Counterclockwise rotation of both the maxilla and the mandible is the only viable option to achieve adequate lower face projection.

Accordingly, the following treatment plan is undertaken: anterior maxillary impaction only of 4 mm to normalize lip–tooth relations and to temporarily create an anterior open bite, mandibular advancement and counterclockwise rotation of the distal segment resulting in a 10-mm advancement at the first molar, and an 18-mm advancement at pogonion and advancement genioplasty of 7 mm providing a total advancement of pogonion of 25 mm.

The results of this approach are quite evident. The patient is left undercorrected in the vertical dimension and has been overcorrected sagittally. There is a marked increase in the overall skeletal volume of the lower face. There is a good soft tissue response as evidenced by the development of a well-defined cervical mandibular angle. The occlusal result has been stable for 4 years.

This patient illustrates the necessity of utilizing counterclockwise rotational vectors of movement to achieve adequate degrees of projection of the lower face. This same principle can be utilized in other patients with enhanced aesthetic results.

PREOPERATIVE

POSTOPERATIVE

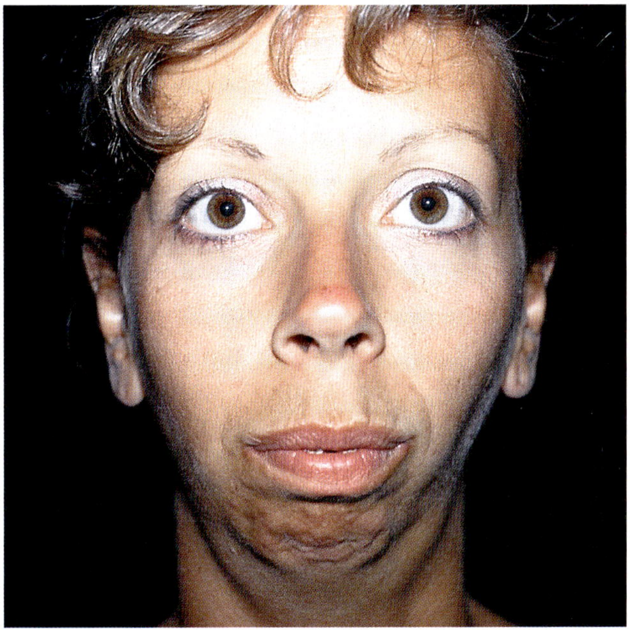

A

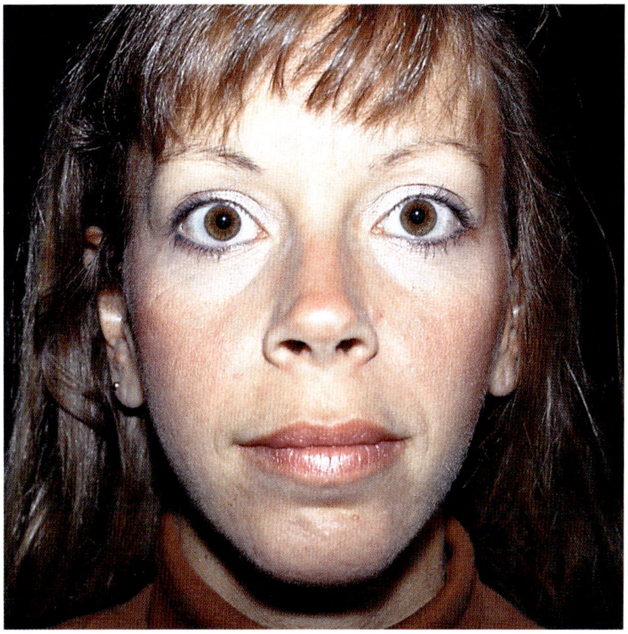

C

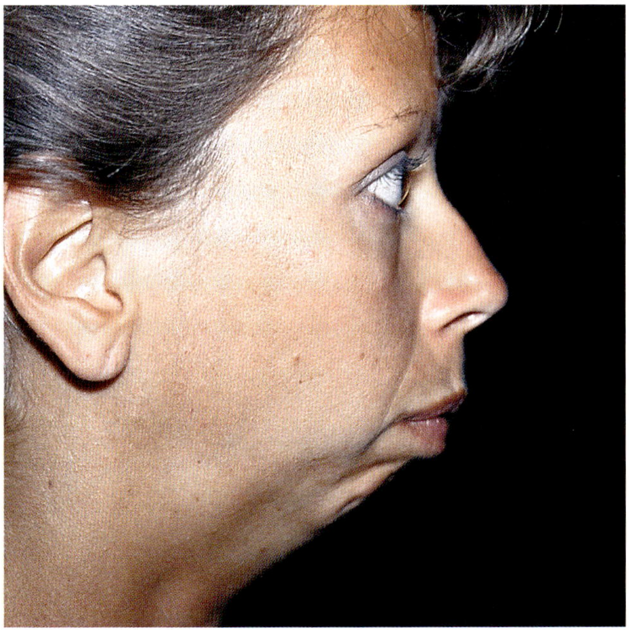

B

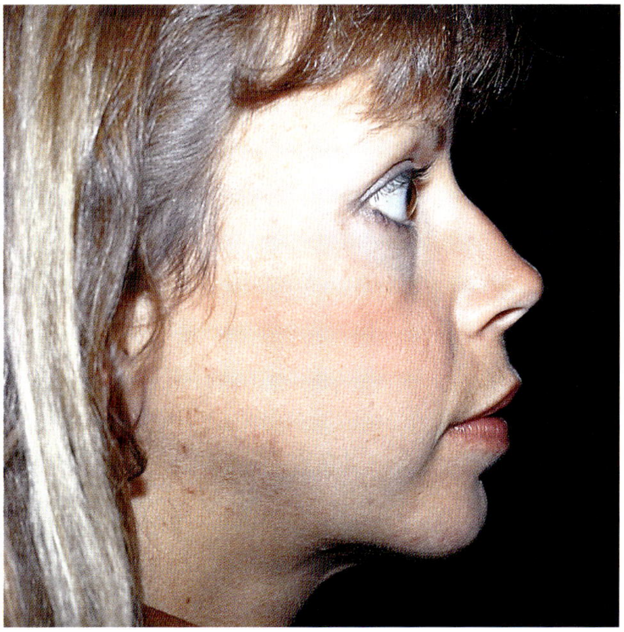

D

Case Study 9-12 A & B: A patient with severe mandibular deficiency involving the ramus and body of the mandible. The maxilla is well positioned sagittally. The surgical correction must rotate the maxilla and mandible counterclockwise to achieve adequate projection of the lower face. C & D: The patient has been undercorrected vertically and overcorrected sagittally with pogonion advancing 25 mm. The cervical mandibular angle is now well defined.

PREOPERATIVE | **POSTOPERATIVE**

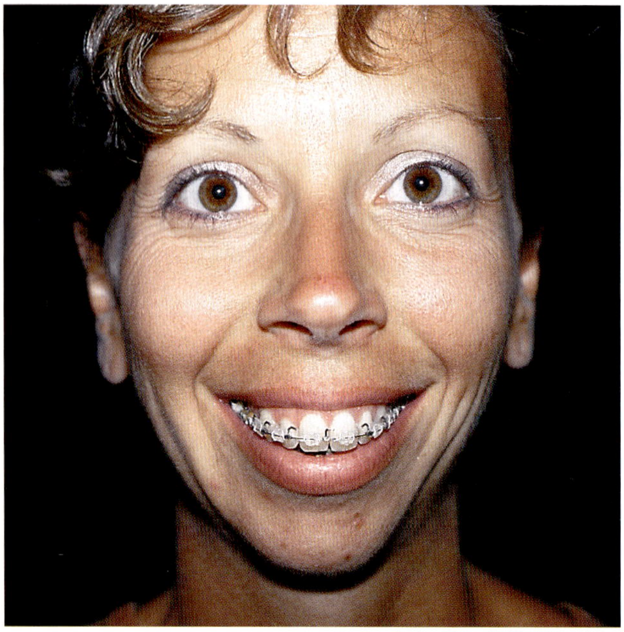

E

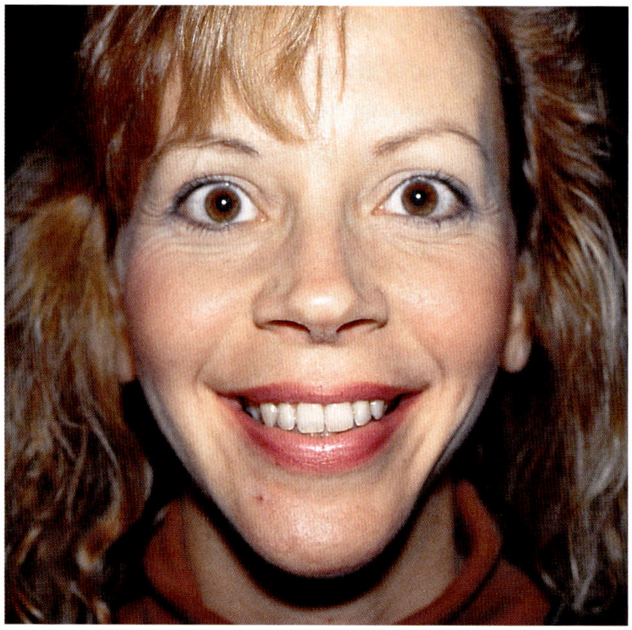

F

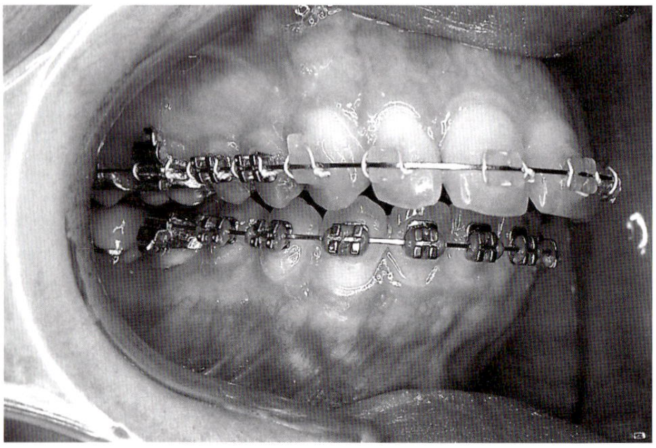

G

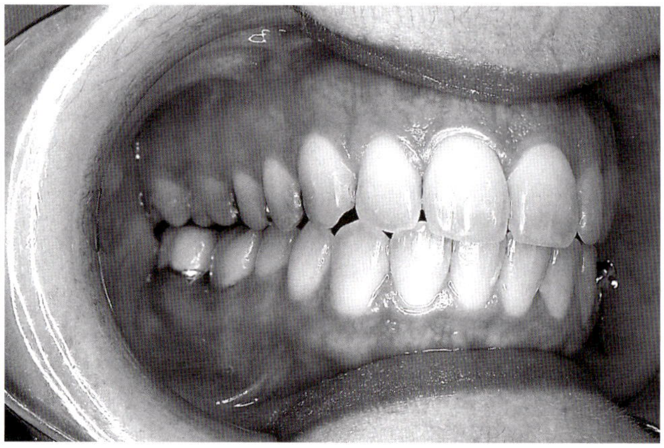

I

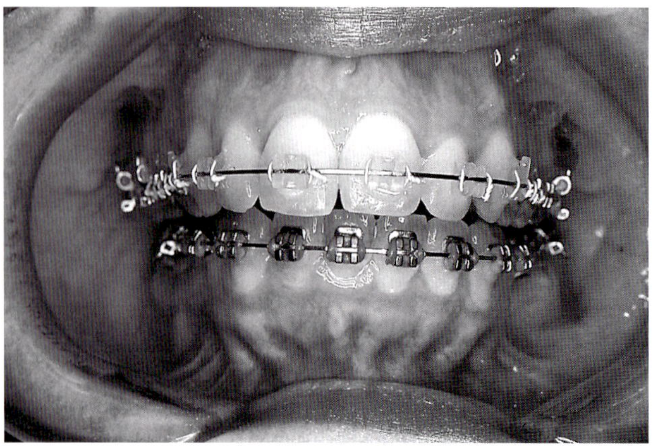

H

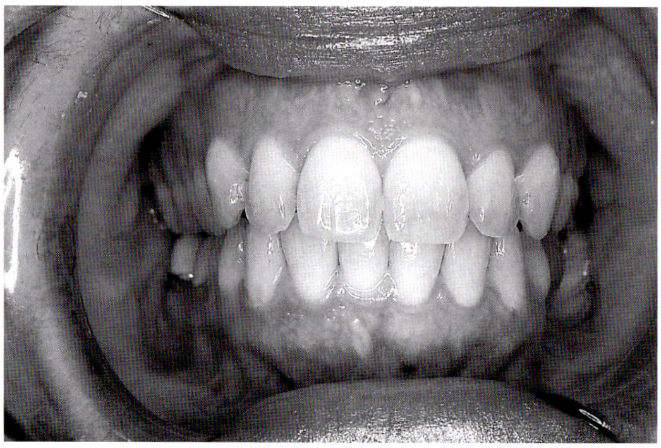

J

FIG. 9-12 (*Continued*) G through J: The occlusal result four years following surgery. The bite has remained closed. Orthodontist: Dr. Gerald Ginsberg

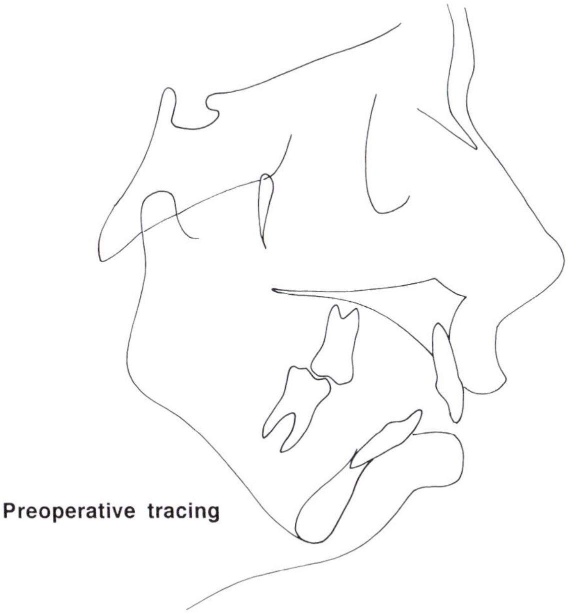

K

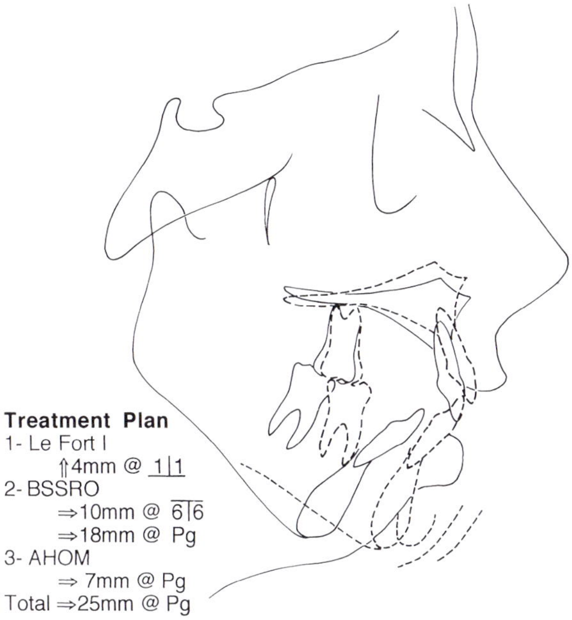

L

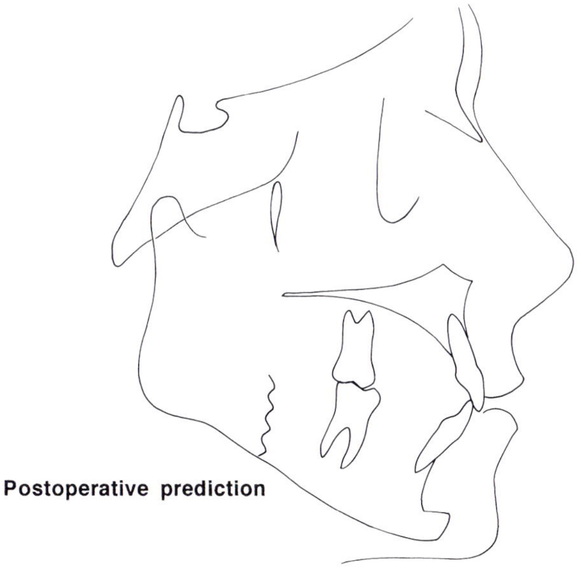

M

FIG. 9-12 (*Continued*) L: Notice the differential degree of mandibular advancement due to counterclockwise rotation of the distal mandibular segment. The mandible comes forward 10 mm at the first molar and 18 mm at pogonion. M: Notice the significant leveling of the occlusal plane and mandibular border.

Case Study 9-13

This patient presents with a long face and a posteriorly divergent, convex profile. There is 5 mm of incisal show at rest and lip strain with attempted closure. She has a component of vertical ramus deficiency, although not as severe as the previous patient. Secondary to the lack of development of the ramus, there is a decrease in posterior face height. This is best detected clinically by a mandibular angle in relatively close proximity to the earlobe (with a normally positioned ear) and a steepened mandibular plane.

Although this patient suffers from vertical maxillary excess and an anterior open bite, counterclockwise rotation of the mandible following bilateral sagittal split ramus osteotomy will be used to close the anterior open bite. The aesthetic rationale for this is to eliminate any further reduction in posterior face height that would occur secondary to maxillary impaction surgery, the standard and more acceptable way to treat anterior open-bite malocclusions.

The overall treatment plan for this patient involves a three-piece Le Fort I osteotomy to advance the maxilla a total of 5 mm and to achieve surgical leveling by intrusion at the premolar teeth. Notice that no anterior or posterior impaction is planned. This will prevent any reduction in posterior face height and maintain a generous 5 mm of incisal show. The anterior open bite will then be closed by advancing and counterclockwise rotating the mandible to achieve an 8-mm advancement at the molar level and a 12-mm advancement at pogonion. Finally, a modest advancement genioplasty will be performed. Overall, the patient will remain undercorrected vertically and will be overcorrected sagittally, at least according to normative standards.

One can see the results of this treatment plan in the postoperative photographs. There is residual quantitative vertical facial excess, but with an absence of lip strain. Incisal show has been totally preserved, and now there is even an attractive, minimal exposure of gingiva with smiling. The profile is characterized by an excessive degree of anterior divergence but with good overall facial convexity, good lip relations, and an adequately projecting nasal tip and dorsum. The occlusal result is excellent.

PREOPERATIVE **POSTOPERATIVE**

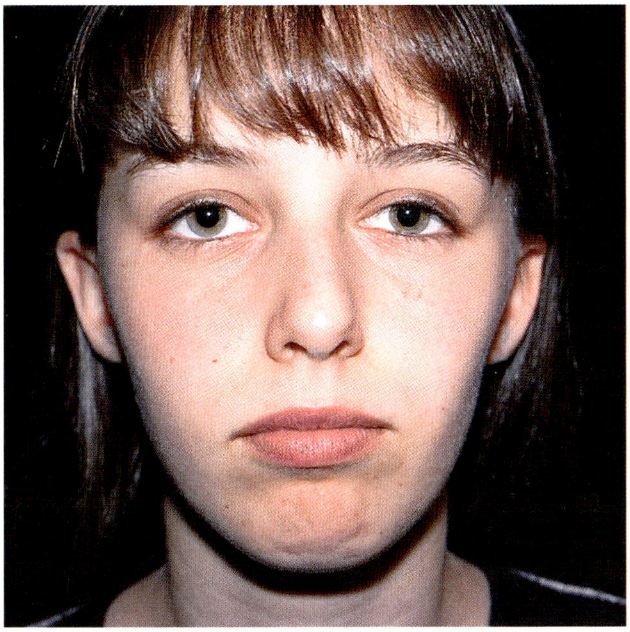

A

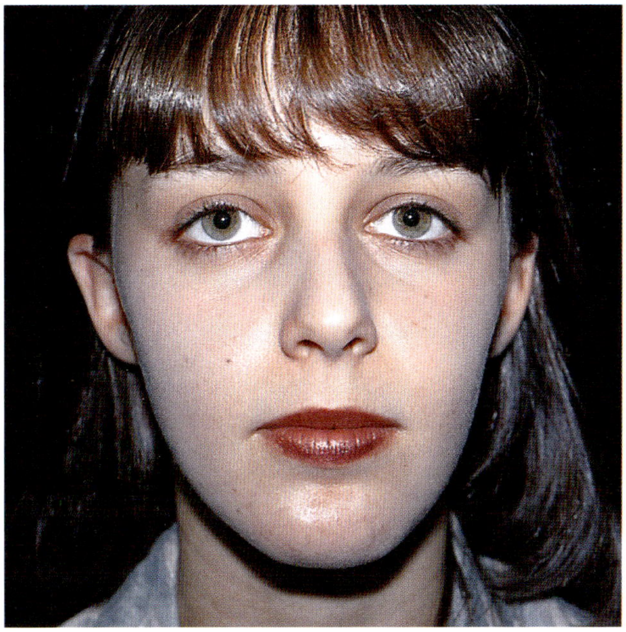

C

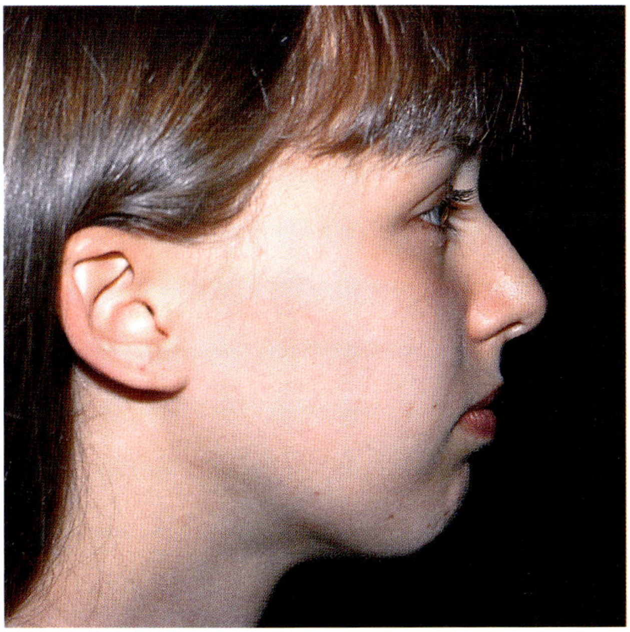

B

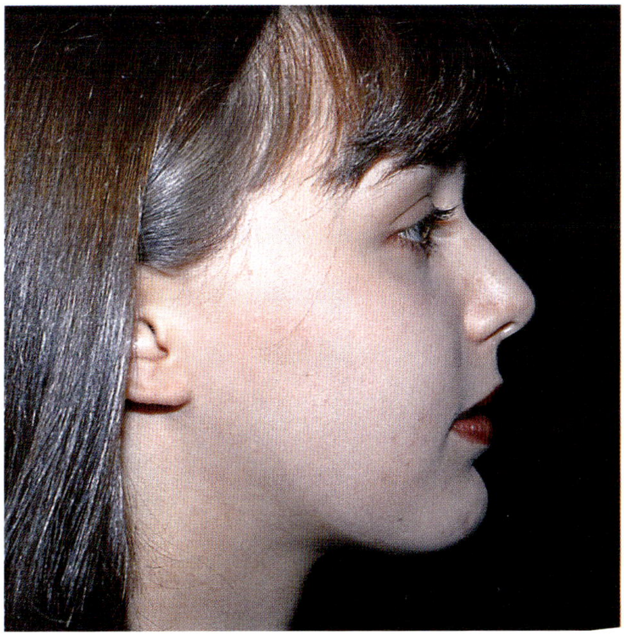

D

CASE STUDY 9-13 A & B: A long face, Class II patient treated with double jaw surgery and counterclockwise rotation of the distal mandible, but with no autorotation of the mandible in an effort to preserve posterior facial height. C & D: The patient undercorrected vertically and overcorrected sagittally, providing for enhanced soft tissue support.

PREOPERATIVE **POSTOPERATIVE**

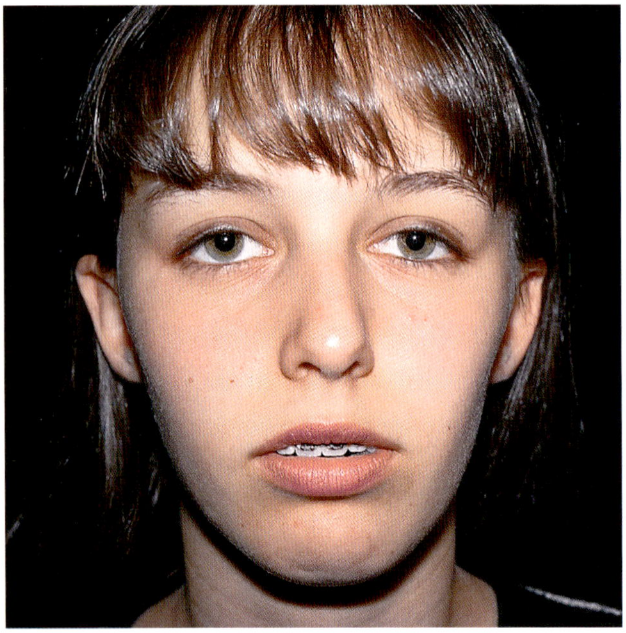

E

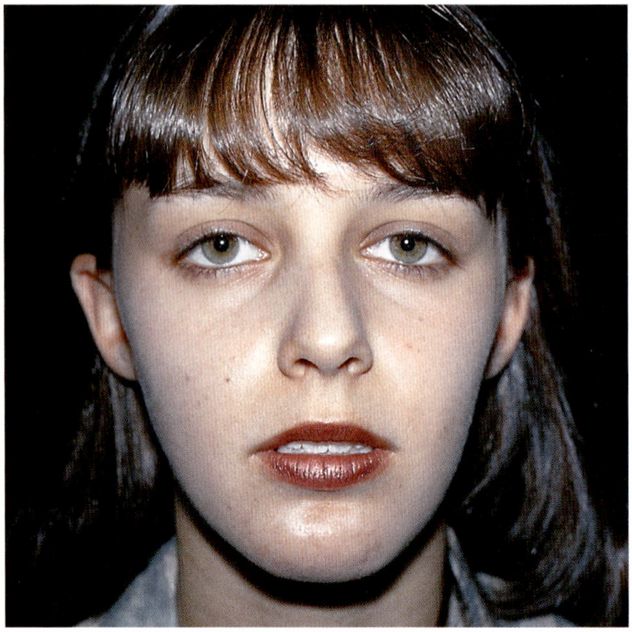

G

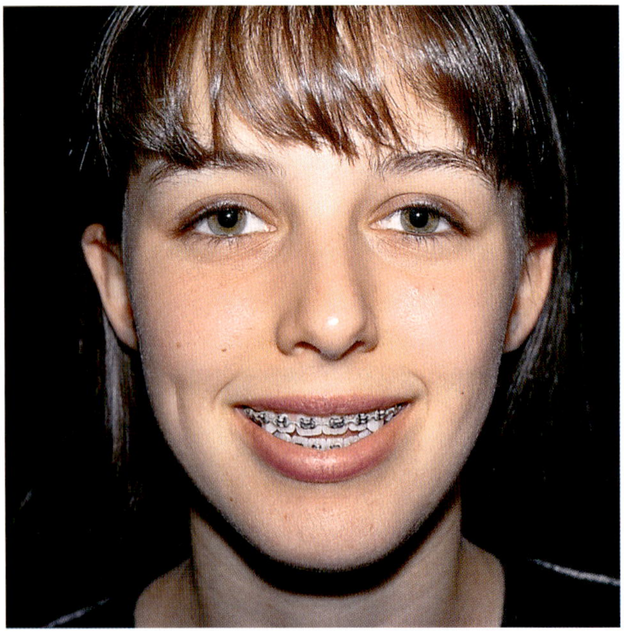

F

H

FIG. 9-13 (*Continued*) E through G: There has been preservation of maxillary incisal show in repose and now there is an attractive display of gingiva with full smiling. The lower face remains quantitatively long. Note correction of maxillary midline.

PREOPERATIVE | **POSTOPERATIVE**

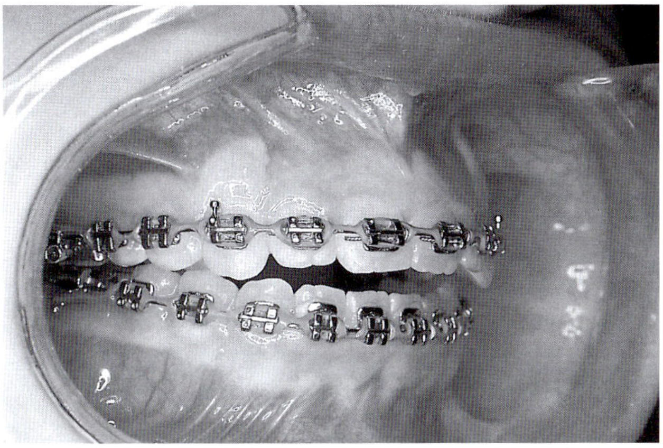

I

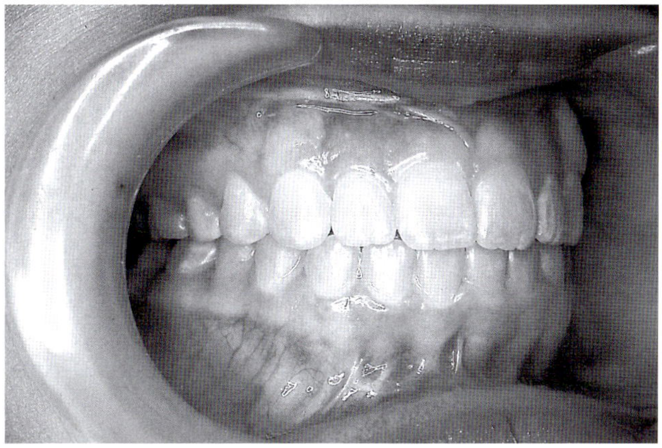

K

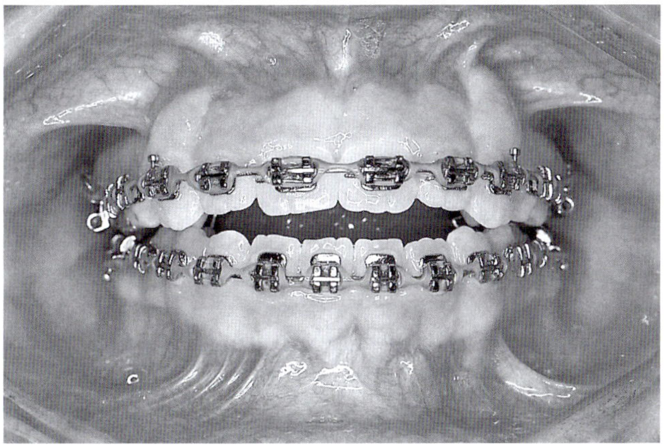

J

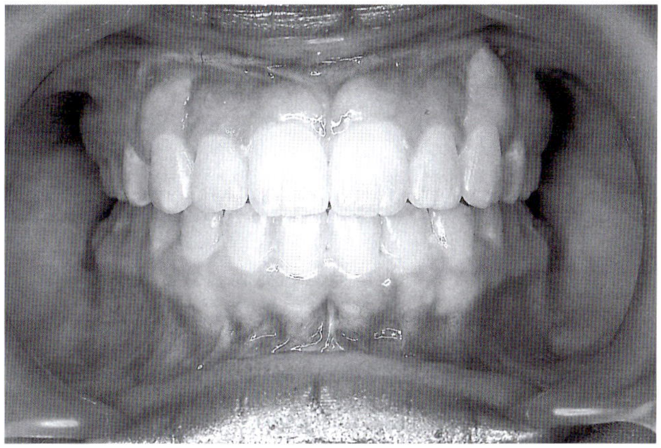

L

FIG. 9-13 (*Continued*) I through L: The maxillary dentition has been leveled surgically in three pieces. The bite remains closed at 20 months following surgery. Orthodontist: Dr. John Steciw

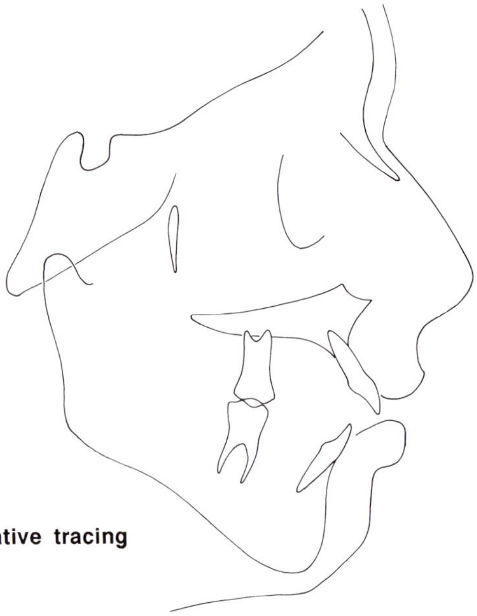

Preoperative tracing

M

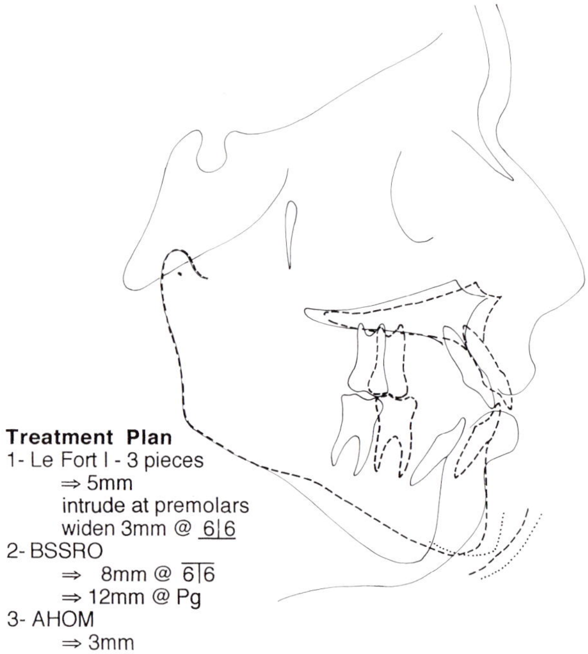

N

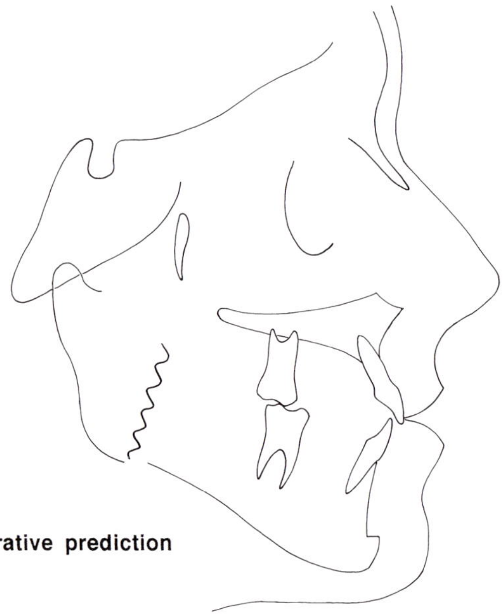

Postoperative prediction

O

FIG. 9-13 (*Continued*) N: There is no posterior or anterior intrusion of the maxilla. Therefore, incisal exposure is preserved and there is no autorotation of the mandible. As a result, posterior facial height is preserved. The open bite is closed by counterclockwise rotation of the distal mandible.

Case Study 9-14

This patient represents still another application of counterclockwise rotational movement of the distal tooth-bearing mandibular segment following sagittal split ramus osteotomy. She is 16 years old and suffers from juvenile rheumatoid arthritis. She has a severe anterior open-bite deformity. Her lower face is quantitatively long relative to the midface. There is severe facial convexity and posterior divergence. A significant morphologic factor that must be considered in her treatment planning is the poor support of the soft tissues. They are excessive and thick and have little definition and angularity. Paradoxically, and quite fortunately as it relates to the soft tissue concerns, there is no aesthetic rationale to vertically contract the facial skeleton. There is no lip strain, the labiomental fold is defined, and there is a total absence of incisal show with the lips in repose and only half a clinical crown exposed with full smiling. This represents an example of the difference between reality and clinical perception in the context of vertical facial excess; that is, a quantitatively long face does not necessarily appear to be so in the absence of the usual stigmata associated with it. As is frequently the case, perception is more important than reality.

Given the soft tissue considerations, everything must be done in this patient to expand the facial skeleton in both the vertical and sagittal planes. Accordingly, the treatment plan calls for converting a quantitatively long face to one that is even longer and a posteriorly divergent face to one with an excessive degree of anterior facial divergence. Soft tissue considerations are dictating the extent and vector of skeletal movement and, therefore, facial proportions. The entire maxilla will be inferiorly repositioned, which will increase incisal exposure and cause a downward and backward rotation of the mandible at the condyle, which will increase posterior face height. This will temporarily cause the anterior open bite to worsen, which will then be closed with counterclockwise rotation of the distal mandibular segment. This will result in a differential mandibular advancement of 7 mm at the first molar and 13 mm at pogonion. Osseous genioplasty will advance the symphysis 7 mm and inferiorly reposition it 3 mm. Overall, pogonion will advance 20 mm, and the lower face height will be increased by 5 mm. In addition, a buccal fat lipectomy and submental lipectomy will be carried out.

Postoperatively, there is a marked increase in the overall skeletal volume of the lower face. There is now a full clinical crown exposed with smiling, and the profile has an excessive degree of anterior divergence, but it is accompanied by good facial convexity and good labial relationships. Note the residual absence of incisal exposure in repose due to upper lip lengthening. Overall there has been an enhancement of soft tissue support, and there is now a moderate degree of facial angularity. The occlusal result is excellent at 3 years.

PREOPERATIVE **POSTOPERATIVE**

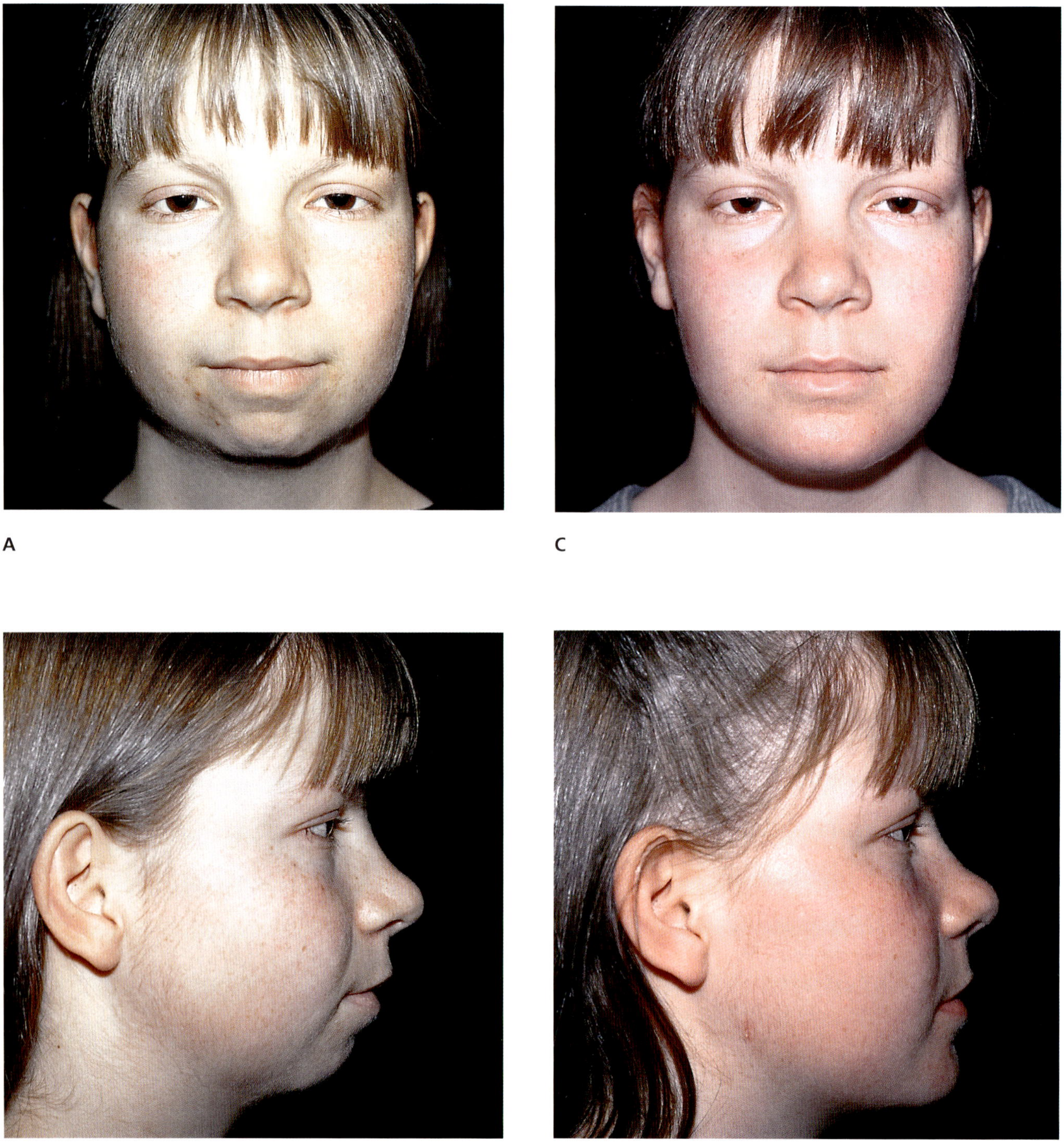

CASE STUDY 9-14 A & B: A patient with severe mandibular deficiency, quantitatively long face, and anterior open bite. There is no aesthetic imperative to shorten her face. Counterclockwise rotation of the jaws must be utilized to achieve adequate facial projection. C & D: Postoperatively, the face has been lengthened and there is now adequate facial projection to support the soft tissues.

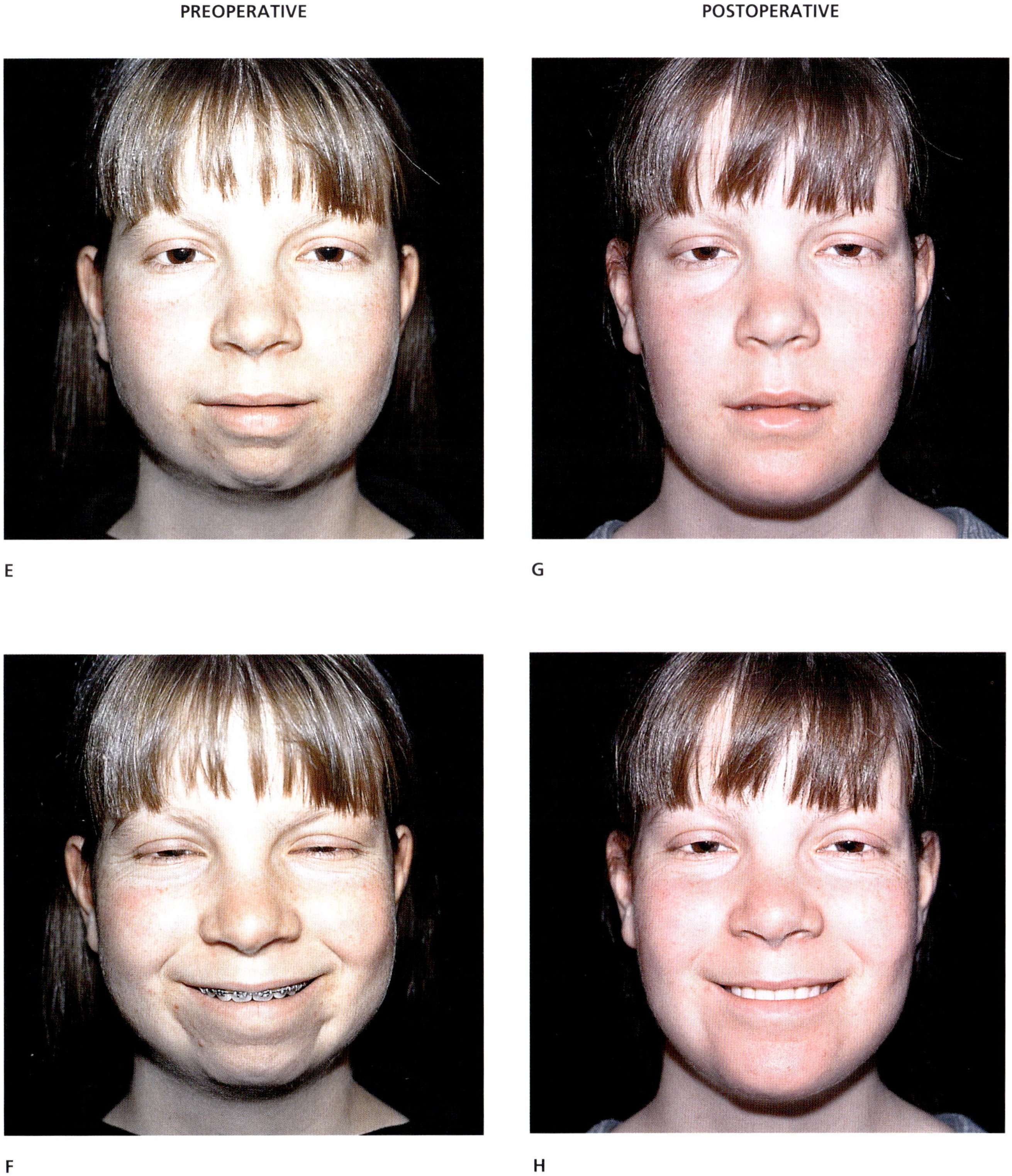

FIG. 9-14 (*Continued*) E & F: Preoperatively, there is inadequate exposure of maxillary incisors in both repose and smiling. G & H: Although this has been improved postoperatively, more incisal show would have enhanced the result.

PREOPERATIVE

POSTOPERATIVE

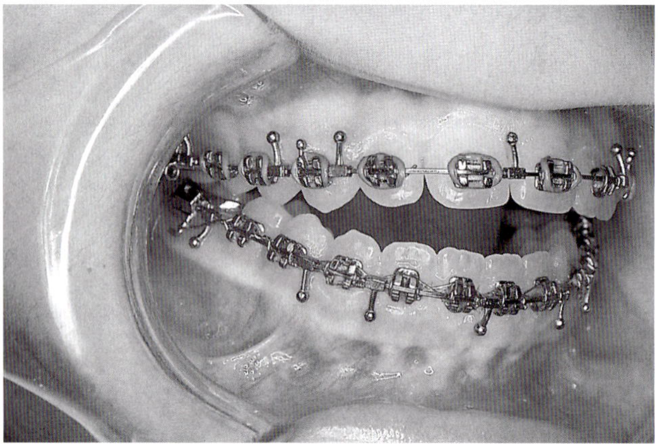

I

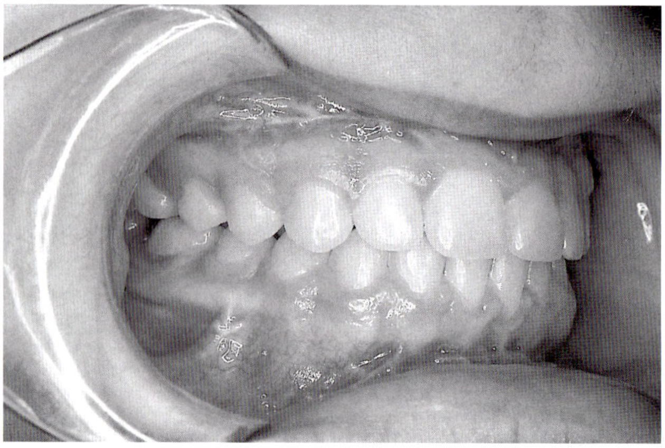

K

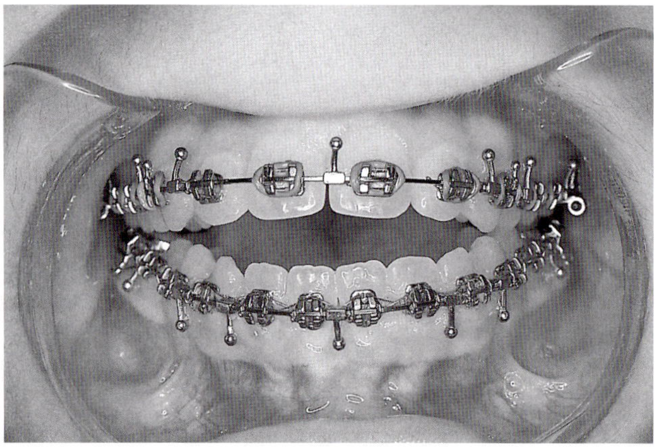

J

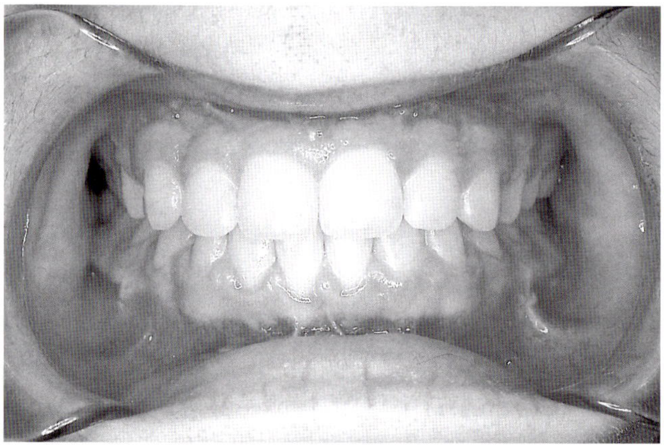

L

FIG. 9-14 (*Continued*) I through L: The postoperative occlusal results. The bite has remained closed for three years. Orthodontist: Dr. James Ackerman

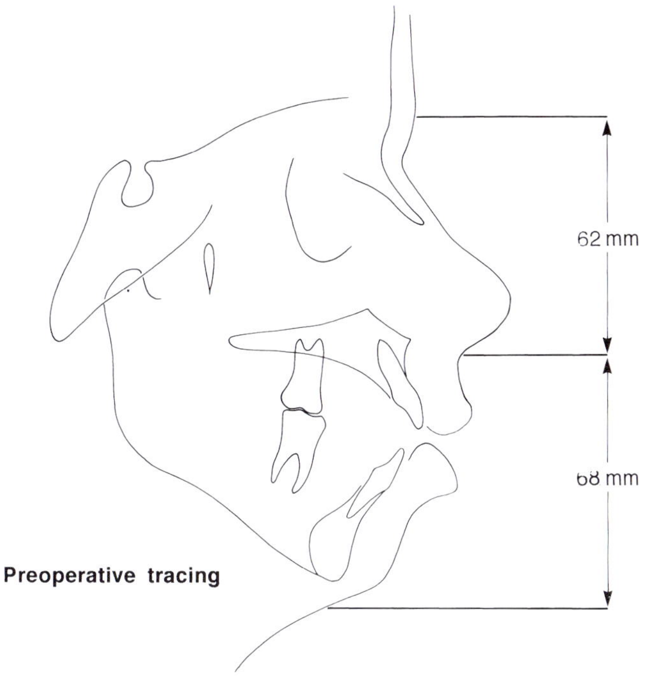

M

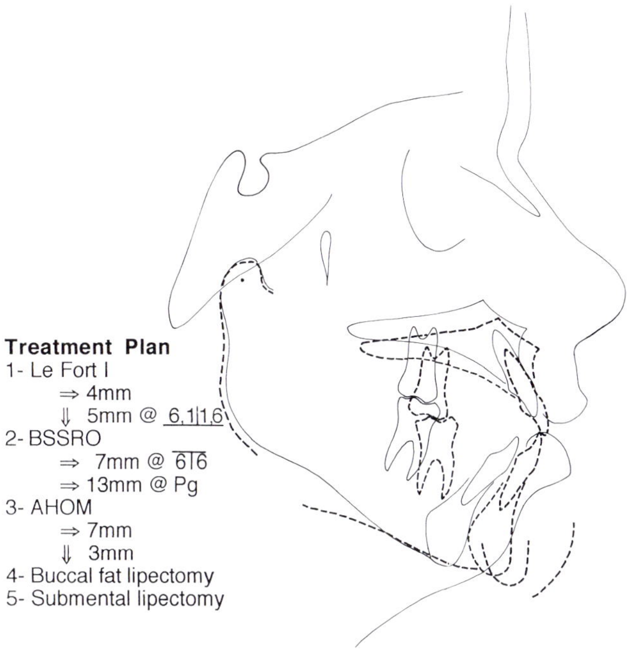

N

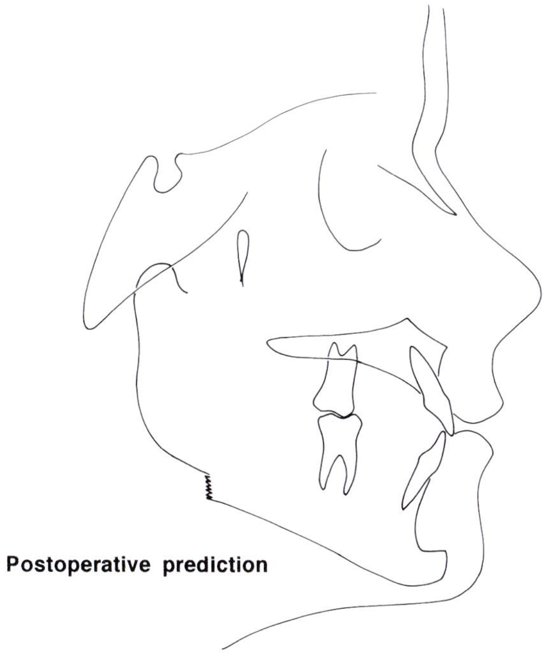

O

FIG. 9-14 (*Continued*) M: Note that the lower face is long relative to the midface. N: The lower face will be lengthened further with inferior maxillary repositioning and lengthening genioplasty. The open bite will be closed entirely with counterclockwise rotation of distal segment. O: This provides for adequate facial projection and subsequent enhanced skeletal support of the soft tissues.

Long-Face, Class III Skeletal Correction

Correction of the long-face patient, when seen in association with a Class III malocclusion, will frequently require some type of posterior repositioning of the mandibular tooth-bearing segment. This is true provided that the skeletal etiology of the Class III malocclusion is an enlarged mandible. This is necessary because maxillary impaction surgery to reduce lower face height will be accompanied by mandibular autorotation that only makes the sagittal discrepancy between the maxilla and the mandible more severe. Even when maxillary impaction is accompanied by a modest maxillary advancement, mandibular setback surgery may be indicated in such individuals. However, one must still be cognizant of over reduction of skeletal volume and, therefore, minimize the extent to which there is posterior movement of soft tissue pogonion. The overall loss of skeletal volume secondary to mandibular setback surgery is compensated, in part, by the accompanying mandibular autorotation and the surgical option of advancement genioplasty.

Case Study 9-15

This patient exemplifies these principles. This patient clearly has a long lower face due to vertical maxillary excess and vertical macrogenia. She is displaying approximately 10 mm of maxillary incisal show and has a rather gummy smile even with minimal elevation of the upper lip. Of importance is the fact that the upper lip is short, being only 16 mm in length. The profile is characterized by an excessively anteriorly divergent profile obviously due to a large mandible. The maxilla is in good sagittal position, although there is some relative paranasal flattening.

The surgical plan for this patient calls for an anterior maxillary impaction of only 4 mm and a somewhat greater impaction posteriorly. The extent of anterior maxillary impaction is purposely undercorrected; an attempt to achieve so-called "normal" lip–tooth relations would reduce the height of the lower face excessively because her short upper lip is, in part, responsible for her toothy appearance. Minimizing skeletal volume reduction should take precedence over the degree of incisal show. This is particularly true because even 6 to 7 mm of incisal show may be attractive provided that it is unaccompanied by excessive lip strain and a very gummy smile. Following mandibular autorotation, a 5-mm mandibular setback will be necessary to achieve a Class I occlusion. Advancement genioplasty of 3 mm as well as the mandibular autorotation that has occurred secondary to maxillary impaction will offset any loss of sagittal skeletal volume. Finally, the chin will be vertically reduced by 4 mm.

The postoperative aesthetic result is a good one because of minimal reduction in overall skeletal volume. There is now approximately 6 mm of incisal show and a full crown exposed with smiling. Although the lower face remains quantitatively long, there is an absence of mentalis lip strain and good definition of the labiomental fold, which deemphasizes residual vertical excess. The profile is still an anteriorly divergent one but with greater facial convexity and good lip relations. There is now a Class I occlusion. A secondary rhinoplasty to narrow the nasal bridge and nasal tip is indicated.

PREOPERATIVE POSTOPERATIVE

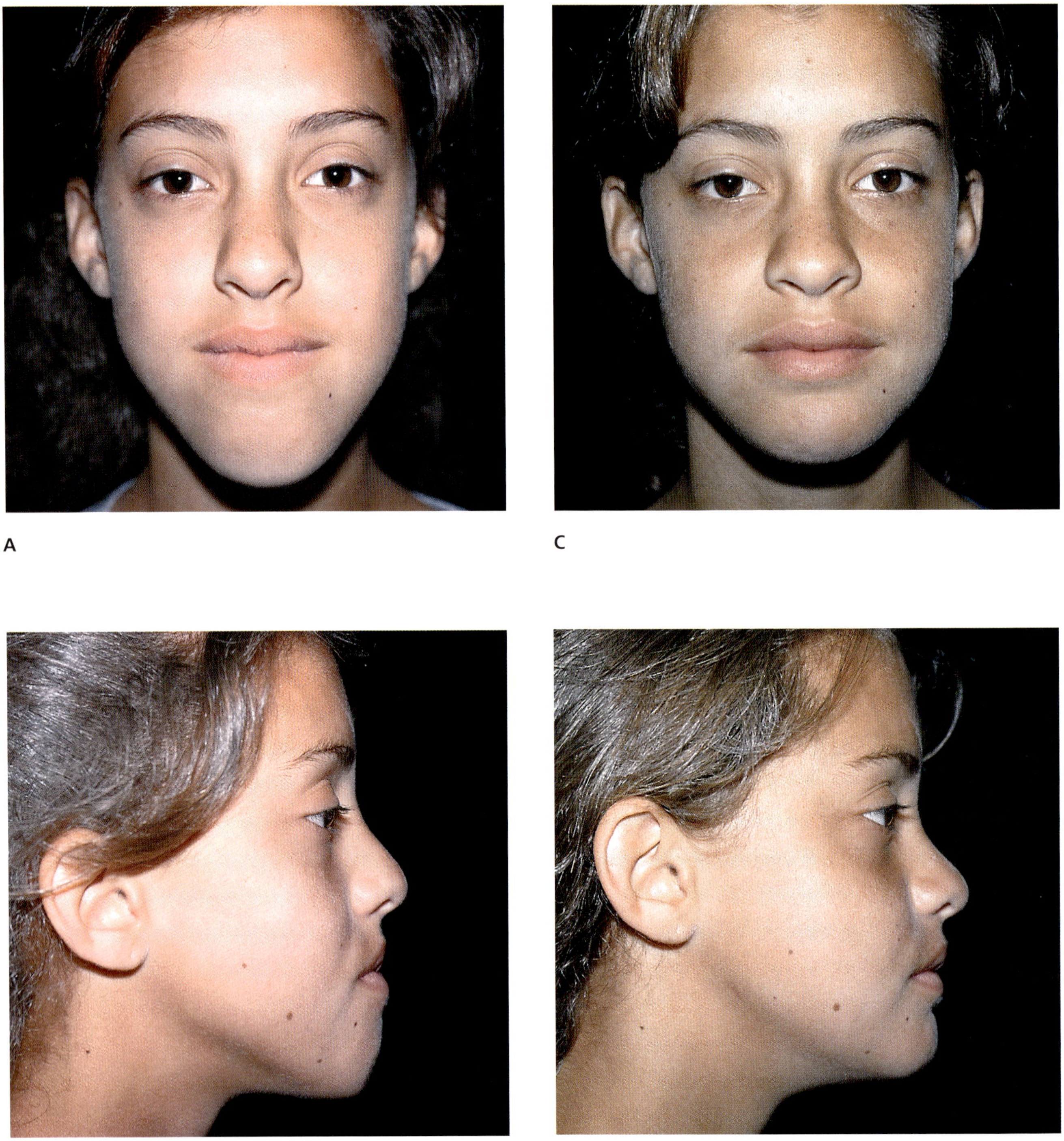

A C

B D

CASE STUDY 9-15 A & B: A patient with a long face and a Class III malocclusion due to mandibular excess. She is to be treated with maxillary impaction and advancement, modest mandibular setback, and advancement genioplasty. C & D: Although the face remains quantitatively long following surgery, the lip strain has been eliminated and the labiomental fold deepened. Anterior facial divergence has been preserved with enhanced convexity.

PREOPERATIVE POSTOPERATIVE

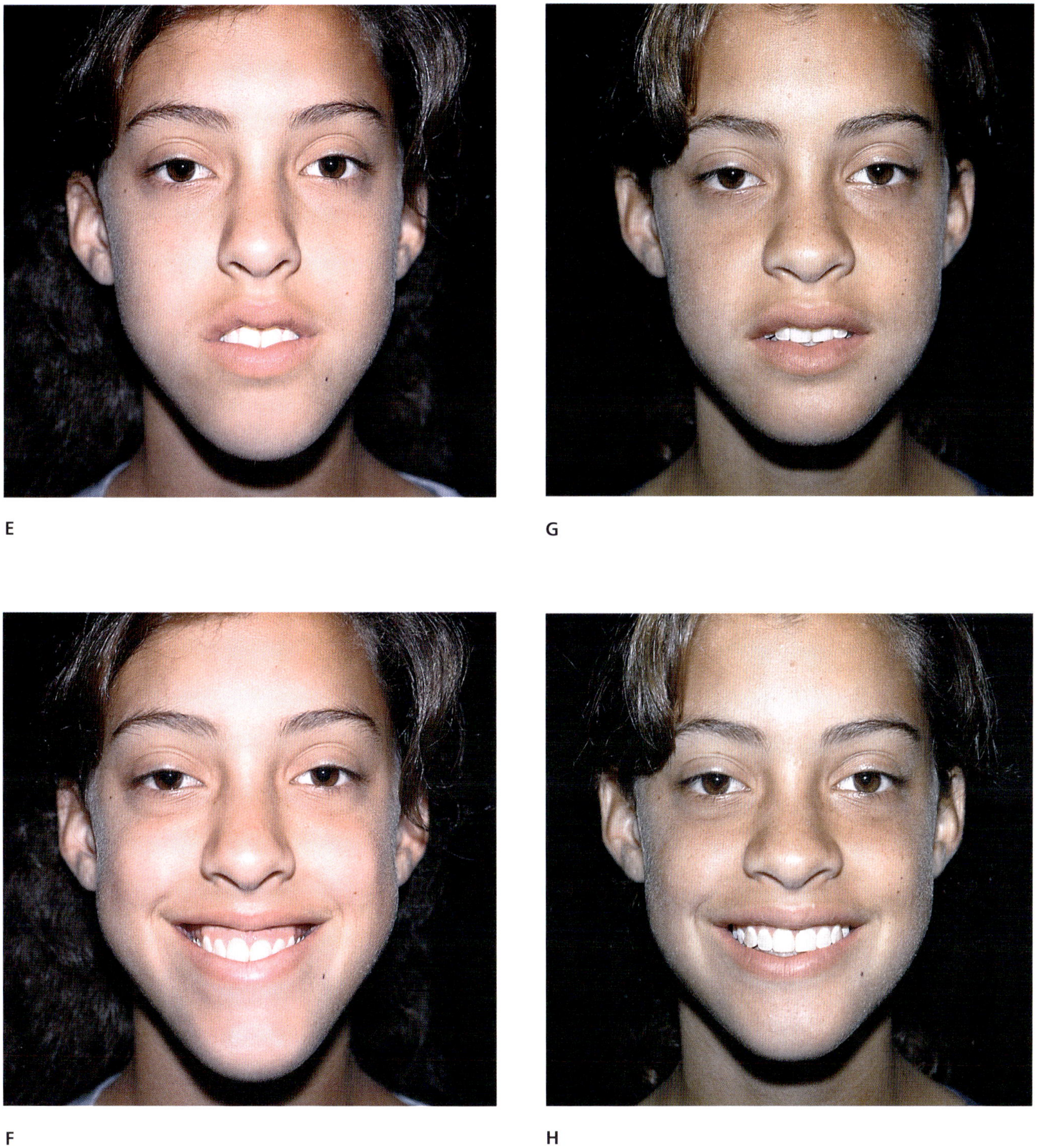

FIG. 9-15 (*Continued*) E & F: Preoperatively, there is full crown exposed in repose and an excessive display of gingiva with smiling. The upper lip is short. G & H: Postoperatively, there is 6 mm of incisal show in repose and an enhanced smile.

PREOPERATIVE

POSTOPERATIVE

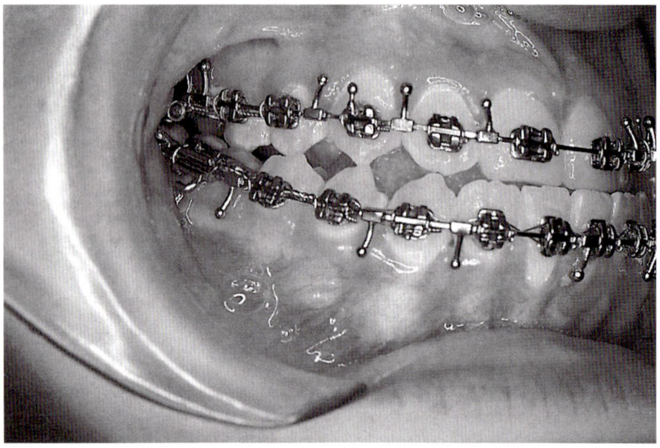

I

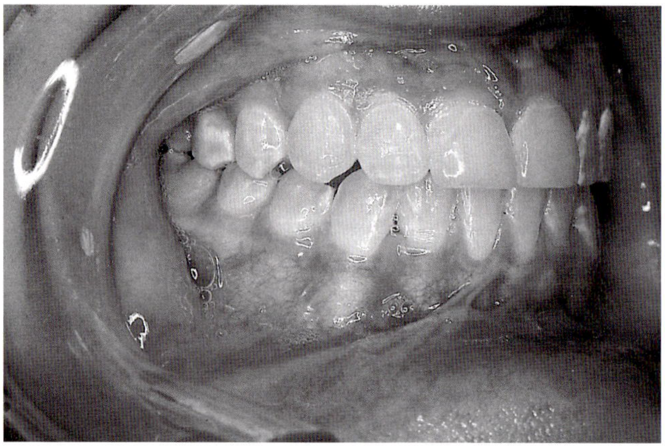

K

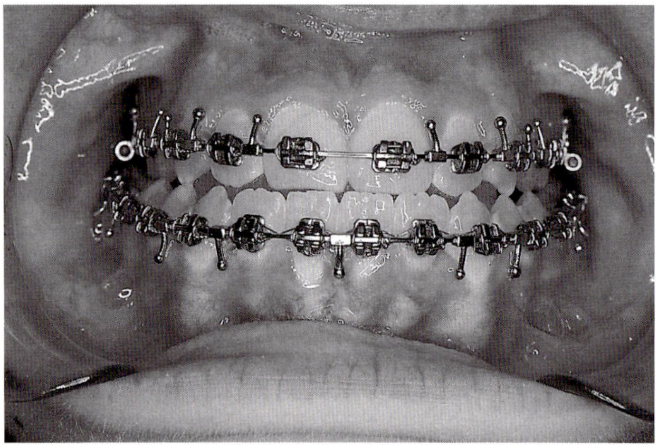

J

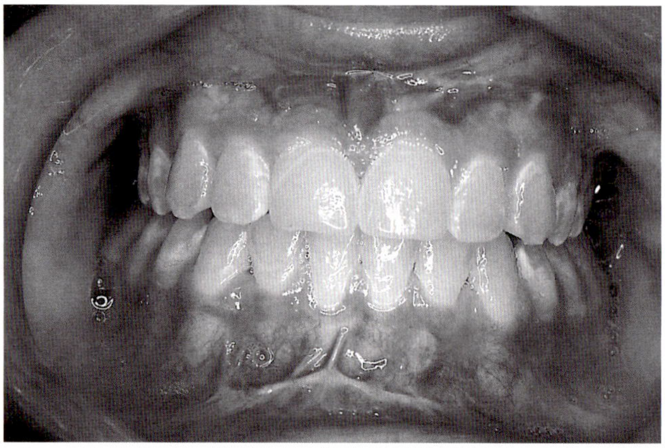

L

Fig. 9-15 (*Continued*) Orthodontist: Dr. James Ackerman

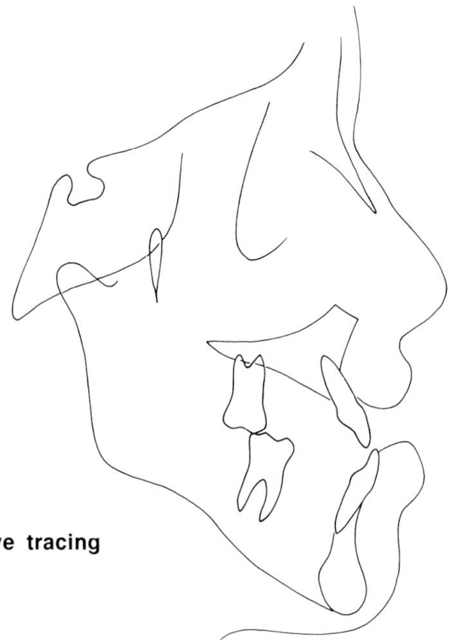

M

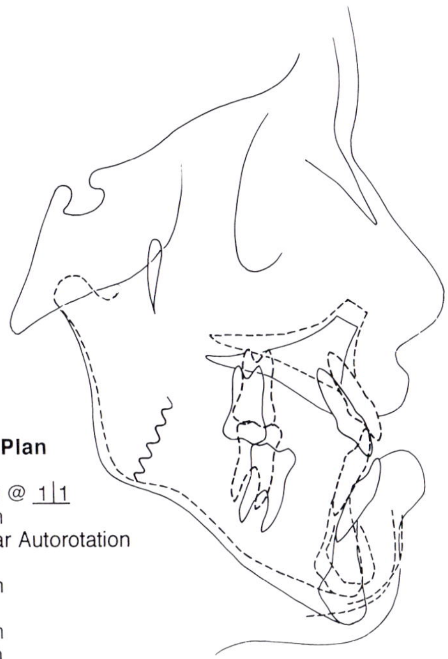

N

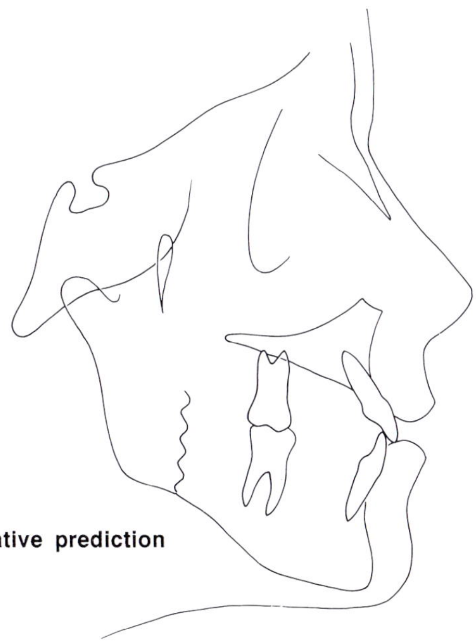

O

FIG. 9-15 (*Continued*) M: Note the short upper lip. N: There is, overall, no loss of sagittal projection of the lower face due to mandibular autorotation and advancement genioplasty.

Case Study 9-16

This patient, perhaps more than any other shown thus far, exemplifies the concept of preservation and augmentation of skeletal volume in an effort to achieve soft tissue goals. The resulting skeletal disproportion in both the vertical and sagittal dimensions can enhance the final aesthetic result.

This patient has a Class III malocclusion due to an enlarged mandible. As a result, there is a modest component of anterior facial divergence with parasagittal flattening. In addition, the nasolabial angle is acutely closed. There is quantitative vertical excess in the lower facial third measuring 73 mm compared to a midface height of 67 mm. However, there is no aesthetic imperative to vertically shorten the face; that is, there is an absence of lip strain, the labiomental fold is present, and there is not a toothy or gummy appearance. On the contrary, there is a lack of incisal show in repose.

One word can best describe the patient's facial morphology and that is amorphous. There is a total lack of definition and angularity due to soft tissue excess, a thick facial mask, and inadequate skeletal support. Paradoxically, skeletal expansion is necessary in this long-face, mandibular prognathic patient to achieve an aesthetic response of the soft tissues.

The surgical correction involves Le Fort I osteotomy to advance and inferiorly reposition the maxilla both posteriorly and anteriorly. This will increase posterior face height through downward and backward rotation of the mandible and exaggerate the anterior open bite. In addition, it will enhance lip–tooth relations. Further augmentation of the maxilla will be performed through cranial bone grafting its anterior surface. Closure of the open bite will be performed with sagittal split osteotomy to counterclockwise rotate the distal segment. Osseous genioplasty will advance the chin 5 mm and inferiorly reposition it 3 mm. The buccal fat pad will be resected. Overall, in this long-face, anteriorly divergent patient, soft tissue pogonion will come forward an additional 7 mm and lower face height will increase by 4 mm.

Examination of the postoperative result demonstrates a favorable soft tissue response. There is enhanced zygomatic highlighting with a submalar hollow, good definition of the inferior border of the mandible, and a well-defined cervical mandibular angle. There is improved incisal exposure. The face diverges anteriorly to an excessive degree, but it is seen in concert with good facial convexity, good labial relations, and adequate nasal projection. The bite has remained closed for 4 years. The patient refused to undergo secondary rhinoplasty.

PREOPERATIVE

POSTOPERATIVE

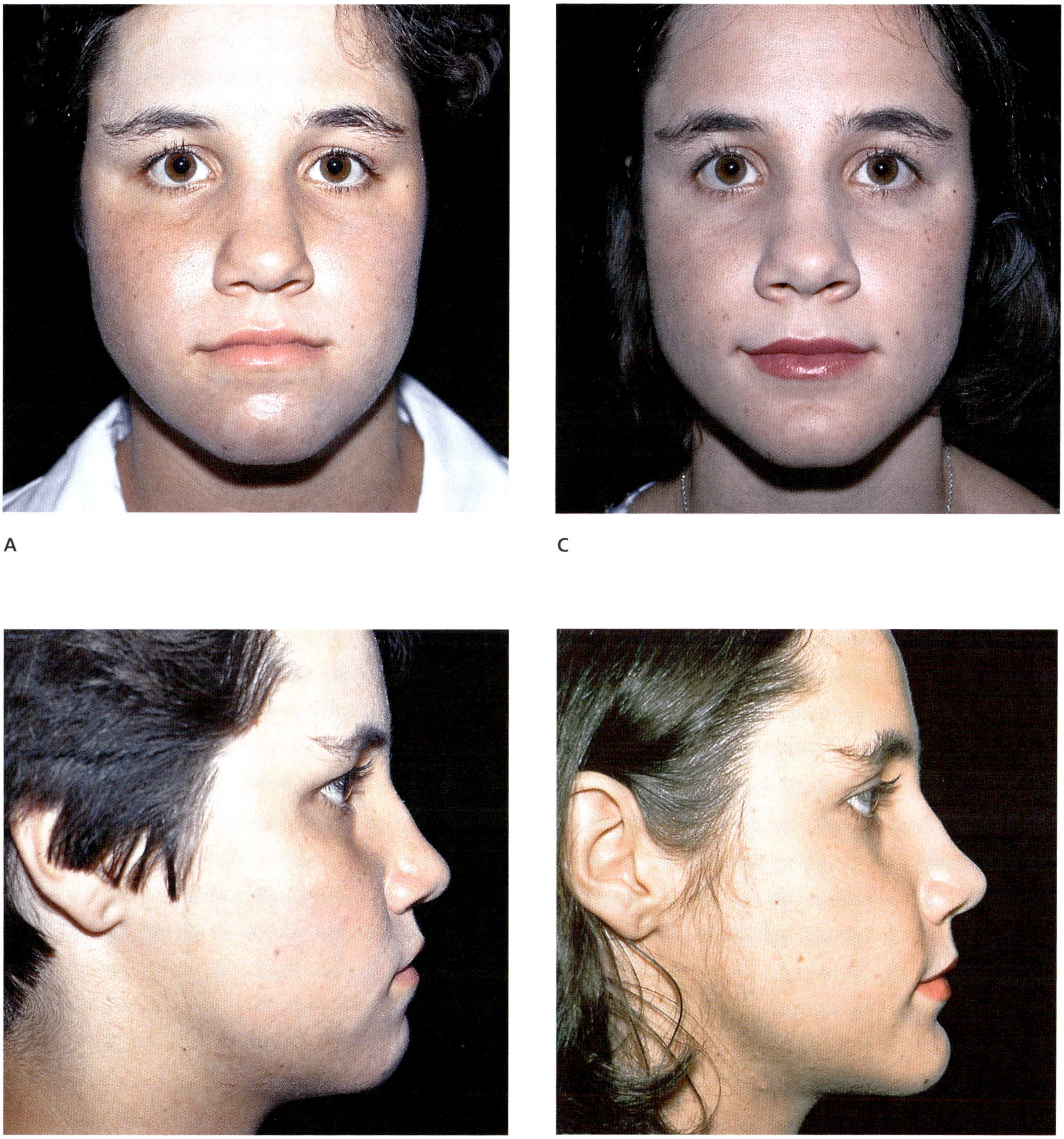

A

C

B

D

CASE STUDY 9-16 A & B: A patient with a quantitative long face, Class III anterior open bite, and anterior divergent profile. The soft tissues are inadequately supported. Double jaw surgery and chin surgery will further lengthen and project the face. C & D: Postoperatively, there is an excellent response of the soft tissues to skeletal expansion with enhanced definition and angularity.

PREOPERATIVE | POSTOPERATIVE

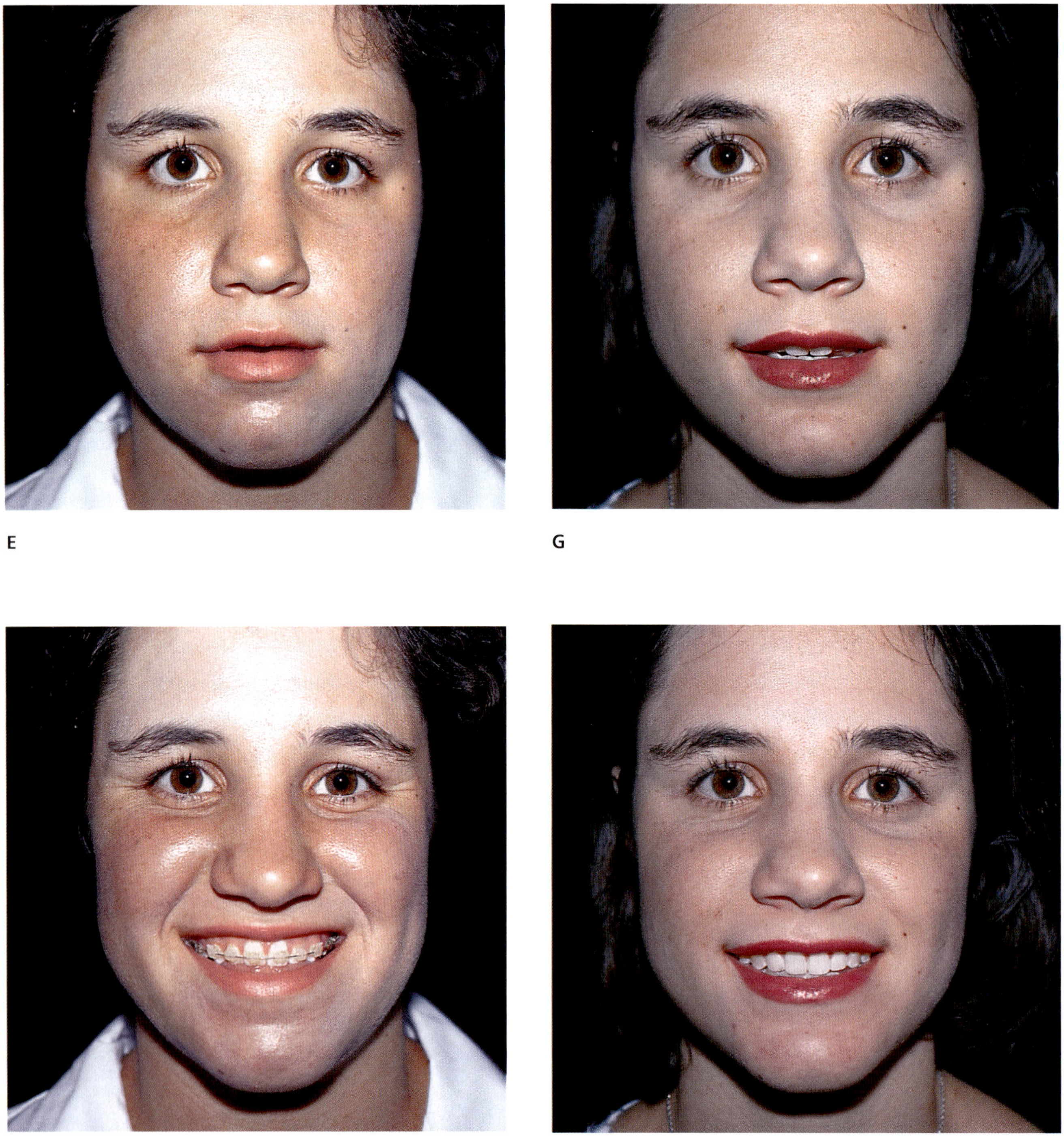

E G F H

FIG. 9-16 *(Continued)* E: There is total absence of incisal exposure with the lips in repose. G: Postoperatively, there is 2 mm of incisal display at rest. Again, note the enhanced definition of the soft tissues.

PREOPERATIVE | POSTOPERATIVE

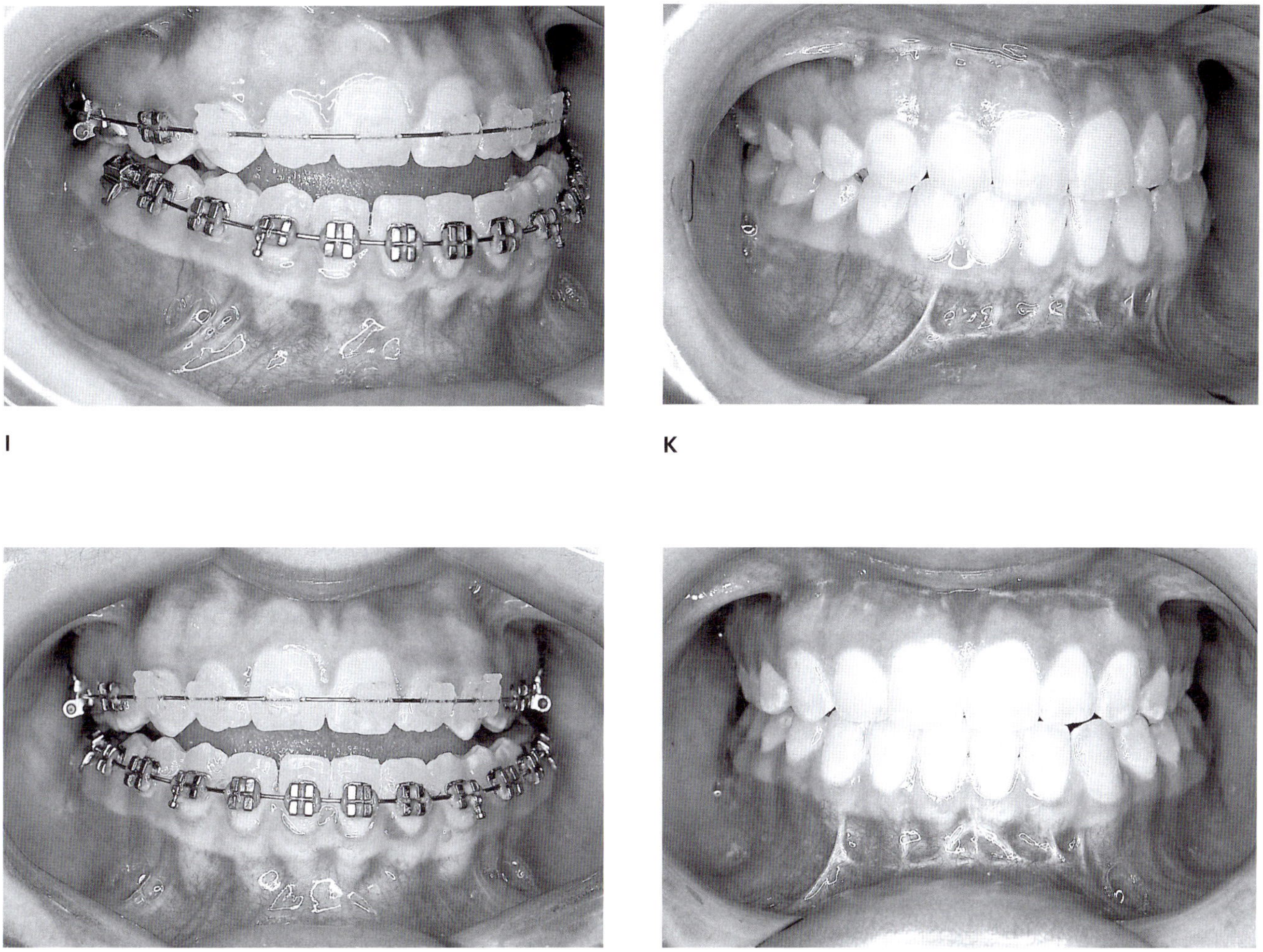

I K J L

FIG. 9-16 (*Continued*) The occlusal result 30 months following surgery. Orthodontist: Dr. Larry Smedley

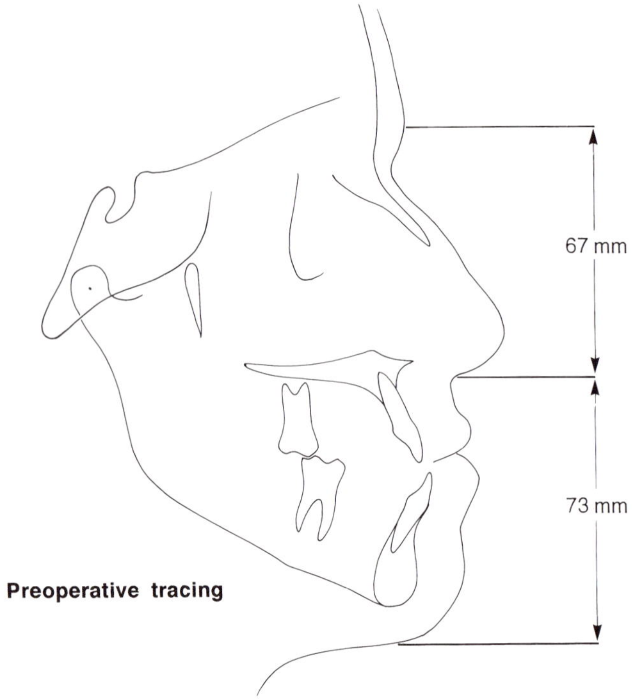

M

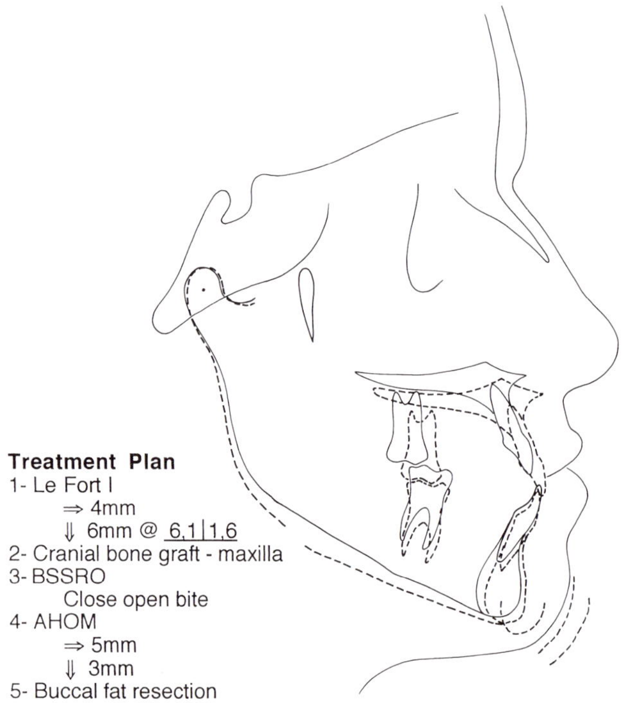

N

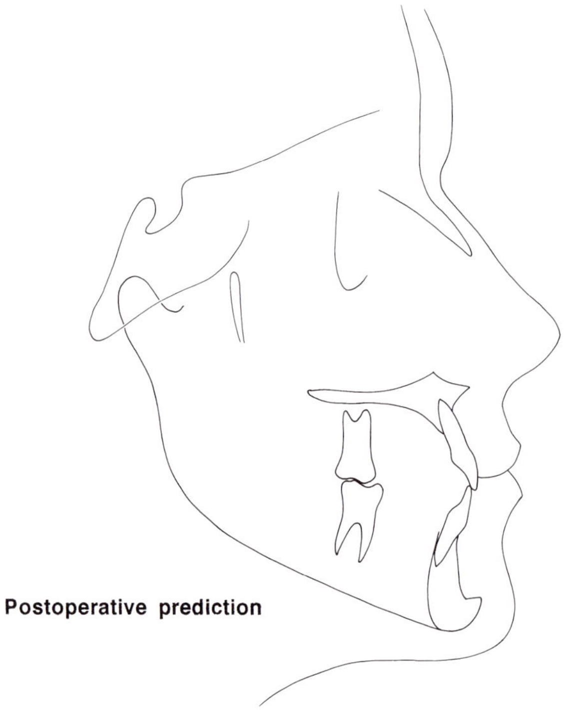

O

FIG. 9-16 *(Continued)* M: The lower face is long and there is overall anterior divergence. N: The treatment plan will further lengthen the face, both anteriorly and posteriorly, and further project the face to create excessive anterior divergence. Soft tissue concerns are dictating skeletal proportions.

Conclusion

The long-face morphologic pattern represents the only one in which surgeons continue to feel obligated to contract the facial skeleton. This results in paradoxical surgical sequelae. On the one hand, it is favorable in terms of restoring vertical proportionality and enhanced lip–tooth relations, but on the other hand, it is detrimental in terms of the unfavorable response of the soft tissues to skeleton contraction. This risks making the patient look significantly older, not only as it relates to the soft tissue response, but also if vertical overcorrection takes place resulting in a short lower face and minimal or no tooth exposure.

The surgeon must keep in mind several principles when dealing with this patient group. First, there must be an aesthetic rationale to shorten a face. In the absence of lip strain, attenuation of the labiomental fold, and excessive incisal and gingival exposure, there is little compelling reason to vertically contract the face. When these stigmata of the long face do exist, it is prudent to err on the side of undercorrecting in the vertical dimension. This undercorrection can then be compensated for by overcorrecting in the sagittal dimension to minimize the loss of skeletal volume. The concept of facial disproportion plays a key role in the management of the long-face patient.

Author's Note: All patients presented in Chapter 9 underwent alar base cinching as part of their initial orthognathic surgery.

Suggested Reading

1. Bell, W.H.: Le Fort I osteotomy for correction of maxillary deformities. *J Oral Surg* 33:412, 1975.
2. Bell, W.H., Creekmore, J.D., and Alexander, R.G.: Surgical correction of the long-face syndrome. *Am J Orthod* 71:40, 1977.
3. Bell, W.H. and Proffit, W.R.: Maxillary Excess. In W.H. Bell, W.R. Proffit, and R.P. White (Eds.): *Surgical Correction of Dentofacial Deformities.* Philadelphia: Saunders, 1980, p. 368.
4. Burstone, C.J.: Lip posture and its significance in treatment planning. *Am J Orthod* 53:262, 1967.
5. Kawamoto, H.K.: Treatment of the elongated lower face and gummy smile. *Clin Plast Surg* 9:479, 1982.
6. Rosen, H.: Lip-nasal aesthetics following Le Fort I osteotomies. *Plast Reconstr Surg* 81:171, 1988.
7. Rosen, H.: Facial skeletal expansion: Treatment strategies and rationale. *Plast Reconstr Surg* 89:798, 1992.
8. Rosen, H.: Aesthetics in facial skeletal surgery. *Perspectives in Plastic Surgery* 6:1, 1993.
9. Schendel, S.A. and Williamson, L.W.: Muscle reorientation following superior repositioning of the maxilla. *J Oral Maxillofac Surg* 41:235, 1983.

Vertical Facial Deficiency

CHAPTER 10

The morphologic designation of short face has several skeletal etiologies. The most common of these is mandibular deficiency with an anterior deep-bite malocclusion. Second in frequency is vertical maxillary deficiency. Last, and least common, is true vertical microgenia, which is rarely seen by itself but usually in combination with mandibular deficiency.

The most severe forms of short face usually involve a combination of vertical maxillary deficiency and mandibular deficiency. Quantitatively, it is defined as a decrease in the height of the lower face, as measured from subnasale to menton, compared to that of the midface, as measured from subnasale to glabella. This results in not only an overall short but also square facial shape. What is more clinically relevant, however, is the qualitative deformity of the soft tissues that results from a decreased vertical dimension of the lower facial skeleton. As a result of this deficiency of skeletal volume, the soft tissues are redundant and poorly supported. The combination of a short face and poorly supported tissues creates an older appearance. Even in younger individuals, there may be jowl formation and submental fullness. In the presence of vertical maxillary deficiency, the maxillary incisor teeth may be completely covered by the upper lip, creating an edentulous appearance with the lips in repose and an unattractive smile. As a result of overclosure of the mandible, the oral commissure is widened, the lips are procumbent, the nasolabial angle narrows, and the labiomental fold deepens. The patients are frequently described as having a saddened or even angry countenance because of the apparent down turning of the oral commissure. Many of these patients are self-conscious about their unattractive smile. As a result, they infrequently smile, which reinforces the perception of sadness. The typical facial morphology of the short-face patient is seen in Fig. 10-1.

The overall treatment planning philosophy in these patients must be to adequately expand the facial skeleton to eliminate the soft tissue

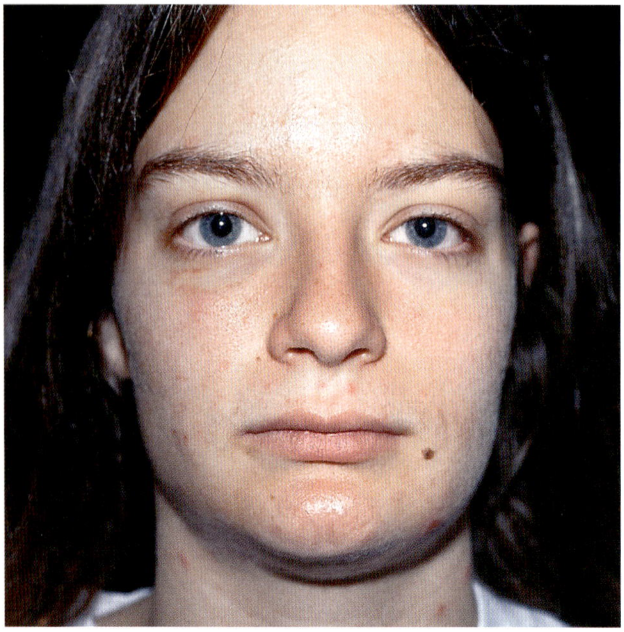

A

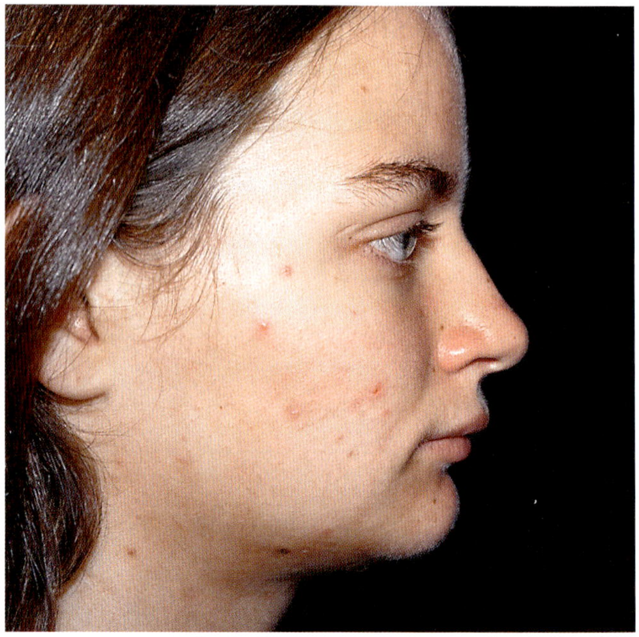

C

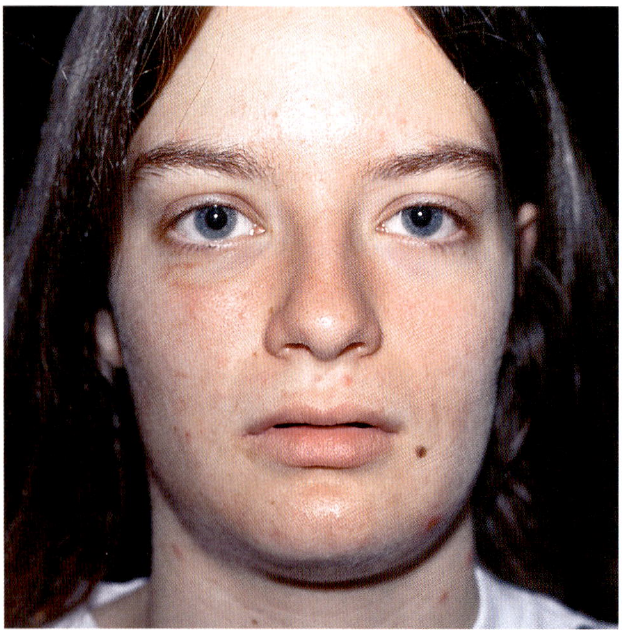

B

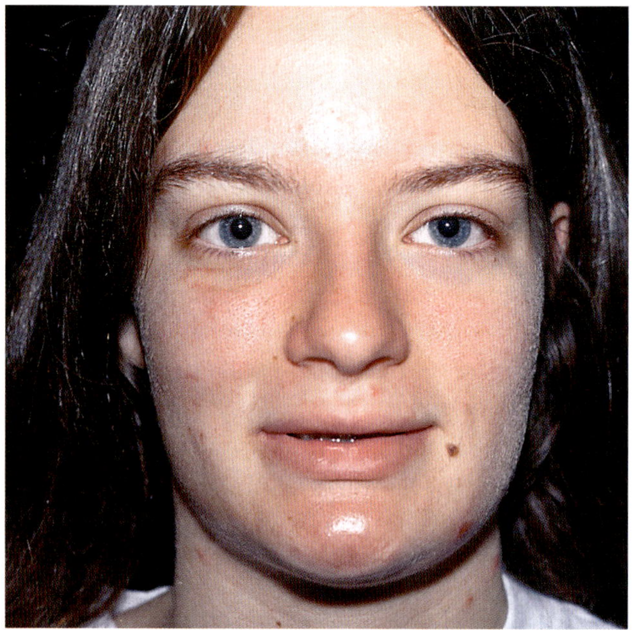

D

FIG. 10-1 A & C: Typical morphology of the short-face patient with a short, square face, inadequately supported soft tissues, and an overclosed appearance. Note the sullen expression. B & D: When vertical maxillary deficiency is present, there is insufficient, if any, incisal exposure in repose and with smiling.

redundancy and provide for a well-supported facial mask. This, in turn, eliminates the aesthetic stigmata of the short face. As the facial skeleton is expanded in the vertical dimension, and frequently in the sagittal dimension as well, the soft tissues take on a more delicate and refined appearance. The oral commissure narrows, the upper vermilion thickens, the labiomental fold becomes less deep, the lips become less procumbent, and the maxillary incisors begin to emerge from behind the upper lip. As the jowls disappear, the lower face narrows, which serves to emphasize its increased height. The shape of the face changes from square to oval (Fig. 10-2). The patient is transformed from having an older, saddened appearance into having a younger, more contented one.

Just as there must be an aesthetic rationale to shorten the long face, there must also be compelling aesthetic reasons to elongate a short lower face. It is important to understand that quantitative vertical facial deficiency does not necessarily correlate with a face that looks too short. It is only when the aesthetic stigmata of a short face accompany the quantitative vertical deficiency that the face, indeed, appears too short. One can then justify vertical elongation. These aesthetic stigmata include an absence of maxillary tooth show beneath the relaxed upper lip (this is primarily an aesthetic issue in female patients); excessive deepening and elongation of the labiomental fold; excessive widening of the oral commissure with procumbent

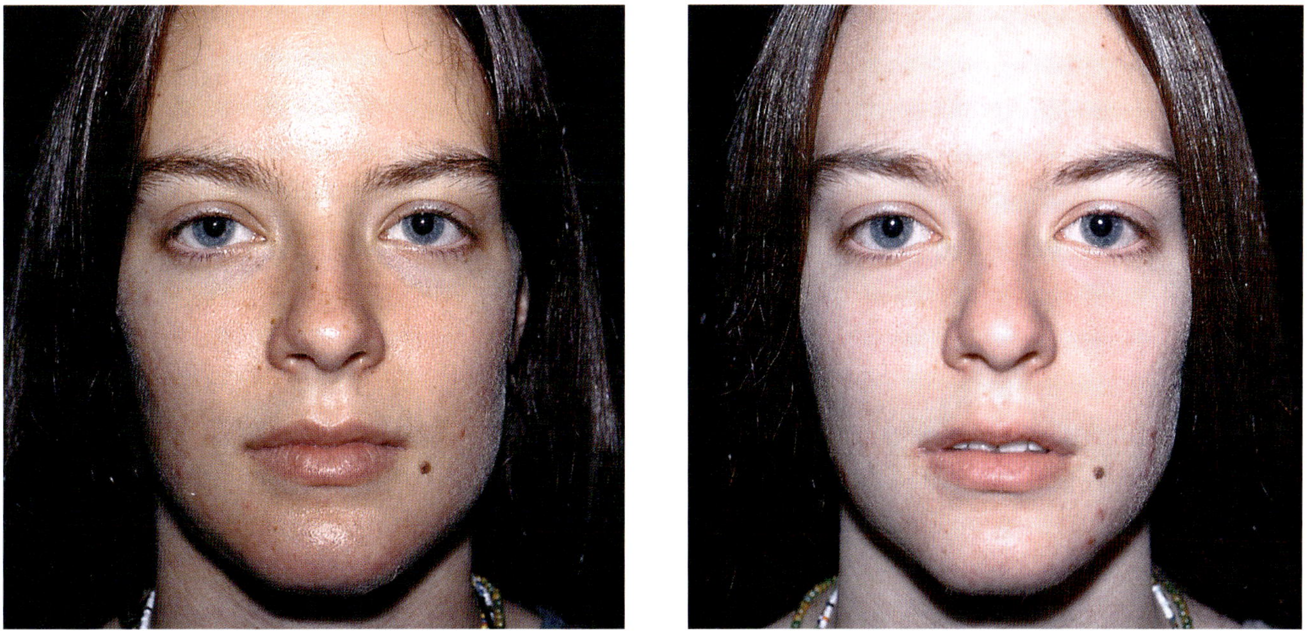

A B

FIG. 10-2 A & B: The same patient as in Fig. 10-1 following surgical correction. As the facial skeleton is elongated, the oral commissure narrows, the lips are less everted, the upper vermillion thickens, the labiomental fold is attenuated, and the jowls disappear, which are all signs of enhanced soft tissue support. The shape of the face changes from square to oval, and the incisors begin to emerge from behind the upper lip when the lips are parted.

lips; and perimandibular soft tissue excess, that is, jowls. It should be obvious why these characteristics make the face look too short. It is the oral commissure and the labiomental fold that run perpendicular to the long axis of the lower face and, therefore, break up the vertical axis of the lower face. When these perpendicular structures are emphasized, the lower face appears shorter. The presence of jowls widens the lower face, which further exaggerates the vertical deficiency.

When one lengthens a quantitatively short face, through either maxillary and/or mandibular surgery, in the absence of aesthetic stigmata of a short face, one risks creating lip strain, excessive attenuation or elimination of the labiomental fold, and an excessively toothy appearance. These can then make a lower face look too long even if it is of normal vertical dimension.

Treatment Planning for Short-Face Correction

The decision to perform single- or double-jaw surgery in the short-face patient should be quite straightforward. If there is any component of anterior maxillary vertical deficiency, defined as minimal or absent maxillary incisal show in the presence of an upper lip that is normal in length, maxillary vertical lengthening is indicated. Double-jaw surgery is almost always indicated when this is seen in combination with a Class II occlusion; that is, mandibular advancement surgery will be necessary because the Class II tendency will only get worse as the maxilla is lengthened and the mandible rotates downward and backward.

When vertical maxillary deficiency is seen in combination with a Class I sagittal relationship, a decision must be made as to whether single- or double-jaw surgery would achieve the best aesthetic result. This primarily relates to the ultimate support of the upper lip and sagittal position of the lower lip and chin. It should be understood that vertical maxillary lengthening in the presence of a Class I occlusion will move the maxilla as well as pogonion posteriorly as the mandible undergoes clockwise rotation, with the point of rotation being the mandibular condyle. If this will result in inadequate midface projection as evidenced by a lack of support of the upper lip, or inadequate sagittal projection of the lower face despite adjunctive osseous genioplasty, a double-jaw procedure is again indicated.

Finally, when vertical maxillary deficiency is found in combination with a Class III malocclusion (typically observed in cleft deformities), isolated maxillary lengthening will improve the occlusal abnormality as the mandible rotates downward and backward. Prediction tracings are helpful in determining the aesthetic advisability of double-jaw surgery and, in particular, adjunctive genioplasty in many of these individuals.

The next decision to be made is the magnitude to which one should vertically elongate the anterior maxilla to achieve adequate degrees of maxillary incisal show. Unfortunately, this is a pure clinical judgment. However, it is a common mistake to err on the side of insufficient elongation as opposed to excessive elongation. It is a fact that the upper lip commonly lengthens as the maxilla is lengthened. Upper lip lengthening is due to less vertical support of the upper lip as the overclosed mandible rotates downward and backward. In addition, as the maxilla is inferiorly repositioned, the nasal spine also moves inferiorly, and this pushes the upper lip downwards.

It has been suggested that one can adequately determine the extent to which vertical maxillary elongation should take place by taking a lateral cephalometric X-ray film with the mandible slightly opened and the lips parted to allow the upper lip to assume a posture uninfluenced by mandibular overclosure. This is a rather naive concept because it is unreasonable to assume that the upper lip will lengthen in the course of several seconds after having been influenced by mandibular overclosure over the course of many years. Given this information, the best rule of thumb is to overcorrect, particularly in the patient with procumbent lips and a narrow nasolabial angle. The tendency to inadequately lengthen the maxilla is illustrated in the patient in Fig. 10-3. Despite a vertical maxillary elongation of 6 mm, there is still an obvious absence of maxillary incisal show with the lips at rest, although there has been marked improvement in the aesthetics of the patient's smile secondary to greater maxillary incisal exposure. It is apparent in this patient that the upper lip has significantly lengthened.

Another clinical decision to be made is the advisability of alar cinching in an effort to avoid excessive alar flaring. This suturing technique should only be undertaken in those patients with preexisting excessive alar base width as vertical maxillary elongation does not widen the alar base or cause excessive alar flaring.

Finally, the extent to which presurgical orthodontics should level the mandibular occlusal plane must be decided. This issue has already been discussed in Chapter 5. However, one consideration bears repeating at this juncture. It is not infrequent to see short-face patients with deep anterior bite, Class II malocclusions. This deep bite is due to supraeruption of the mandibular incisors as a result of a lack of any contact with the maxillary incisors. Orthodontic leveling of the mandibular dentition can take place presurgically with mandibular incisor intrusion or postsurgically with premolar extrusion. Presurgical leveling minimizes the clockwise rotation of the distal mandibular segment as it is surgically advanced. The lower lip and chin will both come forward. With postsurgical leveling, there is significant clockwise rotation that accompanies the mandibular advancement. As a result, the soft tissue chin point is inferiorly positioned at the expense of sagittal advancement; that is, the chin comes forward less than the lower lip. Postsurgical leveling of the mandibular dentition should always take place in a patient with a deep anterior bite that is seen in association with a prominent chin (see Fig. 10-7).

PREOPERATIVE POSTOPERATIVE

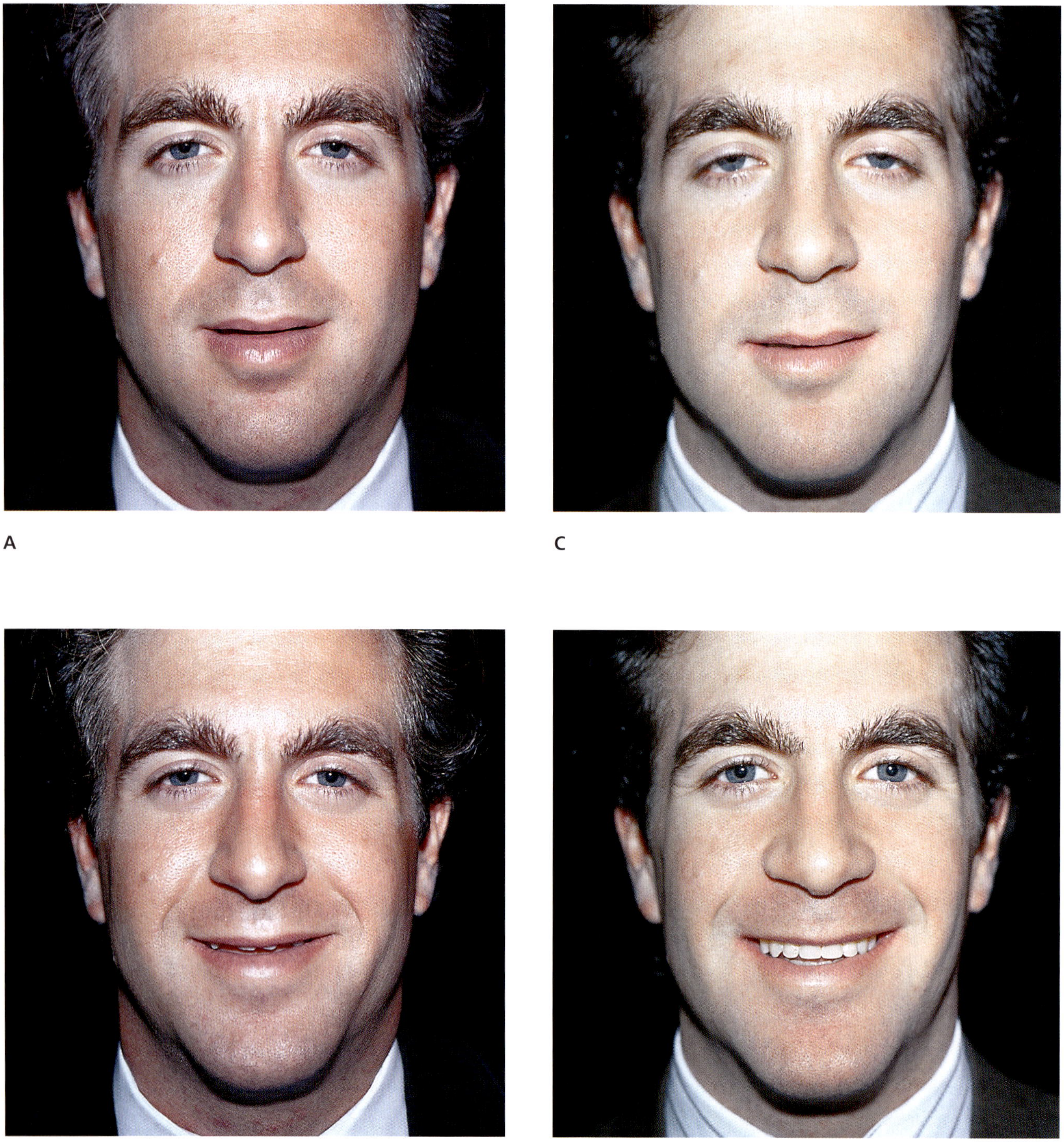

A C B D

Fig. 10-3 A & B: A patient with vertical maxillary deficiency who will undergo inferior maxillary repositioning. C & D: The patient has been undercorrected; that is, there is still an absence of tooth exposure with the lips at rest because the upper lip has lengthened. His smile has been significantly enhanced.

Stability of Maxillary Vertical Lengthening

In the past, conventional wisdom has held that inferior repositioning of the maxilla is a highly unstable procedure. This was initially thought to be due to the resorption of autogenous bone graft material that had been placed into the resulting osteotomy gap secondary to lengthening the maxilla. During the subsequent period of intermaxillary fixation, the constant pumping action of the muscles of mastication would place undue pressure on the grafts, causing resorption. Indeed, reports of total relapse were not uncommon. The use of rigid fixation employing plates and screws improved this situation, but mean relapse rates of 25% were still reported when plate and screw fixation was combined with autogenous bone grafting. It required the combination of plate and screw fixation and nonresorbable bone substitutes, such as porous, block hydroxyapatite, to eliminate skeletal instability following maxillary inferior repositioning.

To understand this enhanced stability, the biomechanical properties of porous, block hydroxyapatite when used in combination with plate and screw fixation must be reviewed. Theoretically, rigid fixation eliminates the need for intermaxillary fixation and, therefore, negates the unfavorable effect of the masticatory musculature that takes place during extended periods of intermaxillary fixation. During the first 6 to 8 weeks, patients are kept on a relatively soft diet to further minimize muscular forces. It is during this time that autogenous tissue, including fibrovascular tissue and bone, is deposited within the interstices of the implant. This increases the compressive strength of the implant so that masticatory forces can eventually be accommodated. As tissue ingrowth proceeds, the compressive strength of the material exceeds those forces generated within the masticatory apparatus. In addition, the elastic modulus of the material prevents adjacent host bone resorption. The final biomechanical property of hydroxyapatite that makes it ideal in this circumstance is that is does not lose volume with time. Data have now been accumulated on patients who received hydroxyapatite implants up to 9 years ago, and the results remain exquisitely stable.

Recently, Major et al have demonstrated good stability of inferiorly repositioned maxillae employing rigid fixation and autogenous cortical bone grafts harvested from the iliac crest. Follow-up was a mean 16 months. Despite these new data, the author still prefers porous block hydroxyapatite because longer follow-up is available in patients treated with hydroxyapatite. In addition, the morbidity associated with harvesting cortical bone from the iliac crest is extremely high.

Questions have been raised regarding the potential problem of free-way space encroachment as the maxilla is inferiorly repositioned. It was postulated that this, too, would lead to instability and/or spasm of the pterygomasseteric muscle complex. It should be realized that the concept of free-way space encroachment was largely generated in the prosthodontic literature. This bears little relationship to the dentulous patient and has certainly not proven to be a concern.

The last consideration regarding the stability of inferior maxillary repositioning relates to performing simultaneous mandibular surgery. It has been theorized that simultaneous osteotomies of the mandibular ramus create less risk of skeletal relapse because there is decreased potential to impinge on the free-way space and because masticatory efficiency will likely be decreased, at least temporarily, with simultaneous mandibular surgery. Both these theories relate to the undesirable masticatory forces transmitted to the grafting material within the maxillary osteotomy gap. I have noted no difference in the skeletal stability of vertical maxillary lengthening procedures performed in isolation or concomitantly with mandibular surgery.

Patient Presentations

Case Study 10-4

This individual is the first, and perhaps youngest, patient in whom short-face correction was undertaken employing rigid fixation and porous block hydroxyapatite. She was 15 years old at the time of her surgery. Her facial morphology is typical of the classic short-face patient. In the frontal view, the overall shape of the face is short and square. There is a clear-cut visual impression of vertical deficiency of the lower facial third relative to the midface. The soft tissues are redundant and poorly supported with jowl formation, everted upper and lower lips, and a deepened labiomental fold. The oral commissure is widened, and there is minimal vermilion exposure. There is a total absence of maxillary incisal show with the lips in repose and minimal incisal exposure with full smiling. The profile demonstrates lip eversion, closure of the nasolabial angle, and deepening of the labiomental fold, all secondary to mandibular overclosure. There is remarkably poor soft tissue support throughout the mid- and lower face. There is overall facial convexity and posterior facial divergence. The occlusion is Class II, Division 1 with significant maxillary spacing. Notice the vertical deficiency of the anterior mandible on the lateral cephalogram. Also note that a repeat cephalogram taken with the jaw slightly parted demonstrates that the maxillary incisal edge is level with the upper lip.

The treatment plan for this individual was designed to achieve an overall increase in the vertical dimension by inferiorly repositioning the maxilla 7 mm, mandibular advancement with clockwise rotation and subsequent inferior repositioning of menton of 3 mm, and osseous genioplasty to advance and vertically elongate the chin 5 mm and 6 mm, respectively. Overall total lengthening of the face would be 16 mm. The magnitude of anterior maxillary lengthening was totally arbitrary and based on the assumption that the upper lip would lengthen a significant amount.

The postoperative result is remarkable for a dramatic increase in the height of the face and a change in the overall shape of the face from square to oval. There is approximately 3 mm of maxillary incisal show with the lips at rest, and a full clinical crown is now exposed with full smiling. The profile is now anteriorly divergent with overall facial convexity. Note the dramatic enhancement of the skeletal support of the soft tissues; that is, the jowls have been eliminated, there is a marked reduction of the soft tissue redundancy in the neck, the lips are no longer everted, the nasolabial angle has opened, the labiomental fold has been attenuated, and the oral commissure has been narrowed in its transverse dimension. The vermilion exposure is also increased. The radiograph taken 8 years following surgery, when compared to one taken 1 day following the surgery, demonstrates exquisite stability of the vertical facial augmentation. The occlusion is now Class I.

This patient represents a dramatic example of the beneficial effect of bidimensional expansion of the facial skeleton on the overall support of the soft tissues. It must be appreciated that the soft tissue response is at least as important as establishing more favorable vertical facial proportions when determining the quality of the aesthetic result.

PREOPERATIVE POSTOPERATIVE

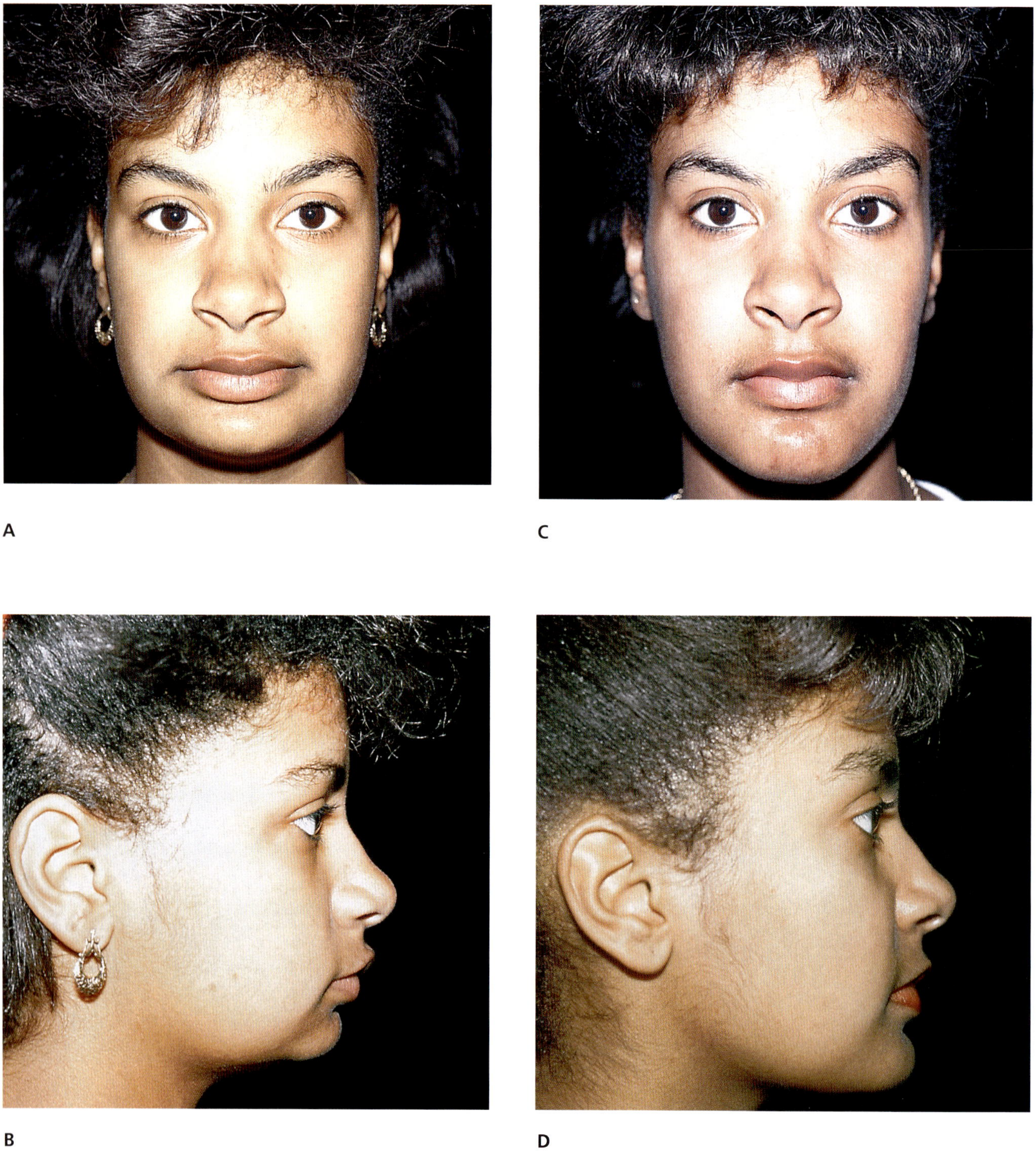

CASE STUDY 10-4 A & B: A short face, mandibular deficient patient, demonstrating the soft tissue aesthetic stigmata of the short face "syndrome." Surgical correction involved total facial lengthening of 16 mm. C & D: Note the obvious bidimensional increase of the facial skeleton and the profound effects on the soft tissue support.

PREOPERATIVE POSTOPERATIVE

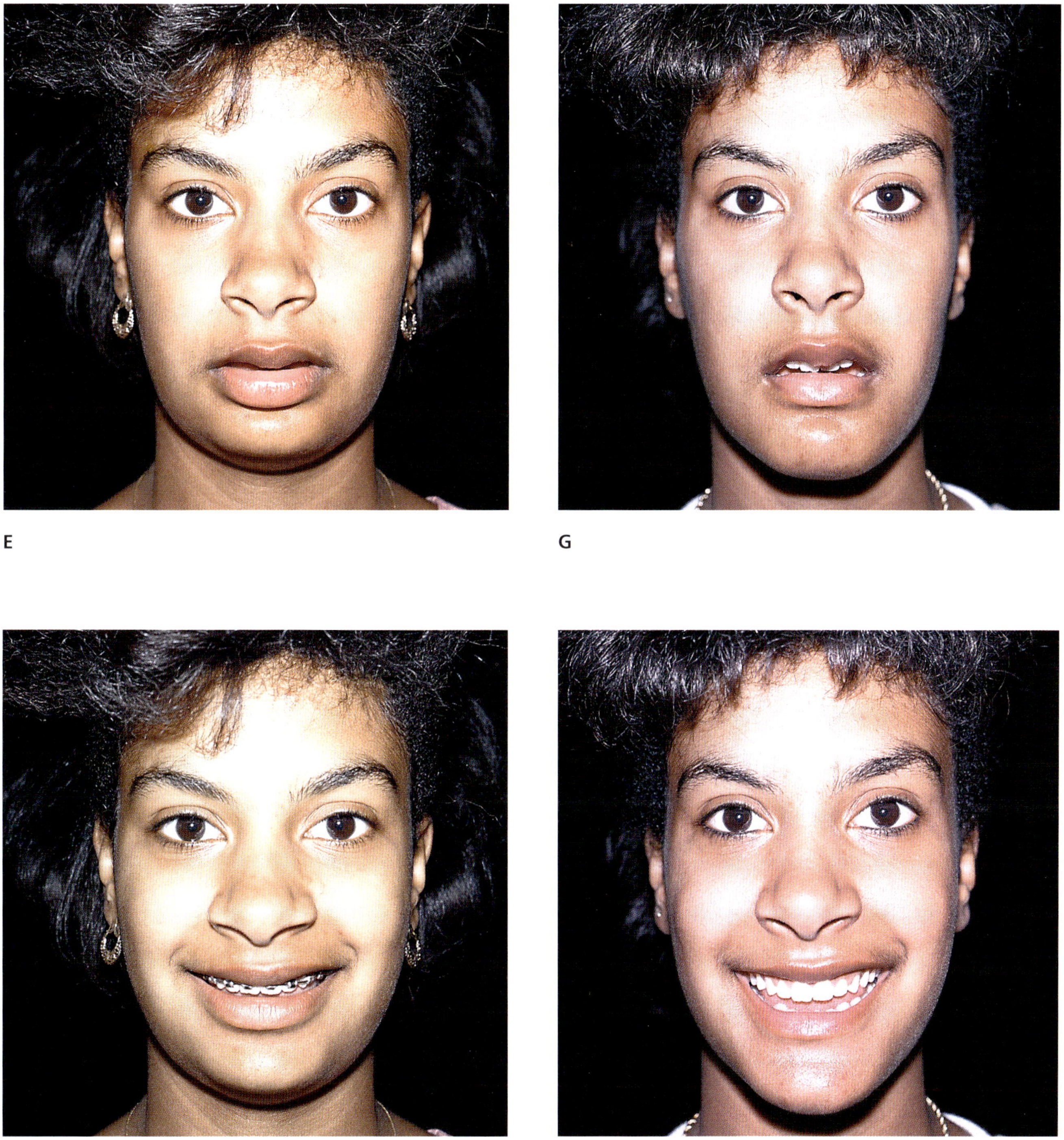

FIG. 10-4 (*Continued*) E & F: Deficient incisal show in both repose and smiling. G & H: Postoperatively, there is 3 mm of incisal exposure in repose and a full crown exposed with smiling. From: Rosen, H.: Definitive surgical correction of vertical maxillary deficiency. *Plast Reconstr Surg* 85:215, 1990. Reprinted with permission.

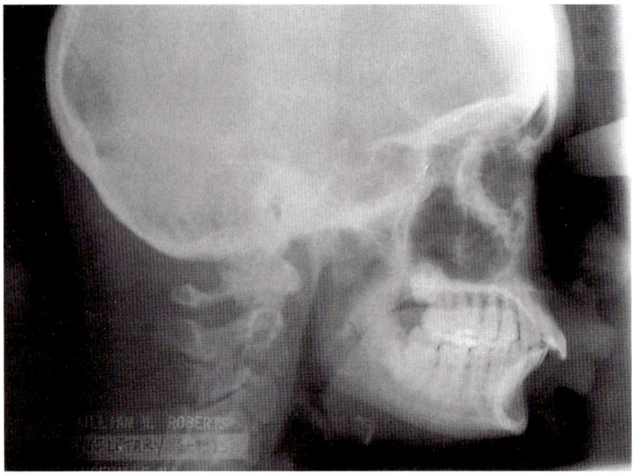

I

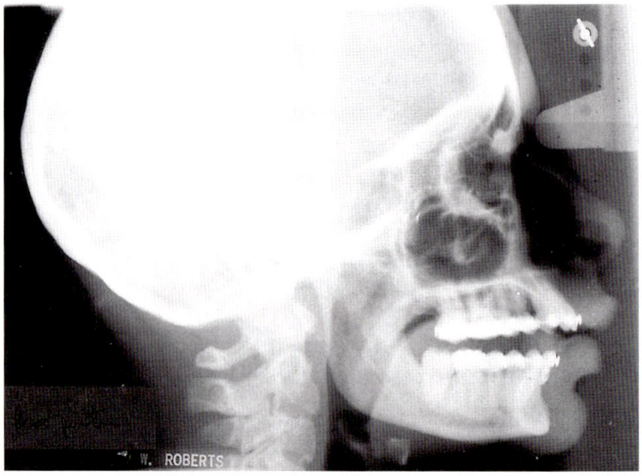

J

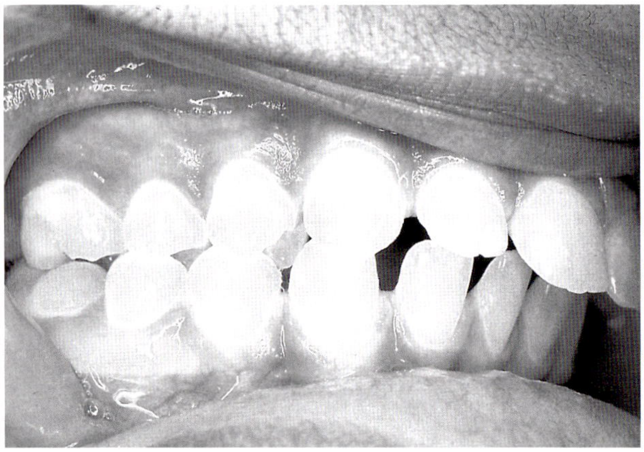

K

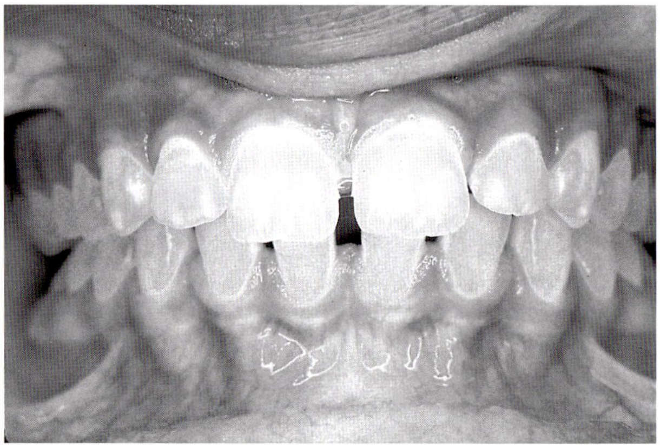

L

FIG. 10-4 (*Continued*) I & J: Preoperative lateral cephalograms in centric occlusion and in physiologic rest position. With the mandible slighly open, the edge of vermillion is even with the incisal edge. Despite this, a 7-mm vertical movement of the anterior maxilla was planned due to the anticipated upper lip lengthening. K & L: The preoperative occlusion.

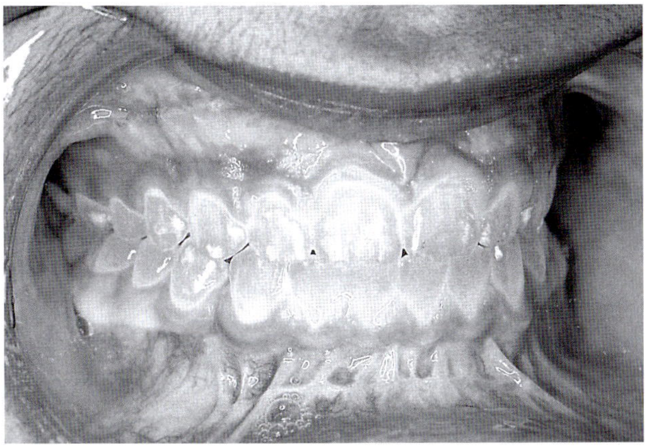

M

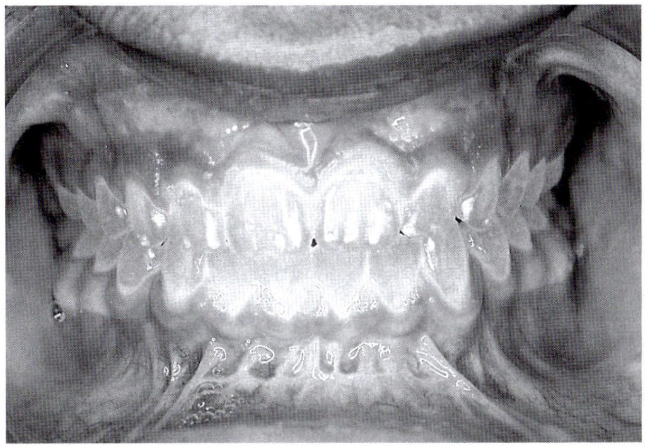

N

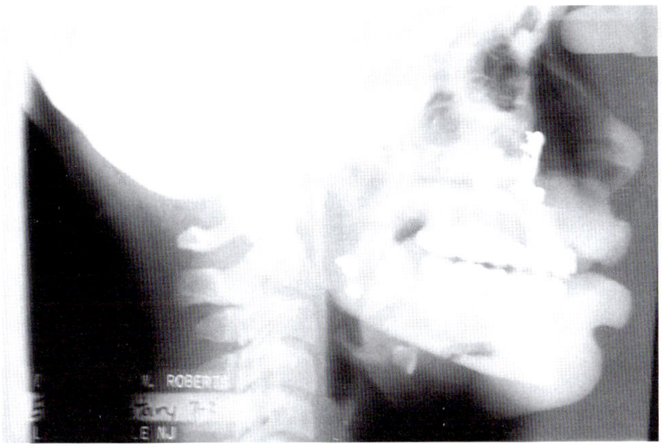

O

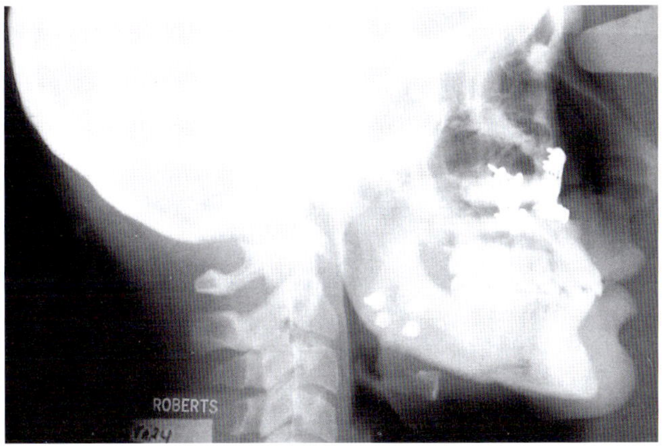

P

FIG. 10-4 (*Continued*) M & N: The postoperative occlusion. O & P: Postoperative lateral cephalograms taken one day and 8 years, respectively, following surgery. There is no relapse in the vertical dimension. Orthodontist: Dr. William Roberts.

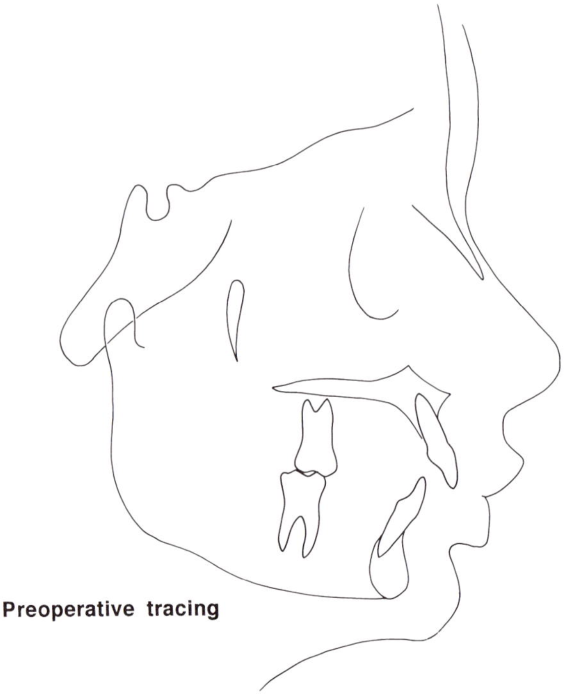

Q

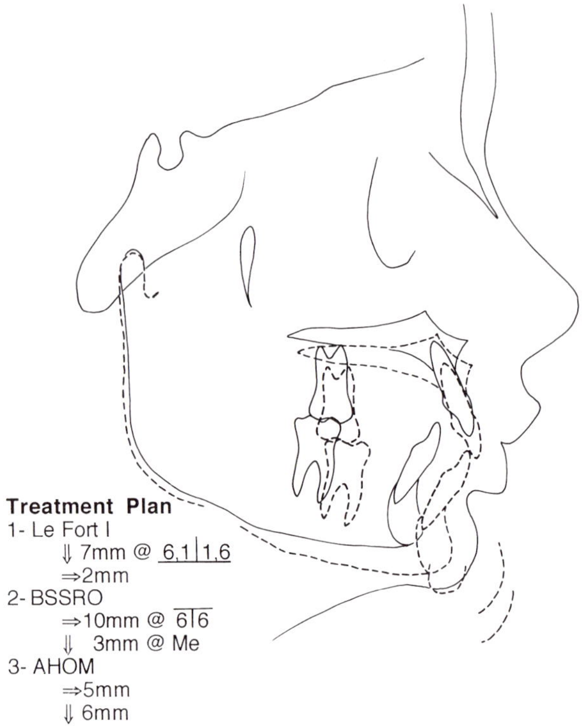

R

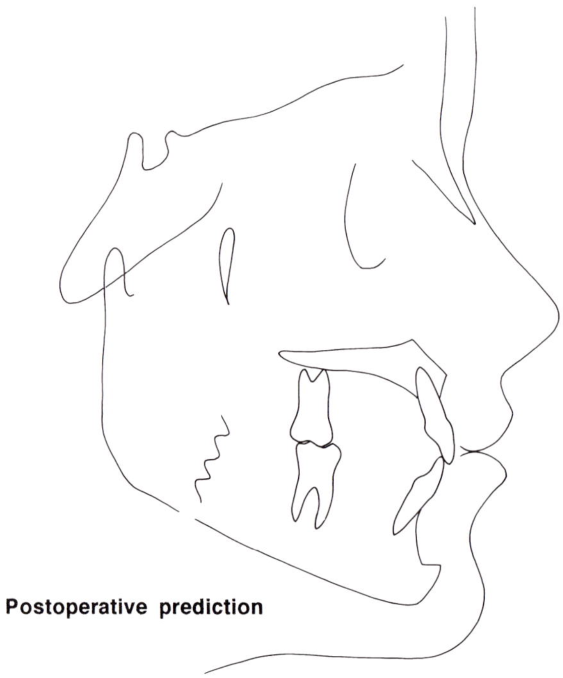

S

Fig. 10-4 (*Continued*) R: Treatment plan proposing a 16-mm increase in vertical facial height.

Case Study 10-5

This individual is clearly not as severely affected as the previous patient presented. The soft tissues are better supported, and there is only a modest reduction in the vertical dimension of the lower face. There is also lower face asymmetry with the chin point deviated to the left. There is, however, the typical sullen appearance seen in many short-face patients. The somewhat overclosed position of the mandible in the lateral view is evidenced by the procumbency of the upper and the lower lips and the closing of the nasolabial angle. There is overall facial convexity and neutral facial divergence. When the patient is asked to relax the lips, there is again absence of incisal show and little tooth exposure with smiling. There is a modest Class III malocclusion with an edge-to-edge anterior bite and a mandibular midline shift to the left.

It was decided to perform maxillary surgery and chin surgery in this patient. This included a 6-mm anterior maxillary vertical elongation and maxillary rotation to the left to restore dental midline symmetry. Chin surgery involved a 5-mm vertical elongation, a 5-mm advancement, and a 4-mm shifting to the right to restore lower face symmetry.

The postoperative result in the frontal view demonstrates a dramatic aesthetic result. Note that the patient has lost her sullen appearance and looks more contented. This is, in part, due to the changes in the oral commissure region, but it is also due to a change in the patient's self-image; that is, she is no longer told she looks sad and so she no longer feels sad. Note also the lack of alar base widening despite the fact that no alar cinch was performed. This is due to the dominant inferior vector of maxillary movement. Also note the increase in vertical length of the upper lip and, as a result, the residual lack of any maxillary incisal exposure at rest. However, the aesthetics of the patient's smile are dramatically improved with fully exposed maxillary incisor teeth. There is also a more natural and less forced appearance to the smile. It no longer looks forced because the patient is more self-confident about her appearance when she smiles. Also note the obvious discrepancy between the maxillary dental midline and her nasal tip. Although there is slight deviation of the nasal tip to the right, this discrepancy is primarily due to the maxillary rotation to the left. The profile is now characterized by a very slight posterior divergence due to the derotation of the mandible at the condyle secondary to maxillary vertical elongation. This is true despite a 5-mm advancement of the chin. We can also observe that there is moderate loss of the overall convexity of the profile, due in part to less eversion of the lips. The other notable findings in the lateral view are the opening of the nasolabial angle and the attenuation of the labiomental fold. The occlusion is now Class I. This is a 4-year follow-up.

If I were to perform this surgery today, a bimaxillary procedure would be undertaken to advance both the maxilla and mandible and rotate the mandible to restore dental midline symmetry. This would have avoided the rather significant discrepancy between the maxillary midline and the facial midline as well as provided a profile with overall anterior divergence, greater facial convexity, and greater support to the upper and lower lips.

PREOPERATIVE **POSTOPERATIVE**

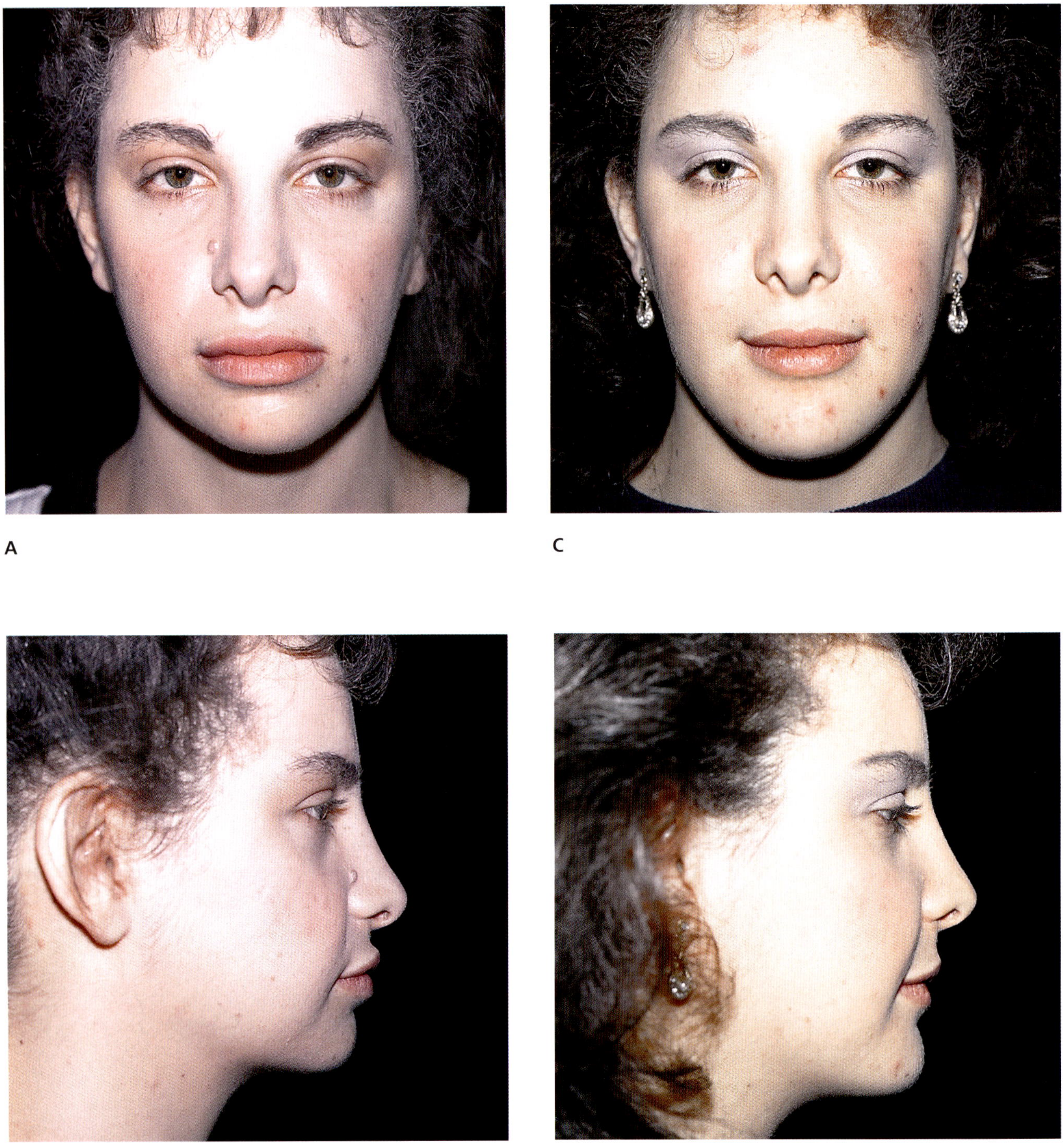

A C

B D

CASE STUDY 10-5 A & B: A short face patient with mandibular asymmetry and an edge-to-edge Class III occlusion. Note the sullen appearance. The patient was treated with maxillary surgery and chin surgery.
C & D: Postoperatively, there is a dramatic improvement in the emotional expression of the face. A better profile result could have been achieved with double jaw surgery.

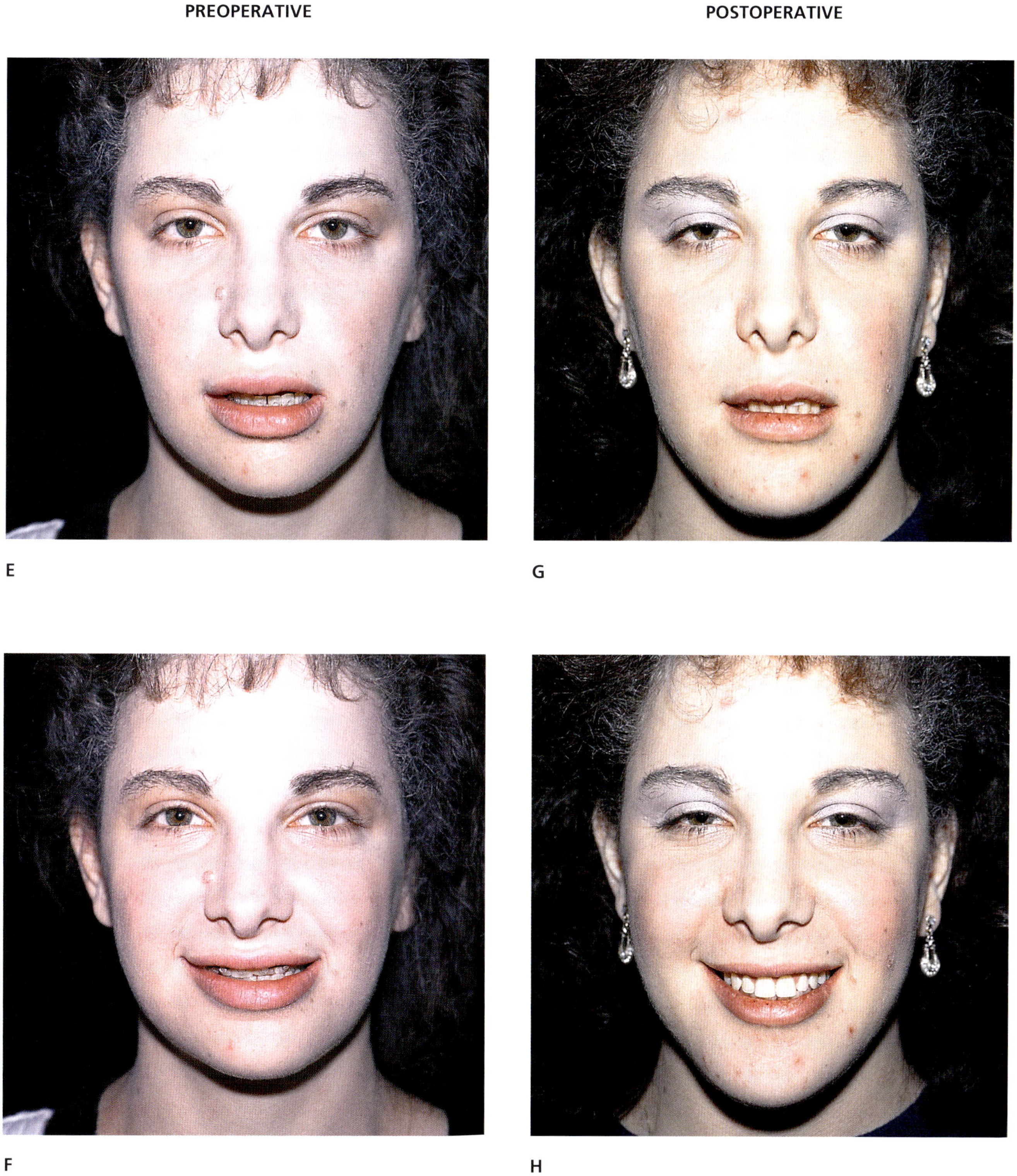

FIG. 10-5 *(Continued)* E & F: Insufficient incisal exposure. G & H: Due to upper lip lengthening, there is still little, if any, incisal show at rest, but the smile has been greatly improved. Note the discrepancy between the facial midline and upper incisors. This would have been avoided if double jaw surgery was performed. From: Rosen, H.: Definitive surgical correction of vertical maxillary deficiency. *Plast Reconstr Surg* 85:215, 1990. Reprinted with permission.

PREOPERATIVE | POSTOPERATIVE

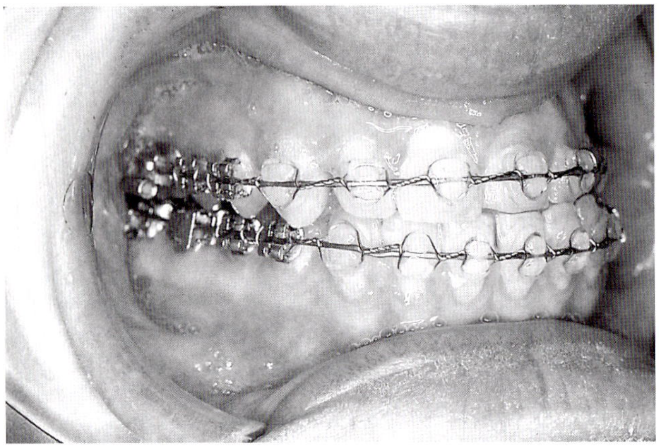

I

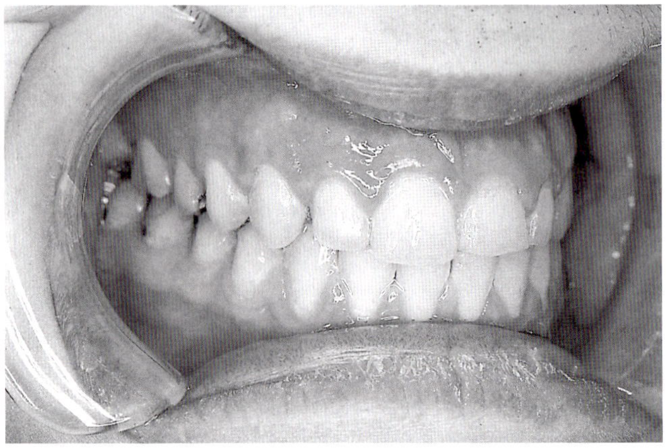

K

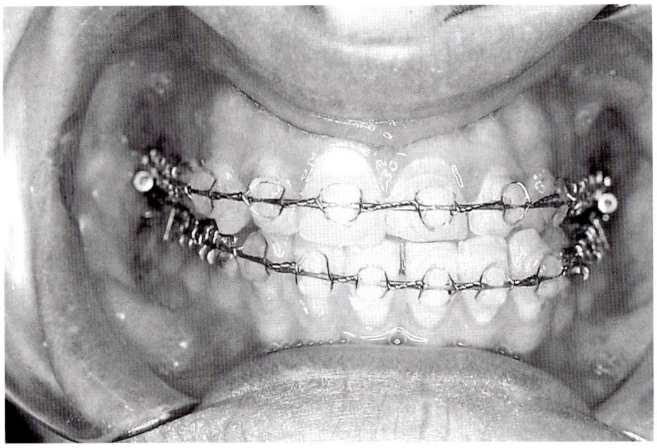

J

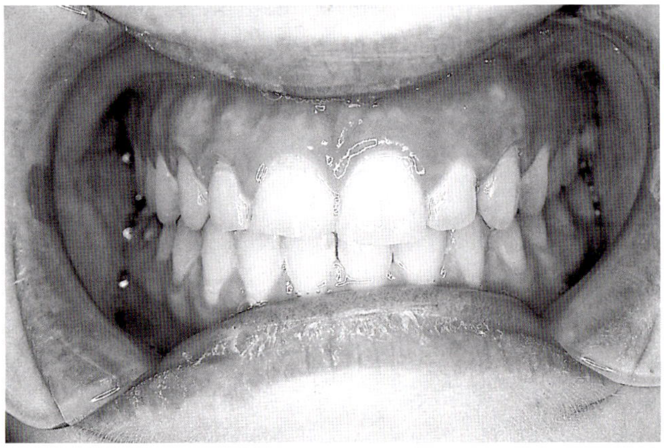

L

Fig. 10-5 (*Continued*) Orthodontist: Dr. James Ackerman

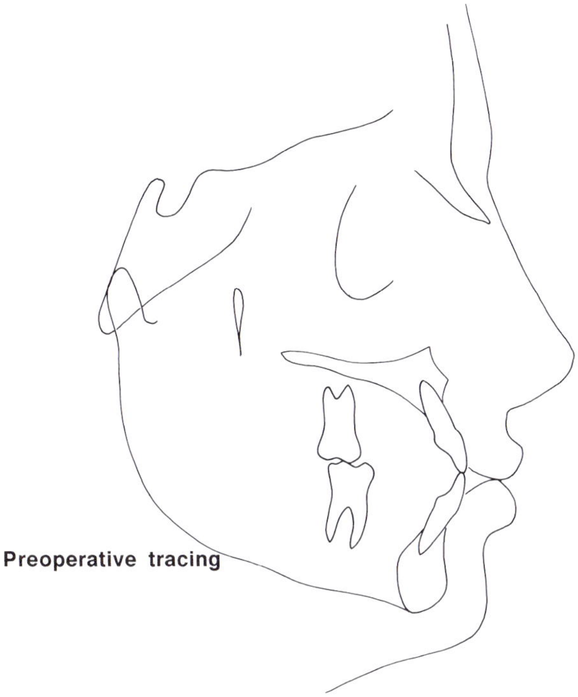

M

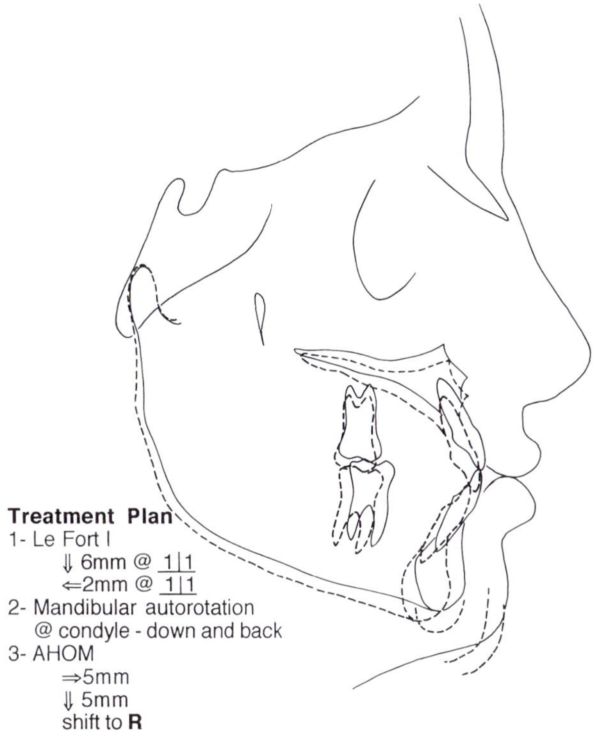

N

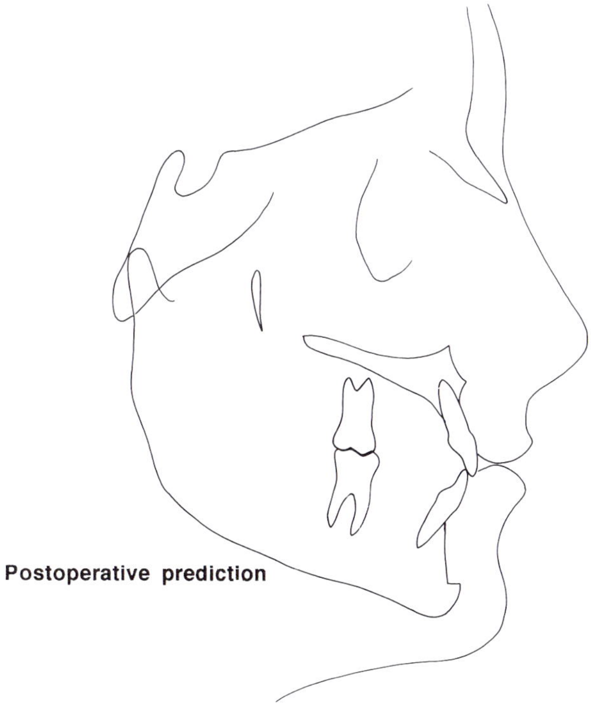

O

FIG. 10-5 (*Continued*) N: The downward and backward movement of the mandible, as the maxilla is lengthened, must be accompanied by posterior movement of the maxilla. Perioral support is, thereby, diminished. Double jaw surgery would have avoided this.

Case Study 10-6

This individual is 42 years old and was referred by her periodontist because of severe impingement of her anterior palatal mucosa by her mandibular incisors. One can see that she has a severe Class II, Division 1 malocclusion with a deep anterior bite. There is supraeruption of the mandibular incisors with the incisors contacting the palatal mucosa. The facial morphology is typical of the short-face patient. Its overall shape is short and square with poor soft tissue support throughout the lower face. Notice the widened oral commissure and deepened labiomental fold. The vermilion is thin. There is an absence of incisal show with the lips parted and 2 mm of incisal show with a full smile. The lateral view is characterized by overall posterior facial divergence and excessive facial convexity. A lateral cephalometric radiograph taken with the mandible in rest position demonstrates the severity of the supraeruption of the mandibular incisors.

It was decided to presurgically level the mandibular dentition by incisor intrusion. This is demonstrated in the immediate preoperative lateral cephalometric radiograph. Photographs of her immediate presurgical occlusion also demonstrate a decrease in the severity of the overbite. The patient desired to undergo a rather significant transformation in her appearance, requesting that she would like a toothy appearance and a broad smile. The treatment plan was designed to achieve this by inferior repositioning of the maxilla 8 mm and advancing it 4 mm, advancing the mandible 9 mm accompanied by clockwise rotation, and advancing and lengthening the chin 5 mm and 6 mm, respectively. Overall, there would be an increase in facial height of 18 mm and advancement of the soft tissue pogonion of 12 mm. This would result in quantitative overcorrection in the vertical dimension of the lower face as well as achieving an excessive degree of anterior facial divergence accompanied by overall facial convexity.

The results of this treatment plan can be seen in the postoperative photographs. There is a marked overall change in the shape of the face from square to oval. She appears less sad. The oral commissure is narrowed, and there is a much fuller upper vermilion. There is less eversion of the lower lip. The soft tissues have improved skeletal support. There is 5 mm of incisal exposure with the lips parted and a full clinical crown exposed with smiling. The profile is now characterized by anterior facial divergence and facial convexity. The occlusal photographs taken at 3 years following surgery demonstrate excellent occlusal interdigitation with 2 mm of overbite. At 6 years following surgery, there is an excellent stability.

This patient underwent intentional overcorrection in the vertical and sagittal dimensions to achieve the stated aesthetic goals. In the right aesthetic circumstances, facial disproportion is a useful treatment planning principle.

PREOPERATIVE **POSTOPERATIVE**

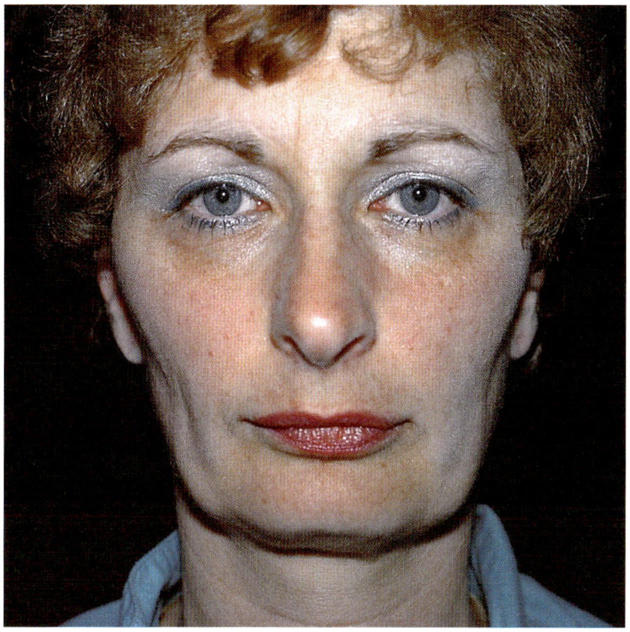

A

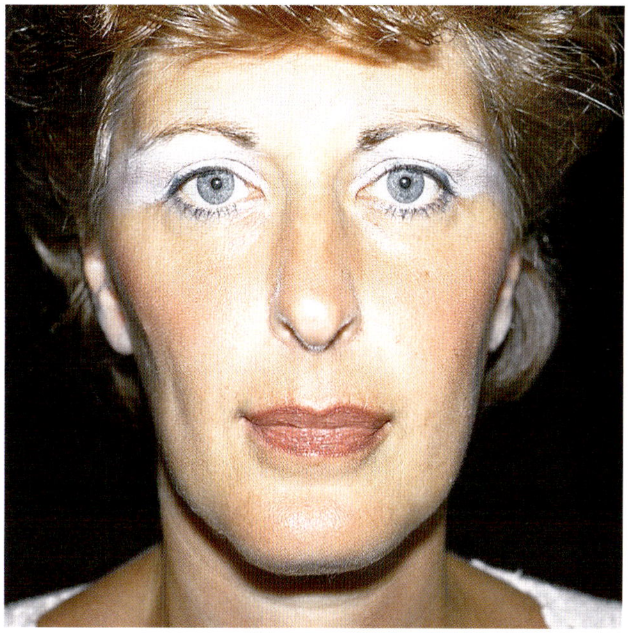

C

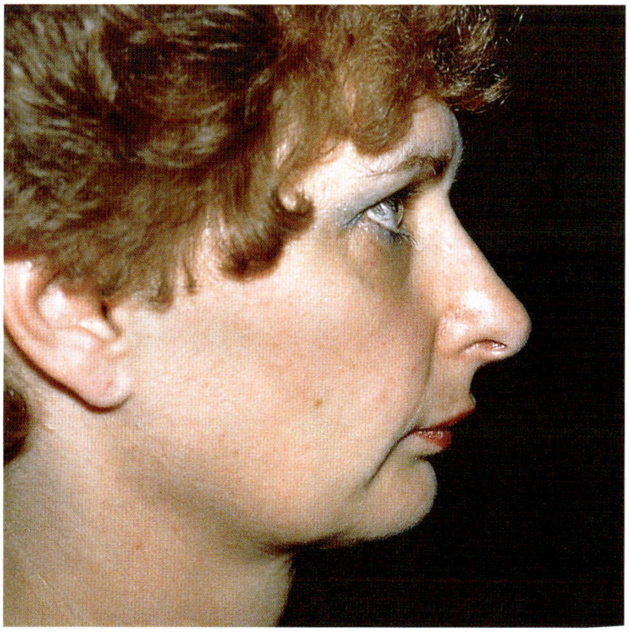

B

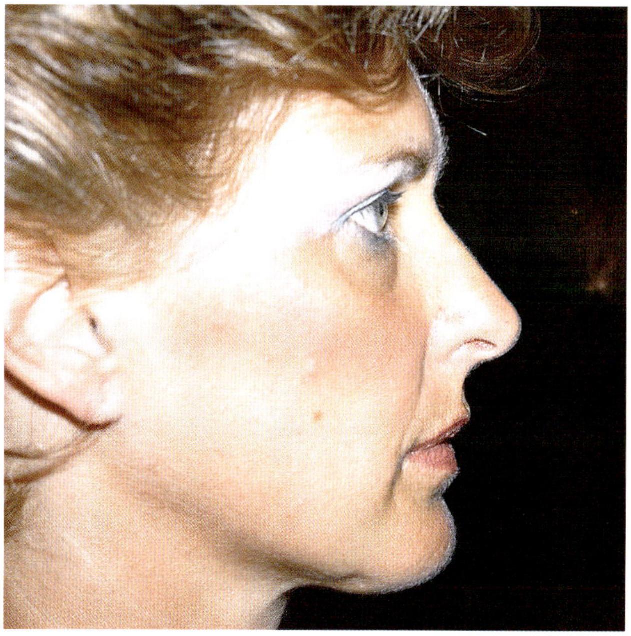

D

Case Study 10-6 A & B: A short-face, mandibular deficiency patient with a Class II deep-bite malocclusion. Surgical correction achieved total facial lengthening of 18 mm and increased facial projection at pogonion of 12 mm. C & D: Note the benefical effects on the perioral soft tissue: narrowing of the oral commissure, vermillion thickening, opening of the nasolabial angle, decrease in lip eversion, and effacement of the labiomental fold.

PREOPERATIVE

POSTOPERATIVE

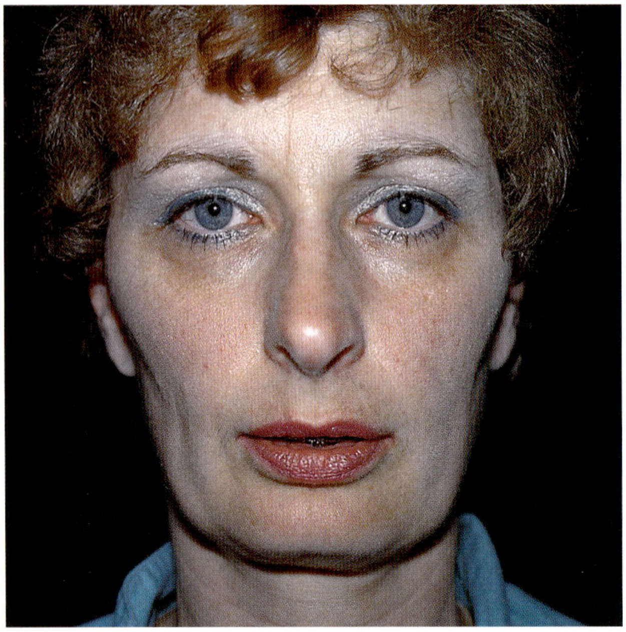

E

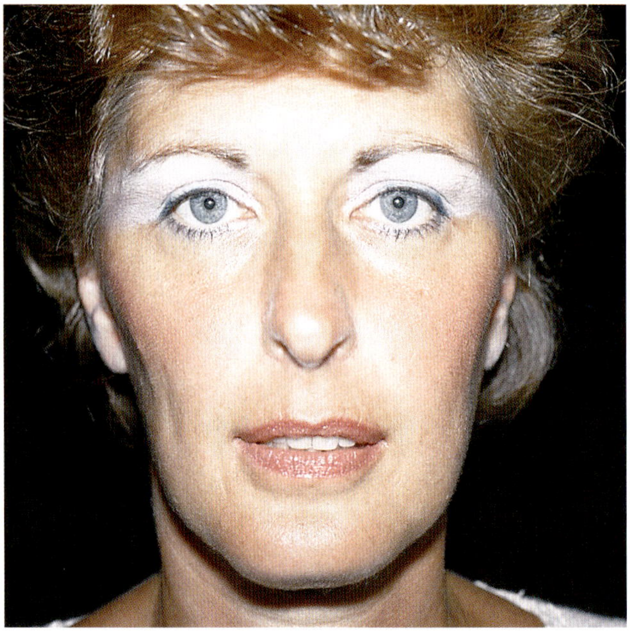

G

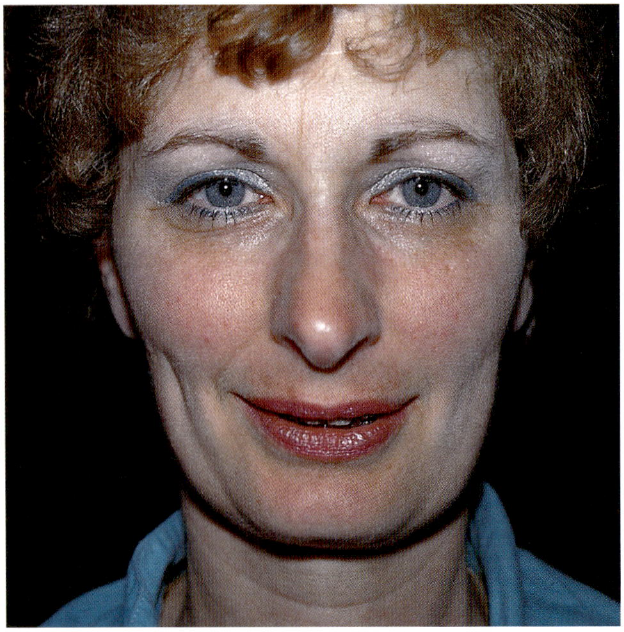

F

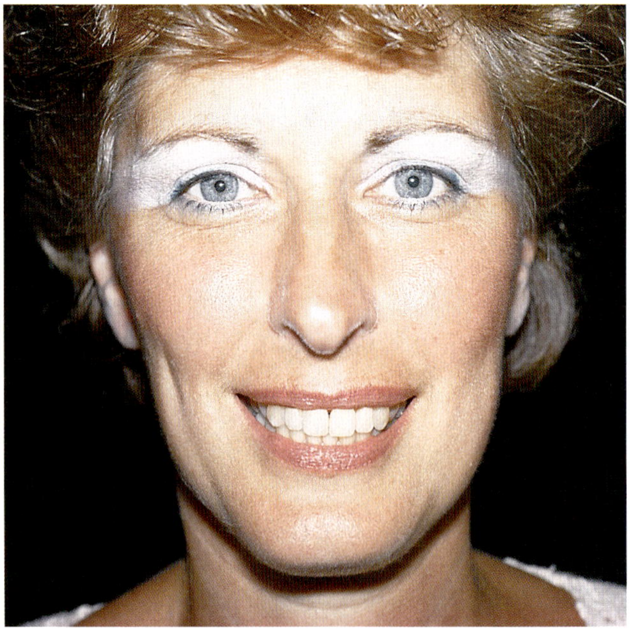

H

FIG. 10-6 (*Continued*) E & F: The patient appears edentulous preoperatively. G & H: Following surgical correction, there is a generous display of the anterior teeth in both repose and smiling. From: Rosen, H.: Facial skeletal expansion: Treatment strategies and rationale. *Plast Reconstr Surg* 89:798, 1992. Reprinted with permission.

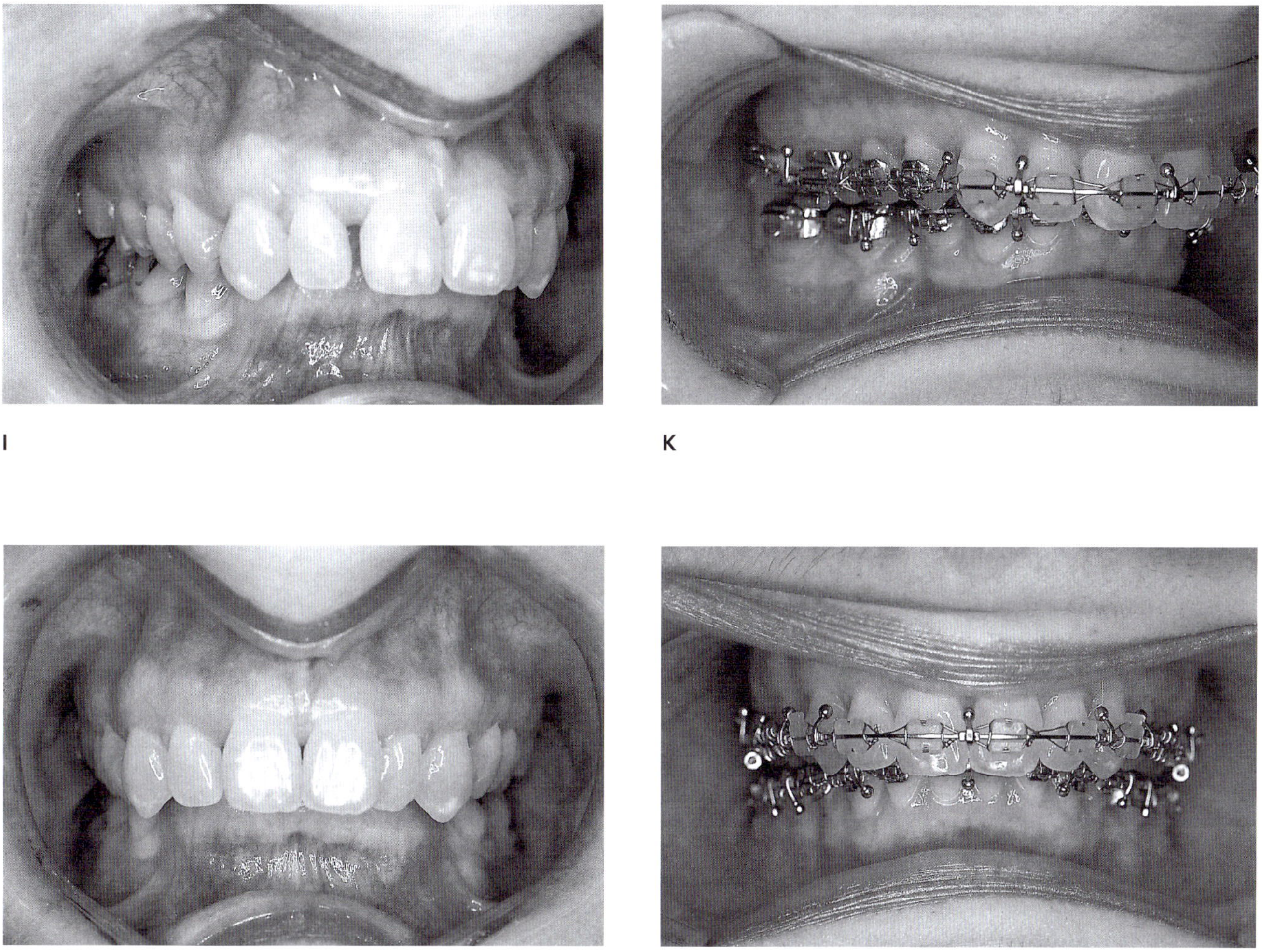

I K

J L

FIG. 10-6 (*Continued*) I & J: There is a Class II, deep anterior bite with the lower incisors impinging on the palatal mucosa prior to presurgical orthodontics. K & L: Following the presurgical phase of othodontics, the mandibular dentition has been leveled, with a decrease in the overbite. This will decrease the extent to which the distal mandible rotates clockwise as it is advanced. Orthodontist: Dr. James Ackerman

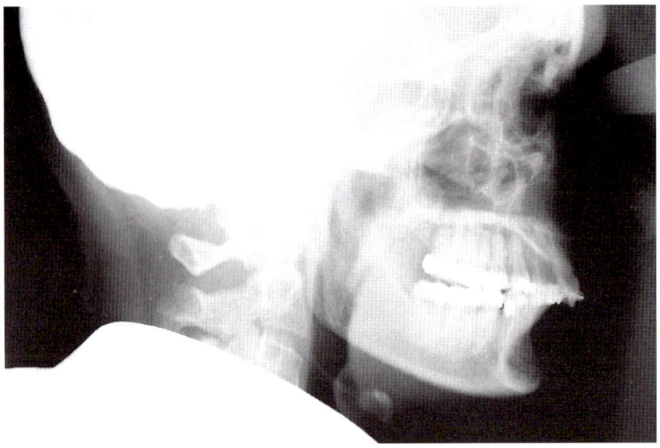

M

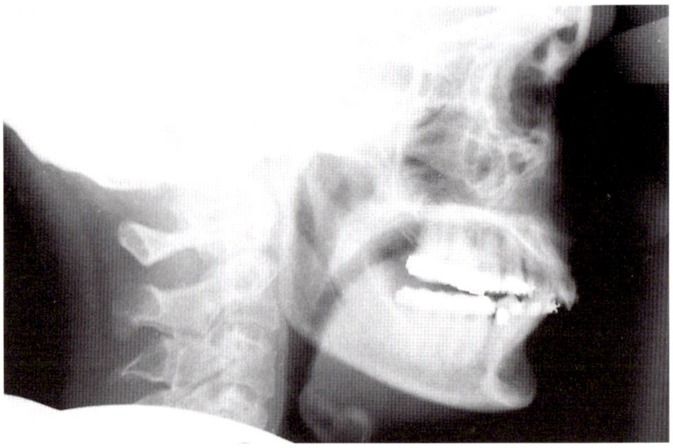

N

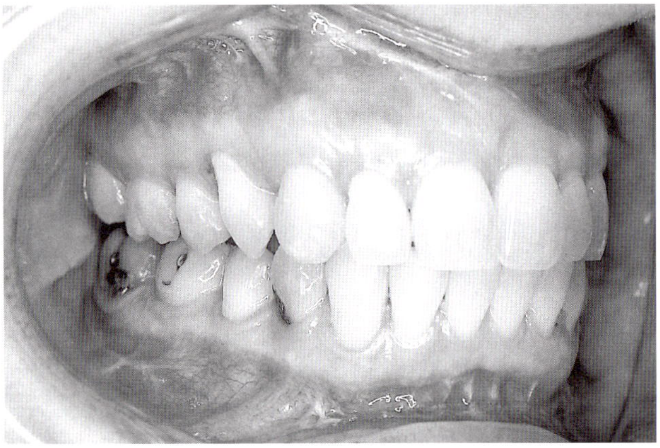

O

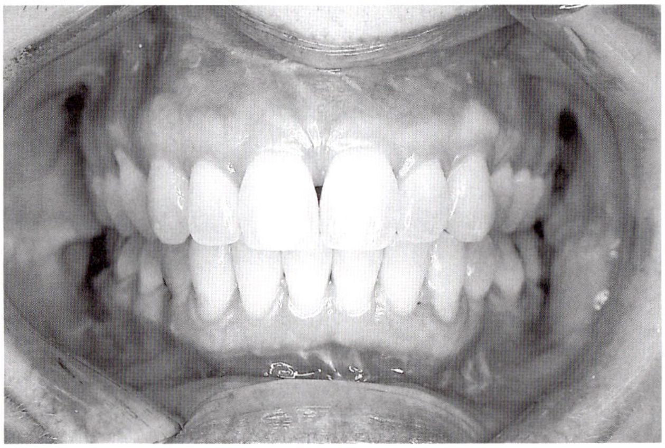

P

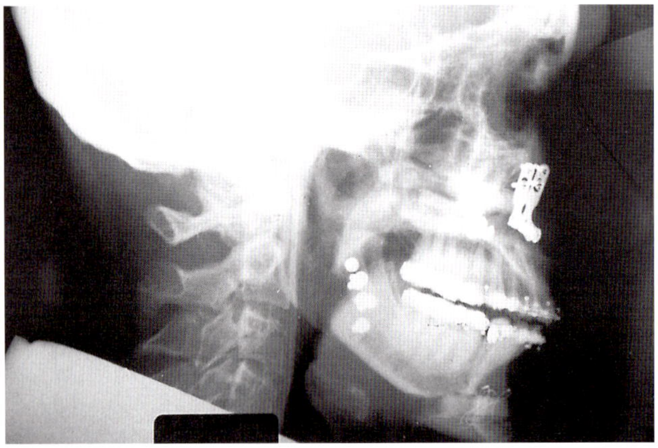

Q

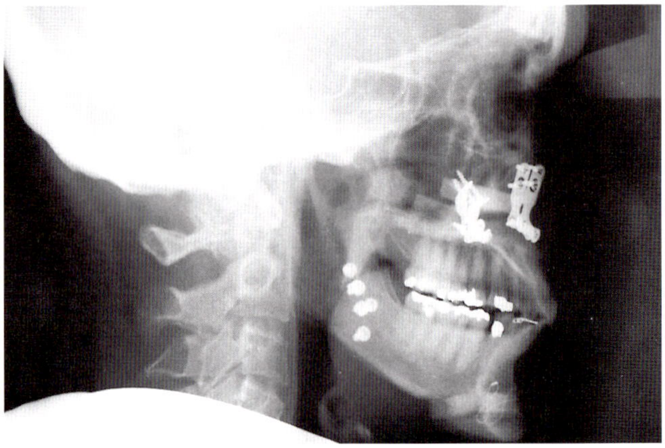

R

Fig. 10-6 *(Continued)* M & N: Before and after orthodontic level of the mandibular dentition. O & P: The postoperative occlusal result at three years. Q & R: Radiographs taken one day and six years, respectively, following surgery. There is excellent stability of the vertical increase of the facial skeleton.

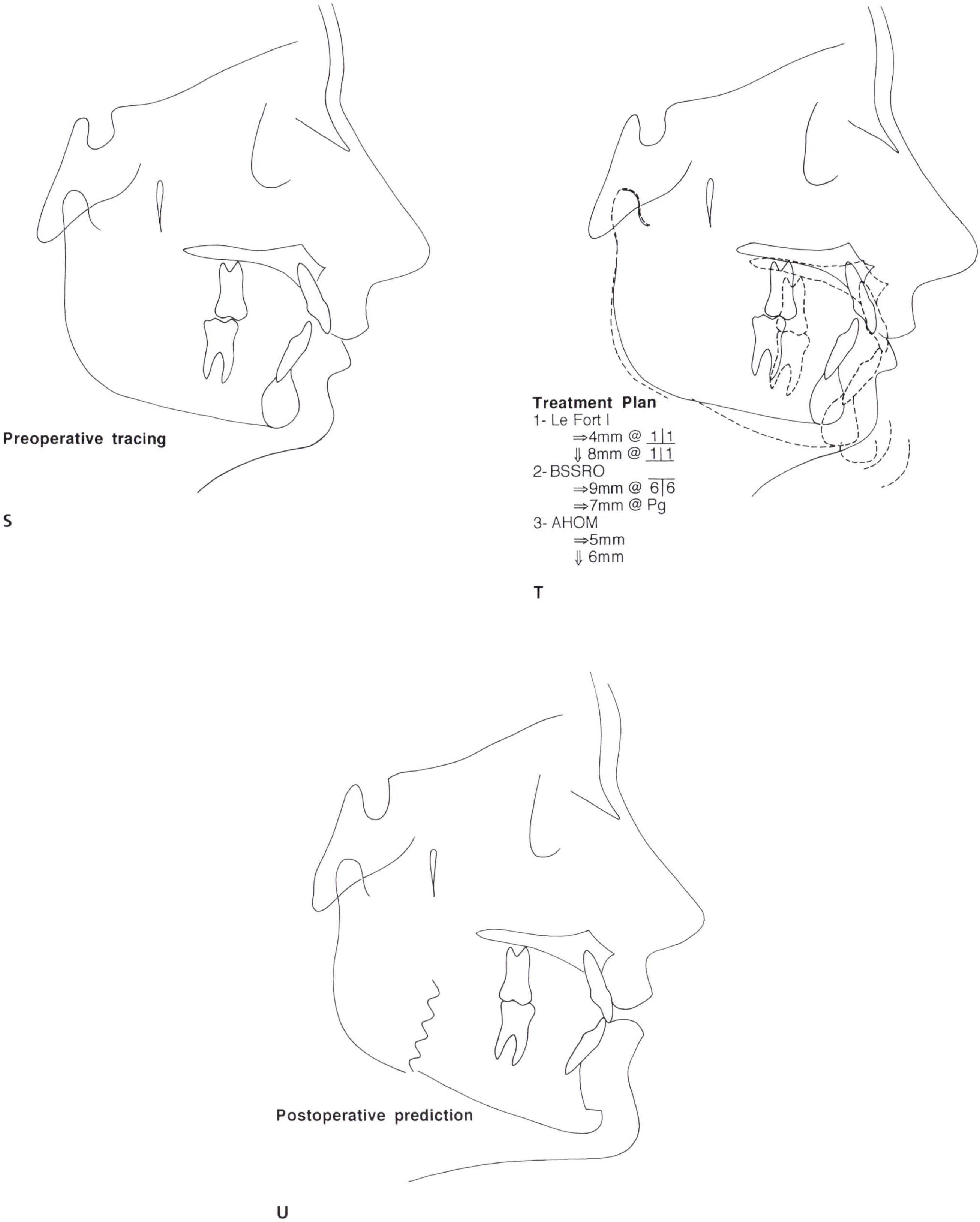

FIG. 10-6 (*Continued*) T: The proposed treatment plan will increase facial height by 18 mm.

Case Study 10-7

This patient demonstrates the aesthetic utility of postsurgical leveling of the mandibular dentition in short-face, deep-bite patients who have relatively prominent chins. Her facial morphology has many of the typical characteristics of the short-face patient. There is a square overall facial shape, a widened oral commissure, an everted lower lip, an absence of incisal show with the lips parted, and the overclosed appearance of the profile. Note the relatively prominent chin and the closed nasolabial angle.

The treatment plan for this patient called for a minimum of presurgical orthodontics to avoid leveling the mandibular dentition and, therefore, maintaining the deep anterior bite. This would maximize the clockwise rotation of the distal mandibular segment as it is advanced, causing the lower lip to come forward but with little forward movement of the chin. A 5-mm vertical elongation of the anterior maxilla would improve the lip–incisor relations. In addition, it would further deepen the anterior bite, resulting in greater clockwise rotational movement of the mandible. There is, thus, little fear of the chin becoming too prominent as the mandible is advanced, and in turn, the need for osseous genioplasty to move the chin posteriorly is eliminated. (This is an undesirable type of genioplasty to perform because it is associated with soft tissue fullness in the submental region.)

Postoperatively, the patient no longer appears angry. There is a narrowing of the oral commissure, fuller upper lip vermilion, an attenuation of the labiomental fold, an attractive exposure of the upper incisors with the lips parted, and a full clinical crown exposed with smiling. The nasolabial angle has softened, the labiomental fold is less deep, and the chin appears to be less prominent relative to the lower lip in the sagittal dimension. The occlusal result demonstrates normal Class I canine relations with 2 mm of overbite. Although it appears that the result could have been improved with a vertical reduction of the chin, this would have unattractively deepened the labiomental fold.

PREOPERATIVE POSTOPERATIVE

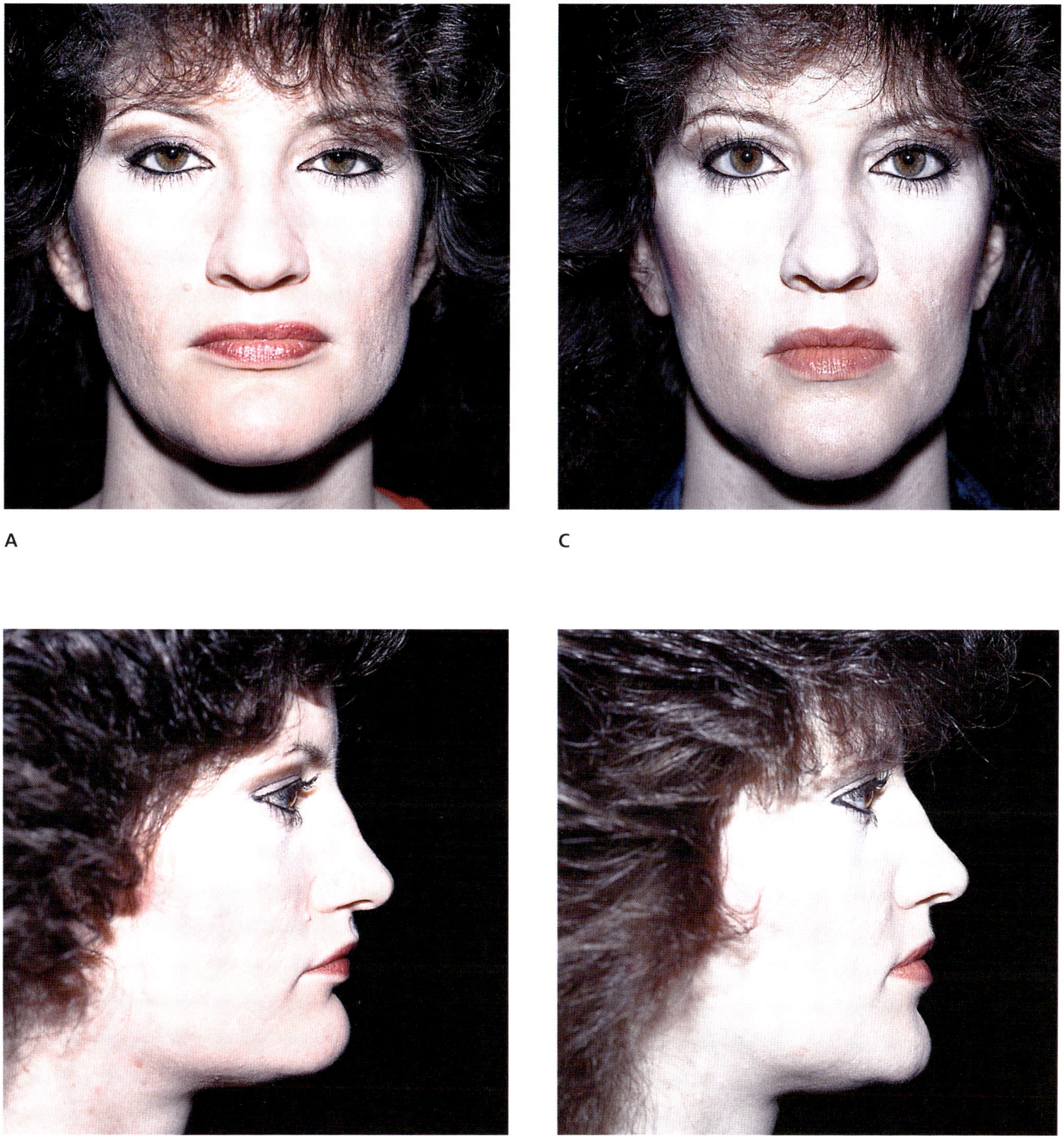

A C

B D

CASE STUDY 10-7 A & B: Short-face patient with a Class II, deep-bite malocclusion. The chin is prominent. Surgical correction must involve clockwise rotation of mandible as it is advanced to keep the chin from appearing too large following surgery. C & D: Postoperatively, the patient displays all of the typical soft tissue changes subsequent to skeletal expansion of the short face. By this time, you should be able to itemize them!

PREOPERATIVE POSTOPERATIVE

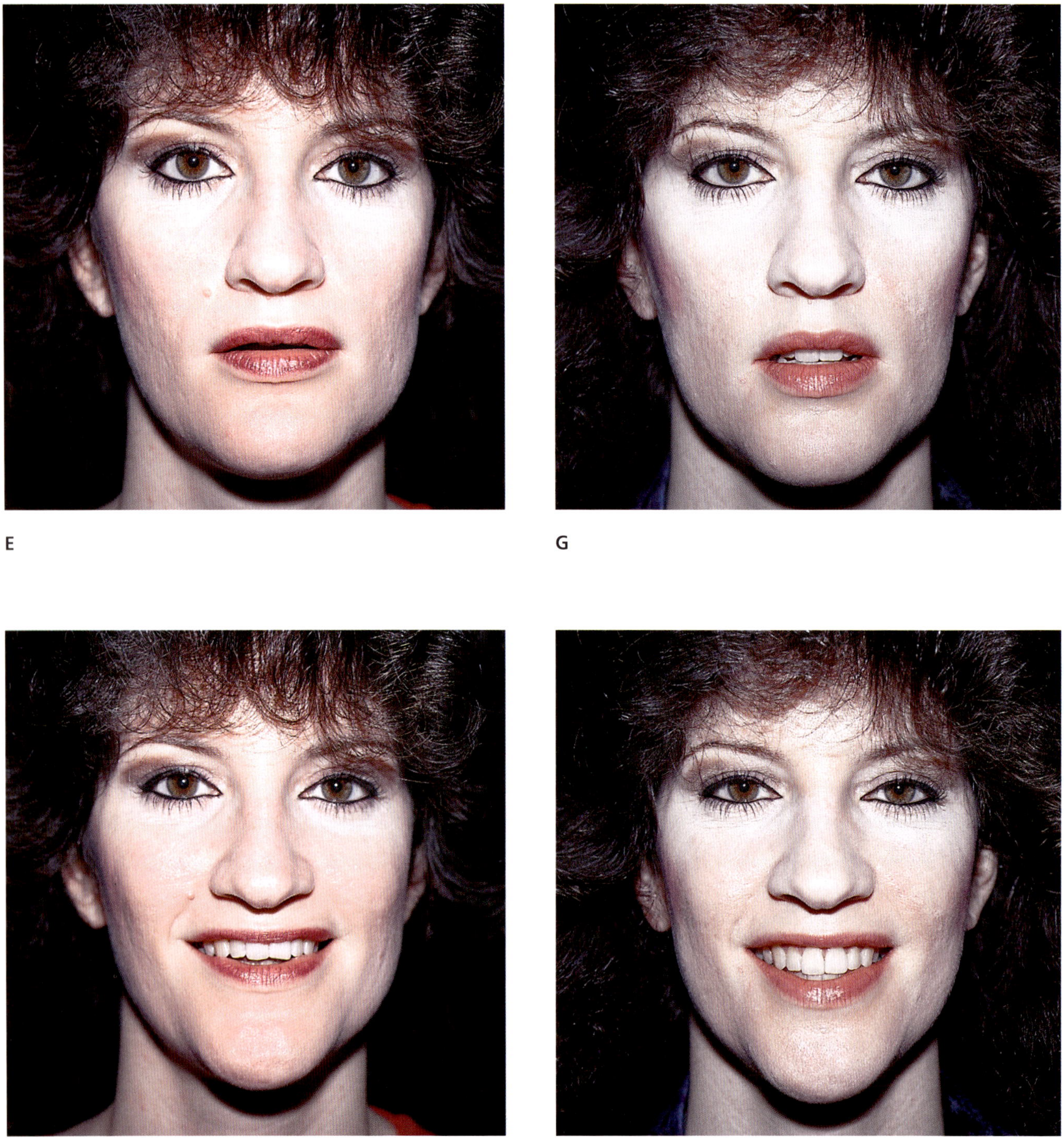

E G

F H

Fig. 10-7 (*Continued*) E through H: The patient displays the typical changes in lip-tooth relations following short face correction. From: Rosen, H.: Aesthetics in facial skeletal surgery. *Perspectives in Plastic Surgery* 6:1, 1992. Reprinted with permission.

PREOPERATIVE | POSTOPERATIVE

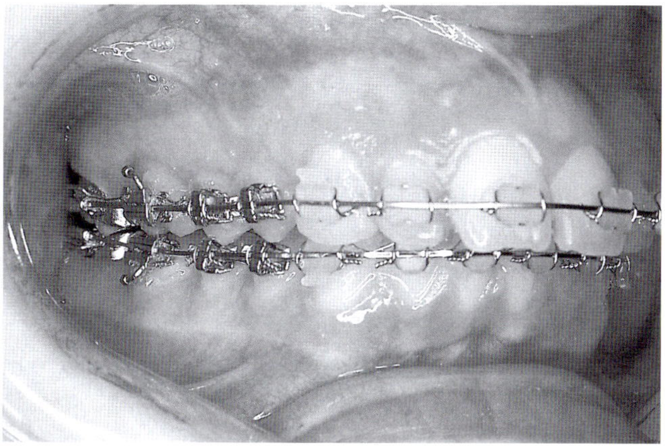

I

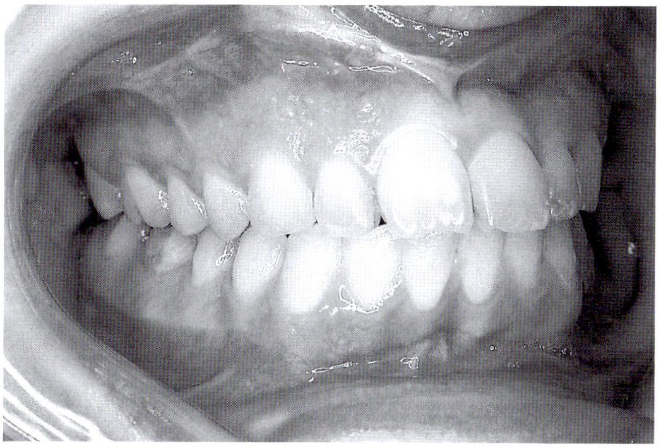

K

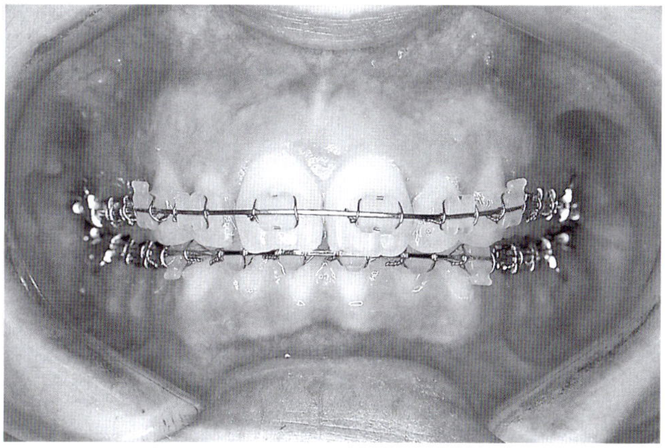

J

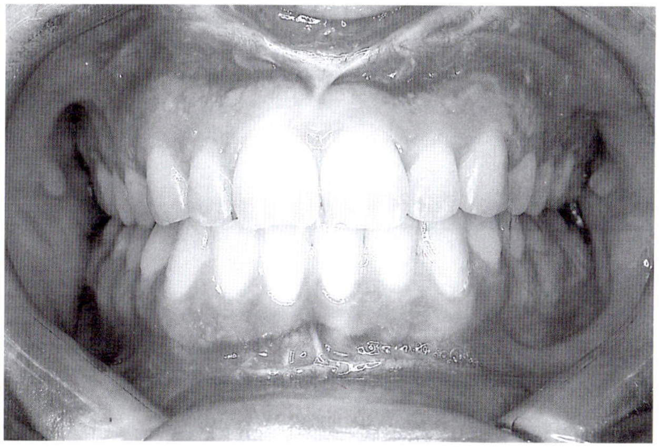

L

FIG. 10-7 (*Continued*) Orthodontist: Dr. Gerald Ginsberg

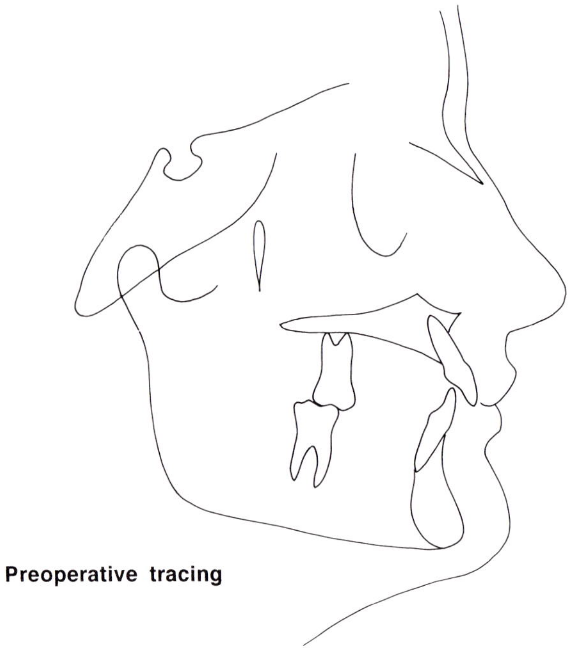

M

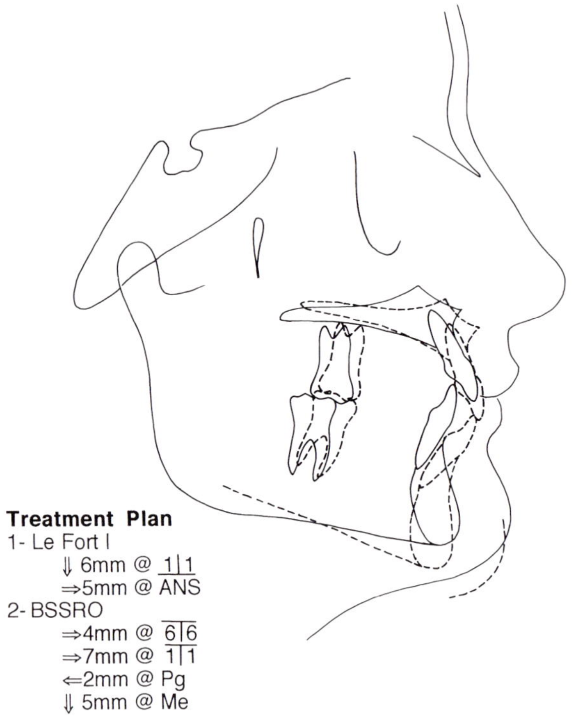

N

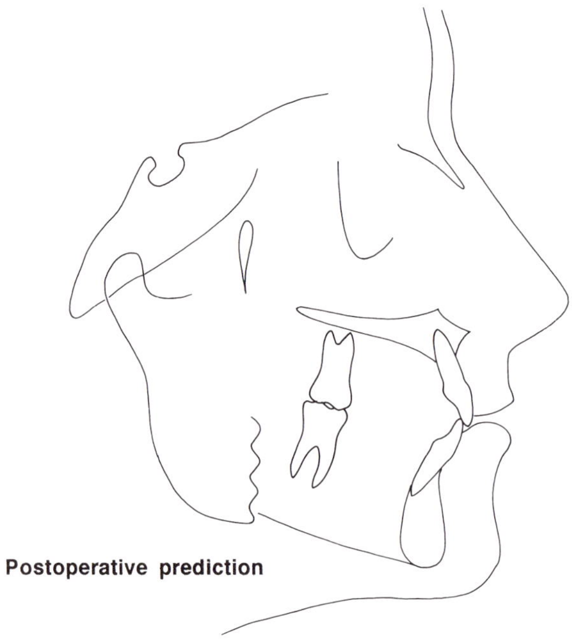

O

Fig. 10-7 (*Continued*) N: Notice the large component of mandibular clockwise rotation as it is advanced. The chin actually moves posteriorly as the lower incisors and lower lip come forward.

Conclusion

It is now possible to adequately treat the short-face patient with excellent stability because of the technological advances of rigid fixation and nonresorbable bone substitutes. Application of these technologies makes one appreciate the dynamic interrelationship between the skeletal foundation and overlying soft tissues. Bidimensional expansion of the facial skeleton in the vertical and the sagittal dimensions creates the skeletal substrate for a well-supported facial mask. This is readily observed not only in the traditional locations of the submalar, perimandibular, and submental regions, but also in the perioral region. The typical response of the perioral soft tissues following skeletal enhancement include a narrowing of the oral commissure, greater vermilion exposure of the upper lip, less procumbency of the upper and lower lips, and an attenuation of the labiomental fold. When this is seen in association with greater degrees of incisal exposure and enhancement of the aesthetics of the smile, the countenance of the individual undergoes a dramatic change. The overall result is a more youthful and contented appearing individual.

Suggested Reading

1. Bell, W.H.: Correction of the short face syndrome in vertical maxillary deficiency: A preliminary report. *J Oral Surg* 35:110, 1977.
2. Bell, W.H. and Scheidman, G.B.: Correction of vertical maxillary deficiency: Stability and soft tissue changes. *J Oral Surg* 39:666, 1981.
3. Hedemark, A. and Freihofer, H.P.: The behavior of the maxilla in vertical movements after Le Fort I osteotomy. *J Maxillofac Surg* 6:244, 1978.
4. Major, P.W., Phillippson, G.E., Glover, K.E., and Grace, M.: Stability of maxilla downgrafting after rigid or wire fixation. *J Oral Maxillofac Surg* 54:1287, 1996.
5. Persson, G., Hellern, S., and Nord, P.G.: Bone plates for stabilizing Le Fort I osteotomies. *J Maxillofac Surg* 14:69, 1986.
6. Rosen, H.: Porous, block hydroxyapatite as an interpositional bone graft substitute in orthognathic surgery. *Plast Reconstr Surg* 83:985, 1989.
7. Rosen, H.: Definitive surgical correction of vertical maxillary deficiency. *Plast Reconstr Surg* 85:215, 1990.
8. Rosen, H. and McFarland M.: Biologic behavior of porous block hydroxyapatite implanted into the maxillofacial skeleton. *Plast Reconstr Surg* 85:718, 1990.
9. Wardrop, R.W. and Wolford, L.M.: Maxillary stability following downgraft and/or advancement procedures with stabilization using rigid fixation and porous, block hydroxyapatite implants. *J Oral Maxillofac Surg* 47:226, 1989.
10. White, E. and Shors, E.C.: Biomaterial aspects of Interpore-220 porous block hydroxyapatite. *Dent Clin North Am* 30:49, 1986.

Adjunctive Considerations

PART 3

This sections deals with the adjunctive surgical options that can be utilized to enhance the final aesthetic result. Surgery of the chin (Chapter 11) can, of course, be performed in conjunction with major jaw surgery or by itself in individuals not interested in undergoing an orthognathic surgical correction. Chin surgery can potentially camouflage a major morphologic deformity or subtly enhance the appearance of a modestly retrusive chin. However, it is a mistake to envision osseous genioplasty as a simple technique. To do so underestimates the potential results that this procedure can achieve as well as the potential problems associated with its use.

The final two chapters deal with ancillary aesthetic concerns in the patient undergoing jaw surgery. These include direct surgical manipulation of the soft tissues that can be safely done simultaneously or at a second sitting. Finally, it addresses those circumstances in the immediate postoperative period that warrant surgical revision for aesthetic reasons.

Chin Surgery

CHAPTER

11

This chapter will focus on the osseous genioplasty or anterior horizontal osteotomy of the mandible. Alloplastic augmentation of the chin will be mentioned only to serve as a point of comparison to osteotomies of the chin.

Aside from rhinoplastic procedures, osseous genioplasty represents the most commonly performed osteotomy of the facial skeleton. Despite this, it is an underutilized procedure because of the ready availability of chin implants and the relative simplicity of inserting them. However, the versatility of osseous genioplasty, as well as its superb aesthetic results and relatively low risk, relegates the appropriate use of chin implants to a very limited number of individuals seeking enlargement of their chins.

Unfortunately, the osseous genioplasty remains poorly understood. Consequently, the technique is abused, and the results become suboptimal, for two basic reasons. The first is that excessive focus is placed on the hard tissue component of the operation with little comprehension of the soft tissue requirements. Secondly, the operation is frequently performed with knowledge of only what it can correct and with a lack of understanding of what it cannot correct. As a result, the chin is advanced to an excessive degree causing an unnatural, operated appearance. The osseous genioplasty is one operation in which one must err on the side of conservatism; that is, undercorrection in the sagittal dimension is far better than overcorrection.

Appropriate Candidates for Chin Surgery

Throughout this discussion, it should be remembered that we are talking about patients whose major aesthetic concern is their perception of a small chin. Consequently, they feel that their request for a larger chin is simple and straightforward.

It must be appreciated, however, that in the overwhelming majority of patients seeking chin enlargement, the reality of the deformity is very different from the patient's perception of it; that is, they have small mandibles, not just small chins, causing the appearance of a recessive chin. If one were to examine the patient's occlusion, more often than not they have a Class II occlusion or, when questioned, give a history of an orthodontic correction of a Class II occlusion. Therefore, it is most commonly the Class II deformity for which these patients seek surgical correction. In a study previously performed by this author, 62 of 68 consecutive patients undergoing osseous genioplasty to enlarge the chin had Class II occlusions or prior orthodontic correction of a Class II occlusion.

The Class II skeletal pattern observed in these individuals has one morphologic characteristic in common: there is sagittal deficiency of the lower face characterized by the soft tissue pogonion being posterior to where it should ideally be. Beyond that, however, the facial morphology may be extremely variable. This variability relates to the height of the lower face. It may be markedly deficient and seen in association with a deep anterior bite and/or vertical maxillary deficiency, or it may be excessive if there is an anterior open bite and/or vertical maxillary excess. As the height of the lower face varies, so too does the morphology of the labiomental fold. Decreased lower face height is almost always associated with an increased depth of the labiomental fold. Increased lower face height is more frequently seen in association with a shallow or attenuated fold. It is much less commonly seen in association with a deepened fold. Of 68 consecutive patients undergoing advancement genioplasty, 40% had decreased lower face height and deepened labiomental folds, 24% had increased lower face height and attenuated labiomental folds, and 34% had lower face height that approximated the midface. A total of 88% of these patients had abnormal labiomental fold morphology.

Correct anatomic placement of an alloplastic chin implant (over the symphysis and not the dentoalveolus) must deepen the depth of the labiomental fold. Osseous genioplasty to advance the chin also deepens the fold. Realization of this operative sequela in the context of the numerical data just presented, it becomes readily appreciated who is an appropriate candidate for chin surgery. It also indicates which is the appropriate procedure, that is, alloplastic chin augmentation or osseous genioplasty.

Alloplastic chin augmentation affords control over the anteroposterior or sagittal dimension only. It cannot effect reliable change in the vertical dimension of the lower face, and it has no control over the aesthetics of the labiomental fold other than to make it appear deeper. As a result, appropriate candidates for alloplastic chin augmentation are only those individuals who have a modest sagittal deficiency of the lower face, who have a symmetric lower face, who have normal height of the lower face, and who have a previously existing modest or shallow labiomental fold. Because this combination of morphologic factors is seen in relatively few individuals requesting aesthetic enlargement of their chins, there are, in fact, few individuals who are appropriate candidates for alloplastic chin augmentation.

On the other hand, osseous genioplasty affords the surgeon direct control of not only the sagittal dimension but the vertical dimension as well. It can easily effect an increase in lower face height or a decrease in lower face height. In addition, one has indirect control over the apparent depth of the labiomental fold. With straight advancement of the chin, the fold always deepens. However, as the vertical dimension of the chin is increased, the tendency is for the fold to become less deep and to have a smoother and less acute transition as one goes from the chin to the lower lip. As a result, one can influence the aesthetics of the labiomental fold. Lastly, the chin can be moved in the transverse dimension to affect improvement in asymmetries of the lower face.

Therefore, it should be evident that the great advantage of osseous genioplasty is its versatility. It can be used to enhance the appearance of the overwhelming majority of individuals that request aesthetic enlargements of their chins because it can also treat abnormalities of lower face height and labiomental fold morphology that are so frequently seen in association with the Class II skeletal pattern.

However, there are aesthetic limitations to osseous genioplasty, and one must appreciate these to select appropriate candidates as well as to minimize the risks of achieving an unnatural surgical result. The first limitation has to do with the extent of improvement that one can obtain with osseous genioplasty. A large number of individuals would be more appropriately treated with orthognathic surgery in terms of extent of improvement. It is the obligation of the surgeon to inform the patient of the true nature of the skeletal deformity and, at least, to present orthognathic surgery as an option. This should be presented not only in terms of its aesthetic result but also in terms of its increased risk and greater overall treatment time. If orthognathic surgery is declined, osseous genioplasty can then be offered as an alternative. An exception to this is in the unusual individual who presents with a retrusive appearing chin in combination with increased lower face height and a deepened labiomental fold. Such an individual is seen in Fig. 11-1. This patient should only be offered an orthognathic surgical correction as her appearance cannot be improved with osseous genioplasty alone. In such a patient, when one moves the chin forward, the fold becomes even deeper. One cannot lengthen the chin in an effort to soften the fold because the lower face height is already increased. Orthognathic surgical correction consisting of vertical maxillary impaction and mandibular advancement with or without genioplasty is the only reasonable surgical option that should be presented to this patient. To reiterate, the combination of increased lower face height and a deepened labiomental fold should be a warning sign to the surgeon contemplating osseous genioplasty to make the mandible appear larger.

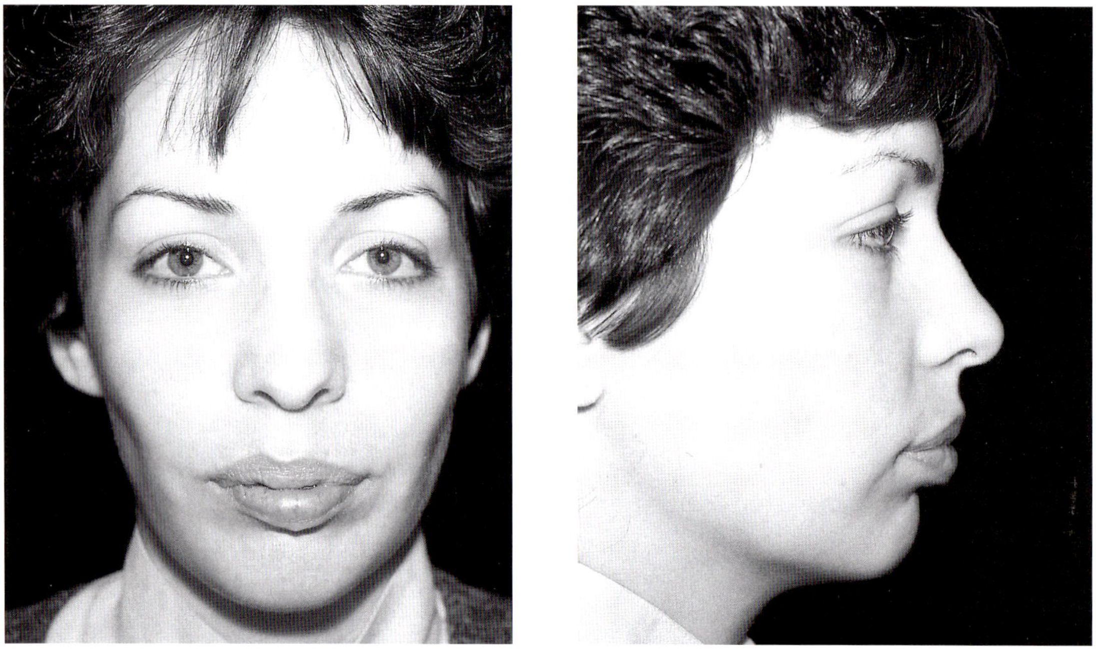

A B

Fig. 11-1 A poor surgical candidate for any type of chin surgery because of the unfavorable combination of a long lower face and a deep labiomental fold. Only double-jaw surgery would improve her appearance.

Treatment Planning for Osseous Genioplasty

After having chosen an appropriate candidate for osseous genioplasty, one must decide upon not only the vector but also the magnitude of symphyseal displacement. Directional changes are dictated by the physical examination of the patient. If the height of the lower face appears to be too long and it is accompanied by lip strain and/or effacement of the labiomental fold, vertical shortening as well as advancement of the chin should take place. Vertical reduction of the chin of more than 7 mm is infrequently indicated. On the other hand, deficiencies of lower face height dictate lengthening of the chin in addition to its possible advancement. Vertical lengthening of the chin of more than 7 to 8 mm is unusual. Lateral shifting of the chin should obviously take place when lower face asymmetry is manifested by deviation of soft tissue pogonion lateral to the midline. Because a retrusive chin may mask a modest skeletal asymmetry, a frontal cephalometric radiograph should be obtained to substantiate this. The extent to which lateral shifting of the chin is necessary to correct facial asymmetries is frequently underestimated. It should be emphasized that the magnitude of vertical as well as lateral symphyseal shifts are best determined from the clinical, qualitative impression and should assume a one-to-one ratio of osseous to soft tissue movement.

Perhaps the most commonly made error in treatment planning is the extent to which one anteriorly advances the chin. Many individuals with Class II skeletal patterns have retrusive lower lips in addition to weak chins. This is clearly demonstrated in the patient in Fig. 11-2, who has a four component deformity to his Class II pattern. He has a retrusive chin, a retrusive, procumbent lower lip, a deepened labiomental fold, and deficient lower face height. It must also be appreciated that it is the retrusive lip component of the deformity over which the surgeon has no control; that is, though the chin can be brought forward, the lower lip remains behind.

Ironically, it is the abnormal position of the lower lip over which there is no control that dictates the extent to which one can move the chin forward. Herein lies the great compromise in advancement genioplasty. One must be willing to accept residual sagittal weakness of the chin if one is to prevent it from being positioned in advance of the lower lip. Standard cephalometric guidelines ignore this aesthetic principle and suggest advancing the chin so that it is positioned in normal relationship to either the forehead, the midface, the nose, or the upper lip. Unfortunately, all of these treatment planning guidelines assume that the lower lip is in normal position or that it can be placed in normal position. The sagittal position of the lower lip, regardless of whether it is normally positioned or retrusive, must dictate the extent to which one can move the chin forward. Residual sagittal deficiency of the lower face is a small aesthetic price to pay in an effort to avoid unnatural, bizarre results in which the chin has been advanced beyond the

A

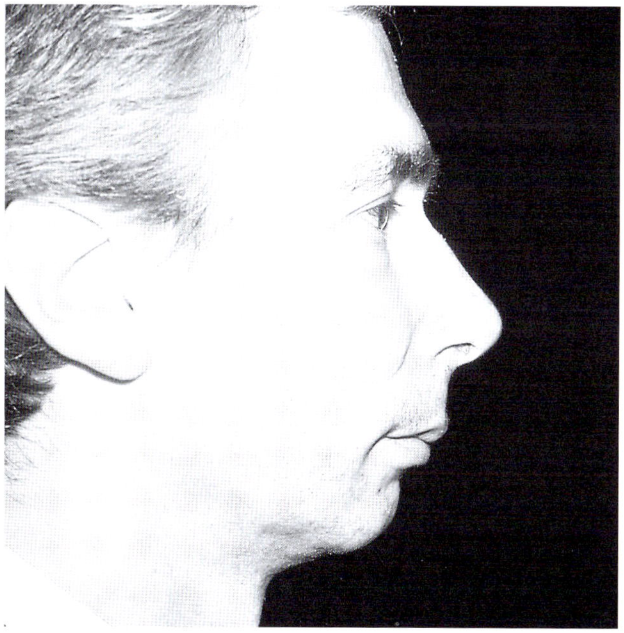

B

FIG. 11-2 This patient requesting chin enlargement has a four component deformity of a Class II pattern: a recessive chin, a recessive lower lip, decreased facial height, and deepened labiomental fold.

lower lip. To reiterate, when moving the chin forward, the extent of advancement should be no greater than the sagittal position of the lower lip, assuming a one-to-one ratio of skeletal to soft tissue movement. Advancing the chin is a very powerful aesthetic tool, and a modest displacement can make an enormous change. It is always best to undercorrect in the sagittal dimension.

Occasionally, there are patients whose lower lips are so retrusive that only minimal advancement of the chin can be accommodated. Provided that there is a decrease in lower face height, significant vertical elongation of the chin can be undertaken to provide the patient with the impression of a significant increase in skeletal volume of the lower face. Paradoxically, overcorrection in the vertical dimension can be planned to compensate for the obligatory acceptance of residual sagittal deficiency. As previously discussed, vertical overcorrection can be an aesthetic benefit provided it is unassociated with lip strain and excessive effacement of the labiomental fold.

A panorex radiograph of the mandible is suggested in all patients prior to undergoing osseous genioplasty. This is obtained to rule out any periapical pathology that may exist in the anterior mandible. In addition, it is helpful in the short-face patient to determine the adequacy of bone that exists in the vertical dimension so that the osteotomy and eventual placement of screws can be located at a safe distance from the apices of the teeth. For inexperienced surgeons, a lateral cephalometric radiograph for the purposes of performing a prediction tracing may be helpful in determining the extent of skeletal displacements.

Technical Considerations

Placement of the incision in the labial mucosa must be far enough away from the buccal sulcus to allow a secure, watertight closure. It should preferably be placed 1 cm anterior to the sulcus. Failure to achieve a secure closure increases the risk of infection in this dependent location. Appropriate incision location is depicted in Fig. 11-3. Note that the lower lip is retracted anteriorly, and the cautery blade incises the mucosa at right angles to it. As soon as the midline raphe and right and left bellies of the mentalis muscle are visualized (Fig. 11-4), the direction of the cautery is changed so that it is now tangential to the mucosal surface and it is, in fact, scything the lower lip. Changing the direction of the cutting instrument prevents perforating the lower lip and also leaves a generous cuff of mentalis muscle attached to the bone so that the mentalis muscle can be reconstructed at the time of closure. Once the bone is reached with the cautery, subperiosteal dissection exposes the minimal amount of bone necessary to perform the osteotomy and to secure fixation. This means that the symphysis is not degloved, which, in turn, helps ensure a close relationship between the extent of the skeletal movement and soft tissue movement and helps avoid ptosis of the soft tissue chin pad following surgery. Dissection must be carried back well pos-

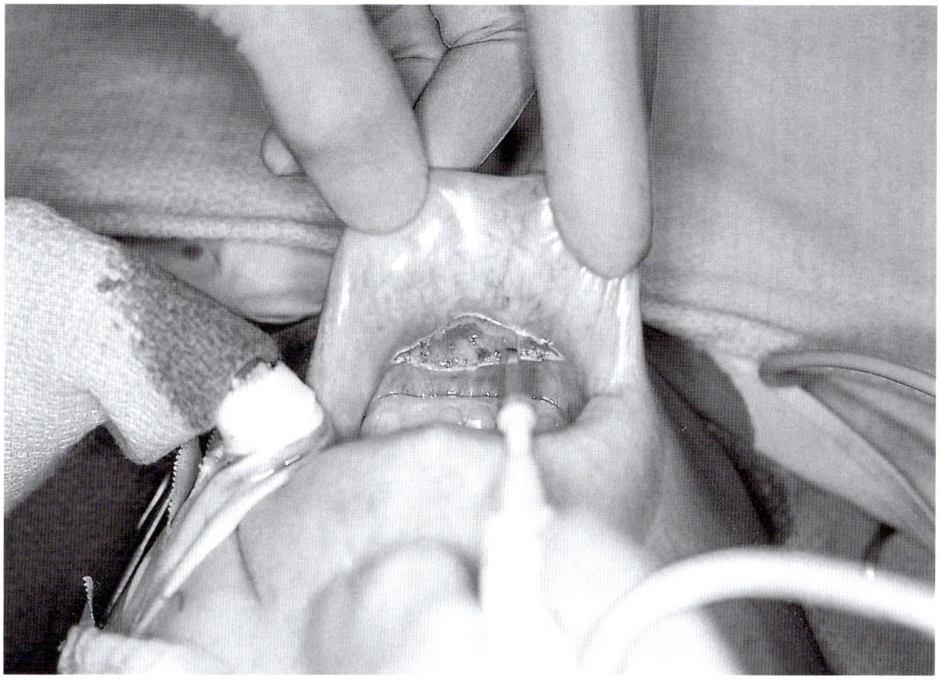

FIG. 11-3 Correct placement of mucosal incision for genioplasty.

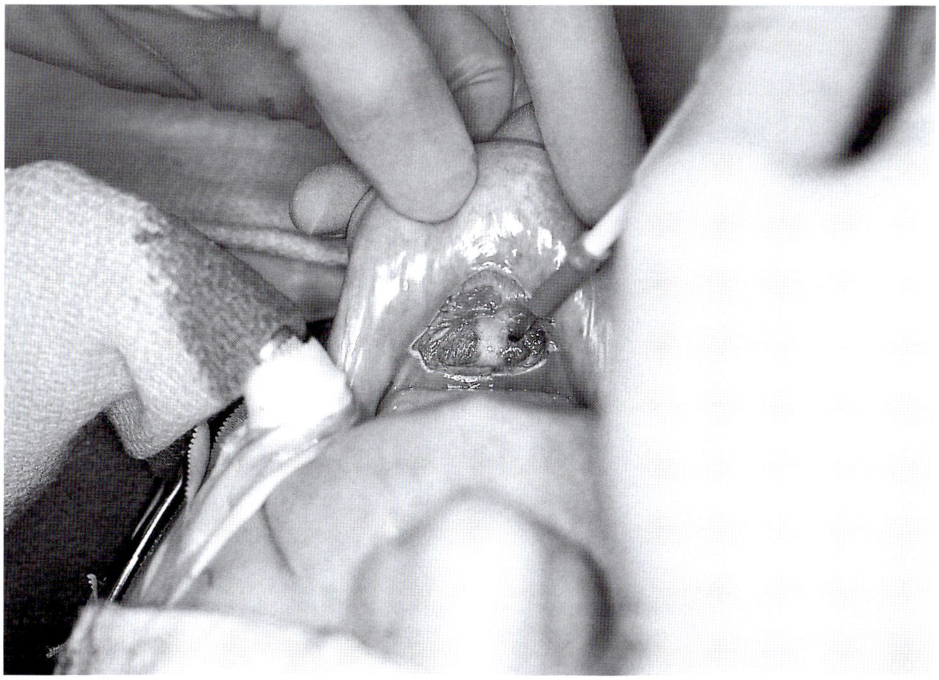

FIG. 11-4 Visualization of mentalis muscle bellies and median raphe.

terior to the mental foramina to the inferior border of the mandible in the vicinity of the first molar. Accordingly, both mental nerves are typically identified. Dissection of the inferior border of the mandible is only done in the region where the osteotomy will be terminated and nowhere else. This type of exposure creates soft tissue tunnels that can be retracted to introduce a reciprocating saw to perform the osteotomy.

Following soft tissue dissection, vertical scoring of the midline of the symphysis is performed as a transverse reference for placement of the repositioned chin. In addition, the horizontal osteotomy is also scored in the bone to ensure symmetry prior to performing the actual osteotomy. The minimal soft tissue dissection off of the inferior aspect of the symphysis, the preliminary scoring of the bone, and the soft tissue tunnels are illustrated in Fig. 11-5.

The actual osteotomy is performed in its entirety with the reciprocating saw. This instrument allows the osteotomy to be extended posteriorly into the molar region with minimal soft tissue dissection. Extending the osteotomy as far posteriorly as possible allows for displacement of a large volume of bone and helps ensure a more natural result. In addition, the inevitable step in the inferior border of the mandible is in a more posterior location where the soft tissues are thicker and, therefore, minimizes the risk of creating an hourglass or "waisting" deformity of the lower jaw. Adequate

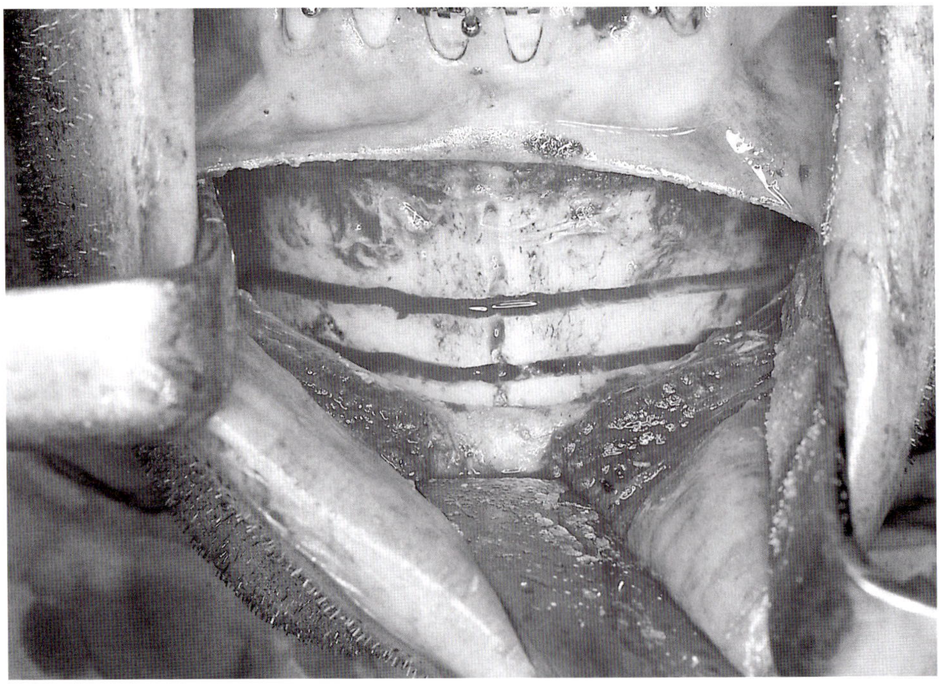

FIG. 11-5 Intraoperative view of genioplasty procedure following soft tissue dissection and scoring of bone. This patient is to undergo a vertical reduction genioplasty.

posterior extension of the osteotomy also minimizes the obliquity of the osteotomy cut so that the anterior mandible is not unintentionally shortened as it is advanced.

A variety of fixation devices including wires, lag screws, or plates and screws can be utilized for fixation of the symphyseal segment. Although theoretic objection has been raised to using a stair-step-shaped bone plate because of its subsequent palpability, I have not appreciated this as being a clinical problem.

Once the bone is repositioned and secured, soft tissue closure is performed. This is accomplished in two layers employing interrupted 3-0 chromic sutures in each layer. The first layer of closure is the mentalis muscle. Reconstruction of this layer avoids postoperative mentalis dysfunction with secondary ptosis of the soft tissue chin pad, mentalis hyperfunction, and unsightly exposure of the lower incisor teeth. Once the muscular layer is reconstructed, the mucosa is closed with interrupted 3-0 chromic sutures.

Proper mucosal and muscular incision, minimal soft tissue dissection, and two-layer closure of the muscle and mucosa are all critical technical aspects of the surgery, which, when properly performed, will minimize complications and unaesthetic results.

Patient Presentations

Case Study 11-6

This 17-year-old has previously undergone orthodontic correction of a Class II, Division 1 malocclusion. This involved premolar extractions. The height of the lower face is relatively normal. The upper lip lacks support, and as a result, the nasolabial angle is excessively opened. There is sagittal deficiency of the lower face, recession of the chin, and a shallow labiomental fold. Although this patient could conceivably be treated with a chin implant (she has relatively normal lower face height and a shallow labiomental fold), in my opinion she will be better served with an osseous genioplasty given her age and the extent of advancement contemplated. The treatment plan calls for advancement of the chin of 7 mm. This distance is intended to bring the chin to the level of the lower lip but not beyond it.

Postoperatively, there is an attractive deepening of the labiomental fold as the chin is advanced. The chin is now in good position relative to the lower lip but is still weak relative to the midface structures. It bears repeating that in individuals with recessive lower lips, residual sagittal deficiency of the lower face is a necessary compromise in advancement genioplasty to avoid an overcorrected, unnatural result.

PREOPERATIVE

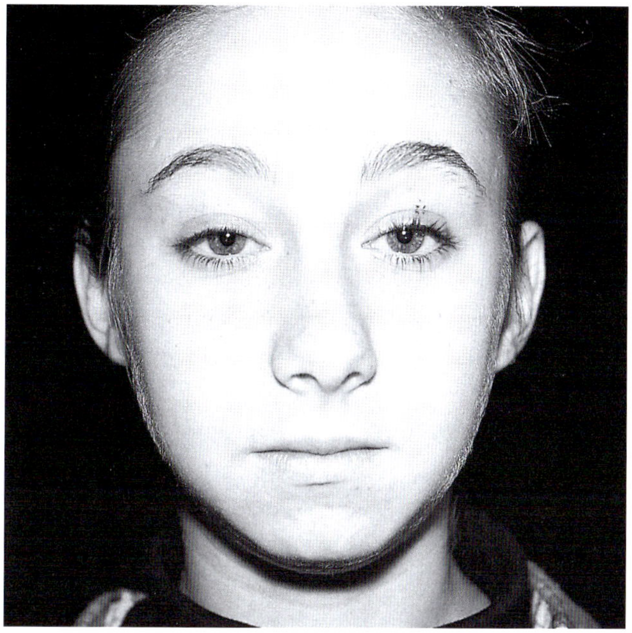

A

PREOPERATIVE

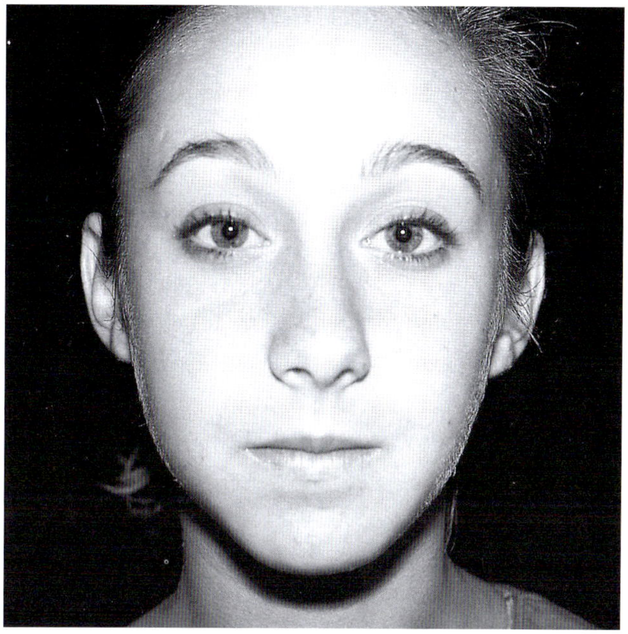

C

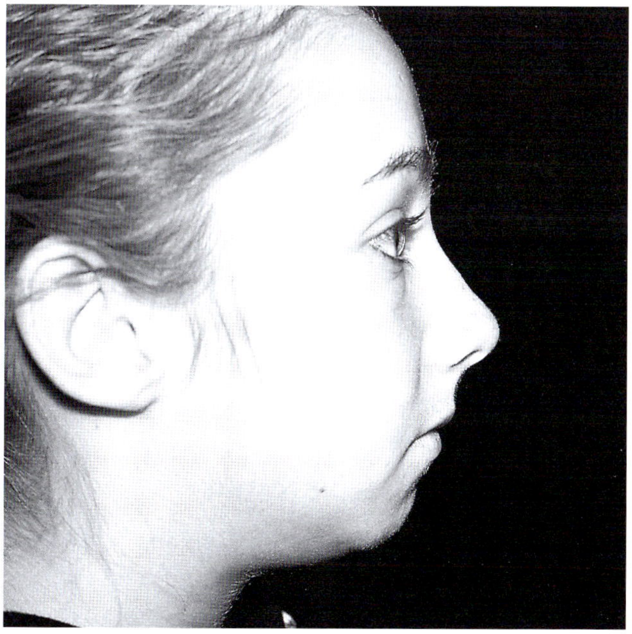

B

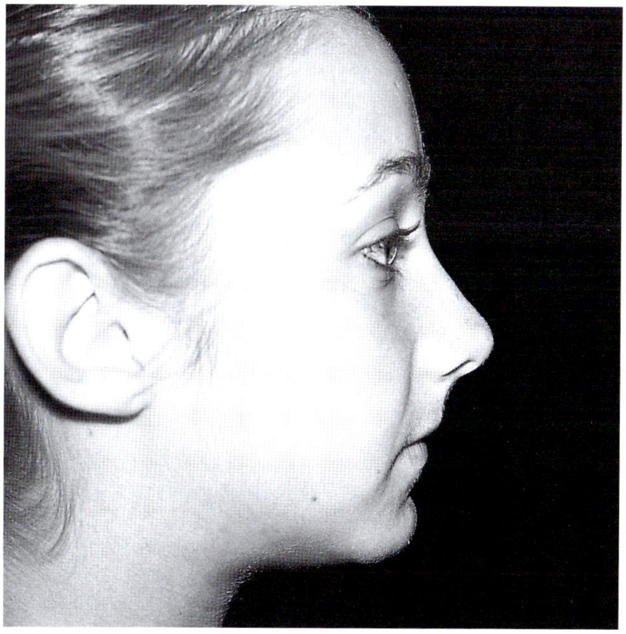

D

CASE STUDY 11-6 A & B: A 17-year-old following orthodontic correction of a Class II maloccusion. The nasolabial angle is obtuse and the lower lip and chin are retrusive. C & D: Following a 7-mm advancement genioplasty. Notice the deepening of the labiomental fold.

Case Study 11-7

This 32-year-old individual thinks his chin is too "weak." He has undergone previous orthodontic therapy, although he is not certain as to what tooth movement took place. His physical examination revealed a lower face height that is modestly elongated relative to the midface, but there is no associated lip strain or excessive tooth or gingival exposure. Although the labiomental fold is somewhat flattened, which further accentuates height of the lower face, anticipated chin surgery will deepen this, which will minimize the impression of excessive lower face height. The soft tissues are rather thick with moderate fat deposits in the jowl and submental region. The treatment plan calls for a 5-mm advancement of the chin and liposuctioning of the jowls and submental region.

Postoperatively, there is better definition of the lower face, and the chin is now in good position relative to the lower lip. The labiomental fold is deepened. This patient could have potentially been treated with alloplastic chin augmentation as well.

PREOPERATIVE POSTOPERATIVE

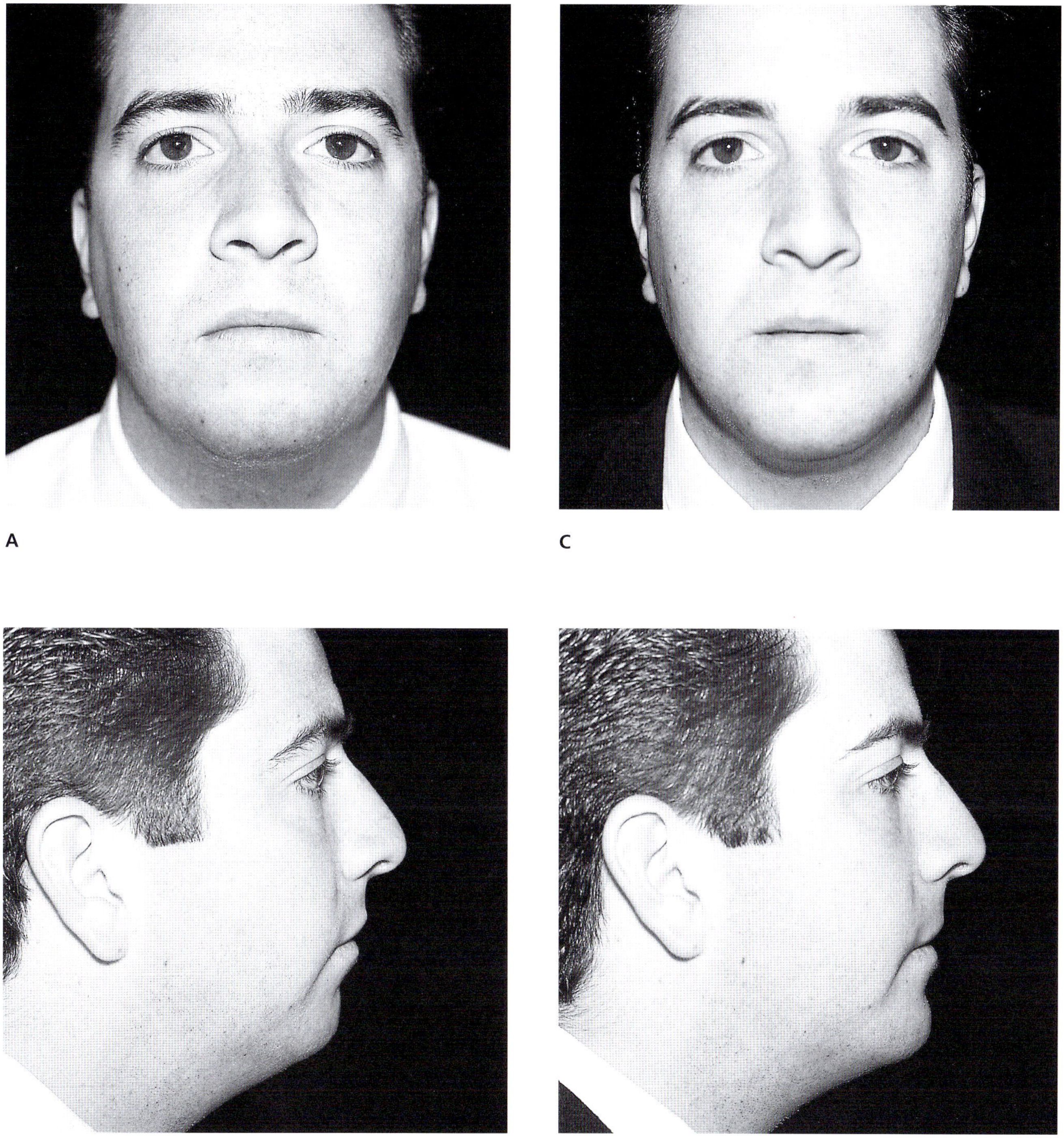

A C

B D

CASE STUDY 11-7 A & B: Appearance of patient prior to advancement genioplasty and liposuctioning. A modest advancement is planned because of the relative retrusion of the lower lip. C & D: The postoperative appearance.

Case Study 11-8

This individual is referred by her orthodontist because of her complaints of a weak chin. She was being treated orthodontically for bimaxillary protrusion and had undergone premolar extractions. At the time that her preoperative photographs were taken, she was at the end of her orthodontic correction. On physical examination, there is good vertical proportionality. The labiomental fold is distinct but somewhat shallow. There is excessive support of both the upper and lower lip from the underlying dentoalvealous. The treatment plan for this patient is for a 9-mm advancement of the chin to allow it to come in line with the lower lip.

In the postoperative, frontal view, one can notice the lower lip already appearing less prominent with a visible deepening of the labiomental fold. The profile demonstrates that the chin is now in perfect alignment with the lower lip but is fairly prominent relative to the upper face. The impression of bimaxillary protrusion has been entirely eliminated secondary to a chin advancement.

These preceding three patients demonstrate the cardinal treatment planning principle of alignment of the chin with the lower lip. Although this may leave the chin in an abnormal position relative to the mid- and upper face structures, it provides good harmony with its most adjacent and clinically relevant anatomic structure.

PREOPERATIVE POSTOPERATIVE

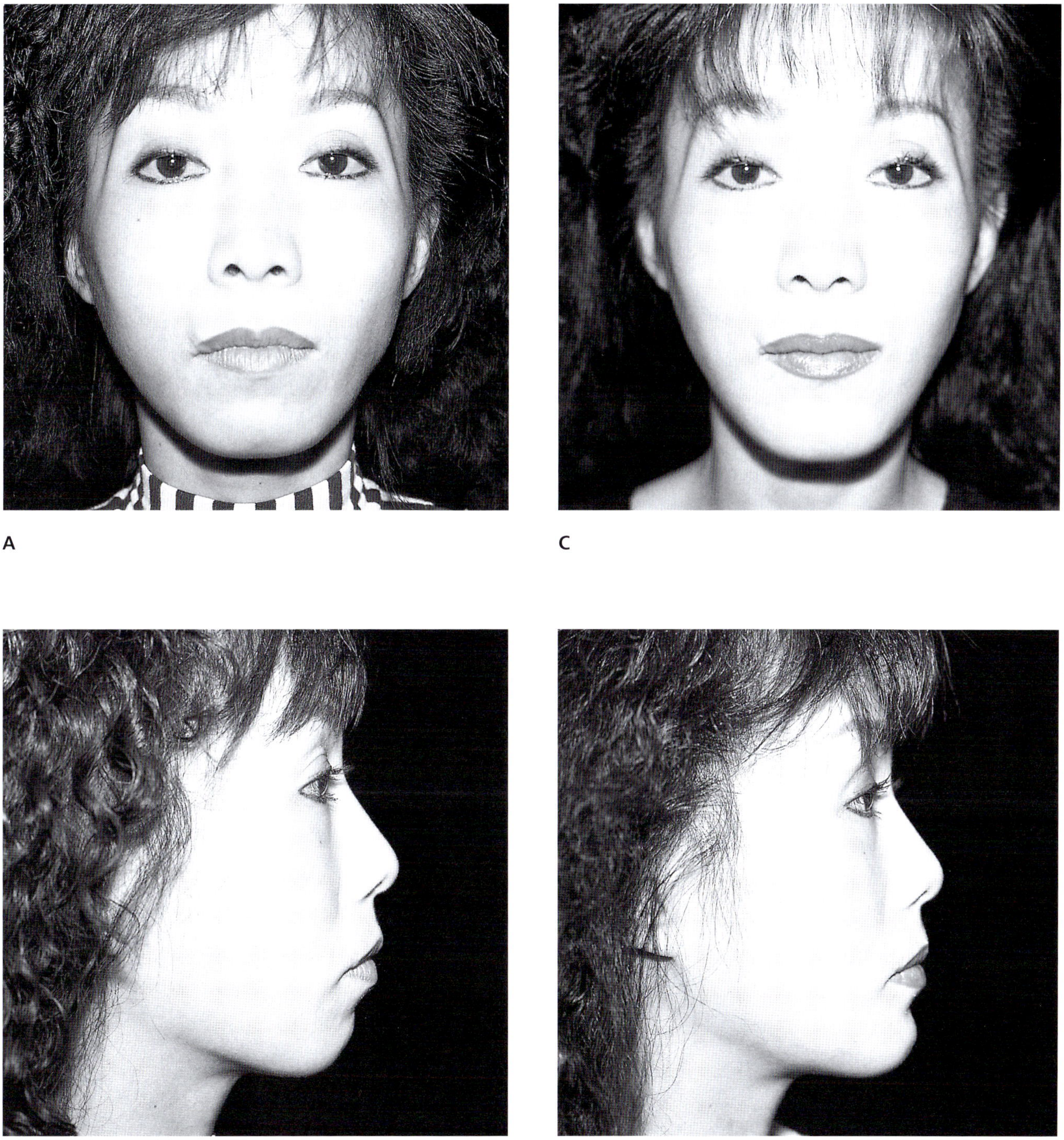

A C

B D

CASE STUDY 11-8 A & B: A bimaxillary protrusive patient following orthodontic correction. The chin remains well posterior to the lower lip, allowing for a planned 9-mm advancement genioplasty. C & D: The postoperative appearance. Note the deepening of the labiomental and alignment of the chin and lower lip.

Case Study 11-9

This patient underwent previous orthodontic correction of a Class II malocclusion. She is concerned regarding the residual impression of a weak lower chin as well as a small dorsal nasal hump. The facial morphology is significant for an increase in lower face height with a significant degree of mentalis and orbicularis oris lip strain in an effort to achieve lip seal. The labiomental fold is shallow and superiorly dislocated due to mentalis hyperfunction. The mandibular plane angle is excessively steepened. Surgical correction involved a vertical reduction of the chin of 6 mm employing two parallel osteotomies and interpositional ostectomy, advancement of the symphyseal segment of 6 mm, and rasping of the dorsal nasal hump.

Postoperatively, several findings are worth noting. This includes the complete elimination of the mentalis and orbicularis oris hyperfunction and a deepening of the labiomental fold as the chin is advanced and reduced in height. Also notice the more inferior position of the labiomental fold due to the elimination of mentalis hyperfunction. The chin has been brought into good alignment with the lower lip, and there is now the visual illusion of flattening of the mandibular plane secondary to superior repositioning of menton.

Alloplastic chin augmentation in this person was absolutely contraindicated because of the requirements for vertical reduction of the chin.

PREOPERATIVE | **POSTOPERATIVE**

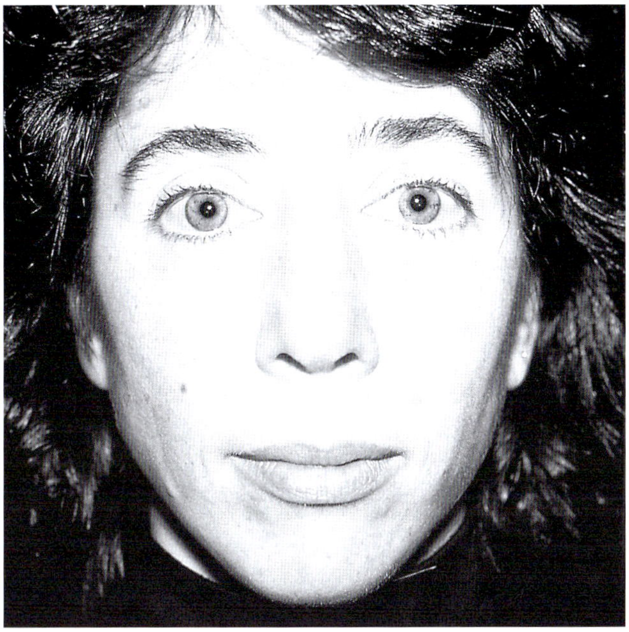

A

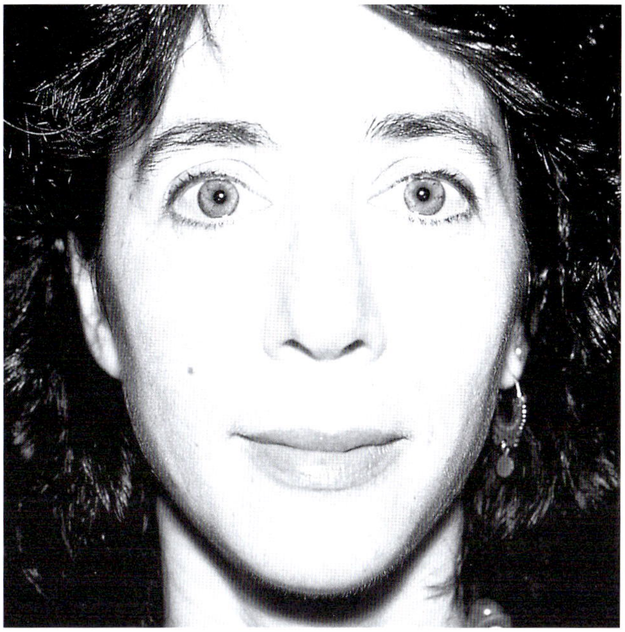

C

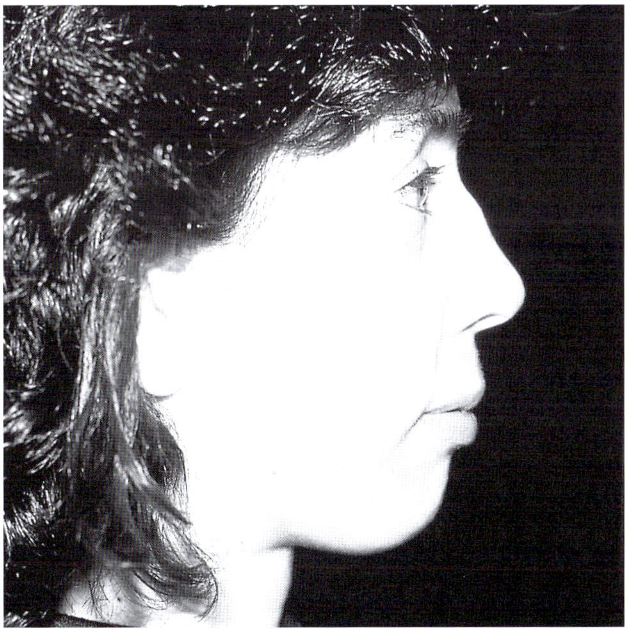

B

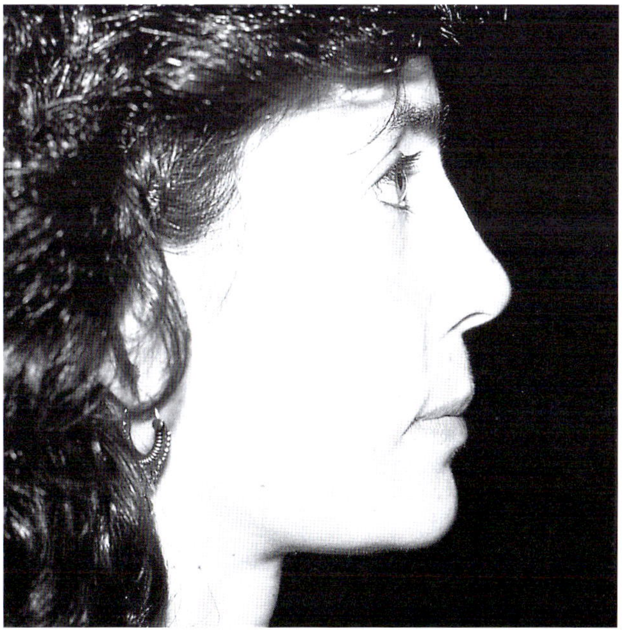

D

CASE STUDY 11-9 A & B: A patient prior to undergoing vertical reduction and advancement genioplasty and rhinoplasty. Notice the lip strain and superior displacement and flattening of the labiomental fold. C & D: Following surgery, the lip strain is eliminated and the labiomental fold is lower and deeper. From: Rosen, H.: Aesthetic refinements in genioplasty: The role of the labiomental fold. *Plast Reconstr Surg* 88:760, 1991. Reprinted with permission.

Case Study 11-10

This patient requested that something be done to enlarge his chin. Historically, there had been a previous orthodontic attempt to correct a deep bite. The facial morphology in the frontal view is characterized by a somewhat sad appearance with a significant deficiency in the height of the lower face and a rather jowly appearance. The labiomental fold is deepened, and the lower lip is procumbent. In the profile view, there is clearly a small mandible with a deepened labiomental fold and a procumbent, recessed lower lip. The patient appears overclosed. His occlusion is a Class II, Division 1 with a deep anterior bite.

The treatment plan involves an osseous genioplasty to vertically elongate and advance the chin 6 mm and 5 mm, respectively. Postoperatively, in the frontal view, there is a rather significant increase in the overall height of the lower face with a marked narrowing of the lower face and elimination of the jowls. The lower lip appears less procumbent, and the fold appears to be less deep. The profile view is characterized by the chin being aligned with the lower lip, a softer transition between the chin and lower lip, and better definition of the cervical mandibular angle secondary to redistribution of the soft tissues in the submental and submandibular regions.

This patient clearly could not have been treated with alloplastic augmentation because of the necessity to effect a significant increase in the height of the lower face. Despite a 5-mm advancement of the chin, the labiomental fold has been attractively attenuated due to the vertical vector of movement.

PREOPERATIVE **POSTOPERATIVE**

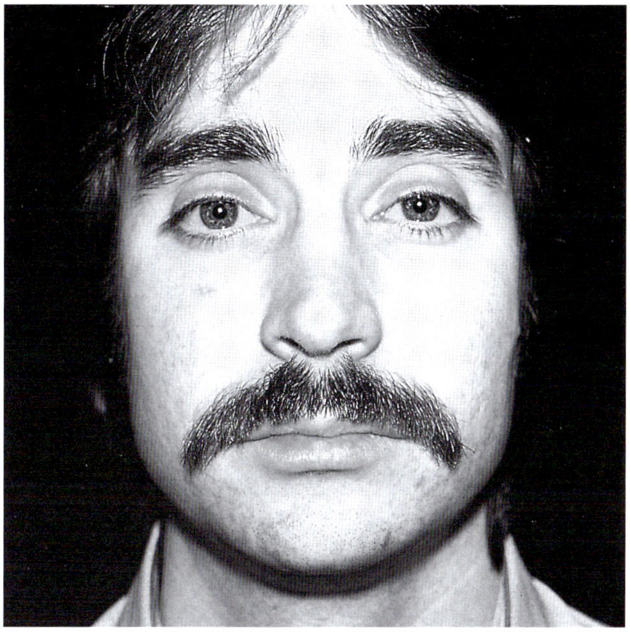

A

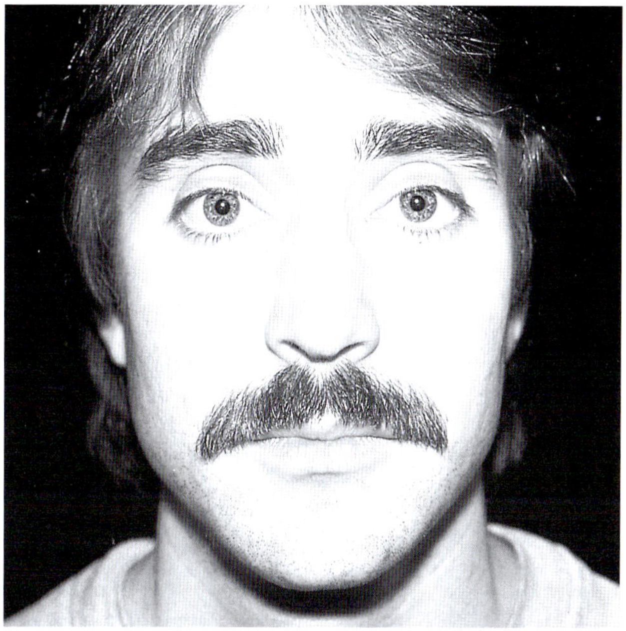

C

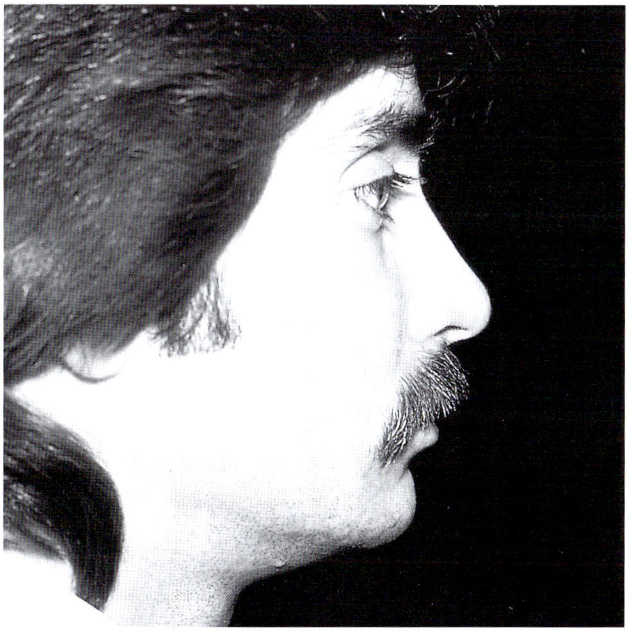

B

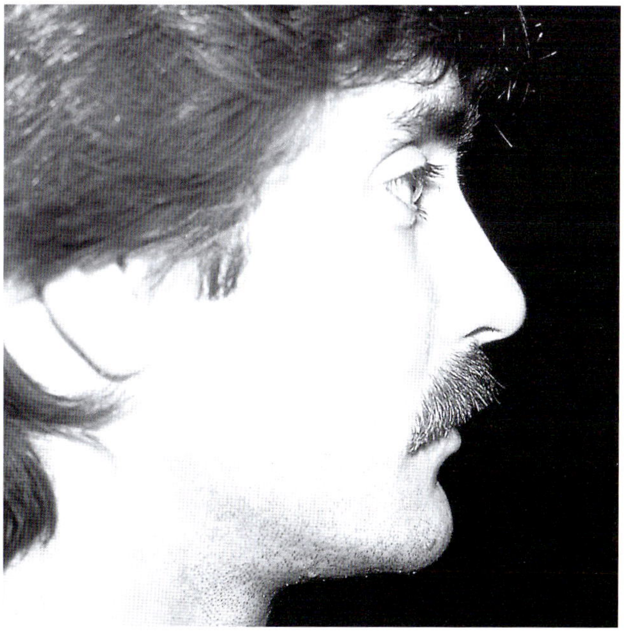

D

CASE STUDY 11-10 A & B: A short-face, Class II deep-bite patient. Notice the deep labiomental fold, procumbent lower lip, jowls, and submental fullness. C & D: Following advancement and vertical lengthening genioplasty, the labiomental fold is less deep and the soft tissues are better supported. No surgery was performed on the soft tissue. From: Rosen, H.: Aesthetic refinements in genioplasty: The role of the labiomental fold. *Plast Reconstr Surg* 88:760, 1991. Reprinted with permission.

Case Study 11-11

This patient presents with a similar history and facial morphology as did the previous patient; orthodontic correction of a skeletally based malocclusion has left him with a residual Class II deformity consisting of diminished height of the lower face, a recessive chin and lip, and a deepened labiomental fold. What makes this patient more difficult to correct is the minimal extent to which one can advance the chin without going beyond the plane of the lower lip. How then does one effect the impression of significant skeletal volume increase in the lower face given this limiting factor?

The treatment plan calls for a rather modest advancement of the chin of only 3 mm to prevent violating the sagittal position of the lower lip. However, the patient was overcorrected in the vertical dimension by lengthening the chin a full 9 mm. In addition, the patient underwent modest suction lipectomy of the jowls and submental region.

The postoperative results are quite dramatic in that there has been a significant enhancement of the overall volume of the lower face due to overcorrection in the vertical dimension. Although there is now quantitative vertical facial excess, the face does not appear to be excessively long because of the absence of lip strain and the presence of a well-defined labiomental fold. However, the fold has been somewhat effaced due to the significant vector of vertical movement. In the profile view, the results are less impressive due to the limited extent to which the chin was advanced. Although there is less sagittal deficiency of the lower face, the overall profile is still characterized by excessive posterior divergence.

This represents a clear-cut example of overcorrecting in the vertical dimension to compensate for the obligatory residual sagittal deficiency that one must be willing to accept when performing osseous genioplasty. To reiterate, the limiting factor over which we have no control in this operation is the position of the lower lip. If this is sagittally deficient, the chin must remain relatively weak.

PREOPERATIVE **POSTOPERATIVE**

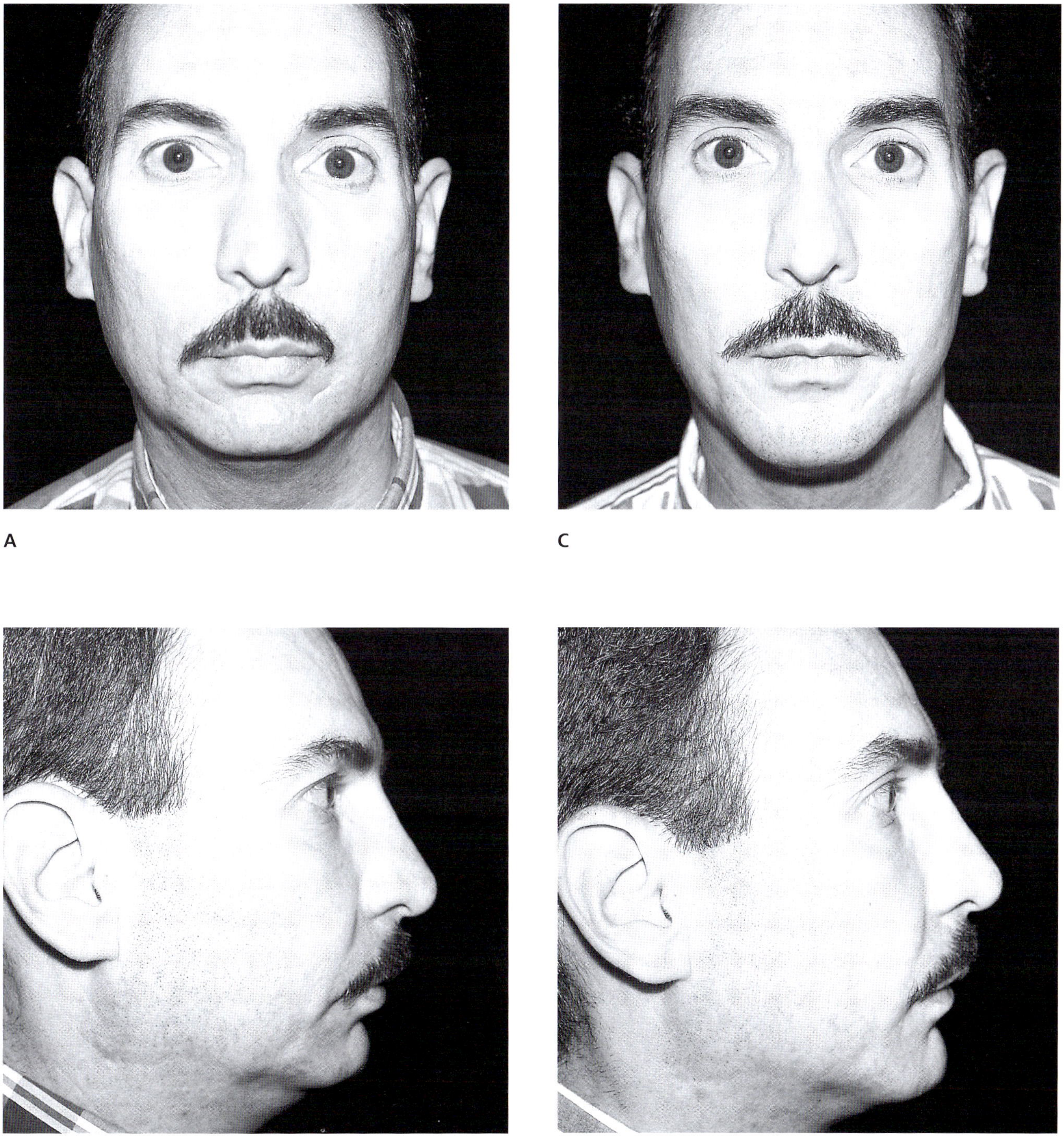

A C

B D

CASE STUDY 11-11 A & B: A patient with a Class II deep bite prior to advancement and lengthening genioplasty. The retrustive lower lip will accommodate only 3 mm of advancement. C & D: The postoperative appearance. The patient has been overcorrected vertically to give the impression of a large increase in skeletal volume of the lower face in the frontal view. The lateral view is, necessarily, less impressive. From: Rosen, H.: Aesthetic refinements in genioplasty: The role of facial disproportion. *Plast Reconstr Surg* 95:463, 1995. Reprinted with permission.

Case Study 11-12

This individual is still another example of an orthodontic compensated Class II patient who is left with a significant Class II deformity. The facial morphology is characterized by a decrease in lower face height, a deepened labiomental fold, excessive facial convexity and posterior facial divergence, and poor soft tissue support throughout the lower face. The nose is excessively large, which makes the chin appear even smaller. Notice the overall posterior inclination of the upper lip secondary to retraction of the maxillary incisors making the nose appear even larger.

Once again, there is a significant limitation placed on the amount one could advance the chin because of the retrusive position of the lower lip. Accordingly, the chin is advanced a modest 4 mm. However, advantage is taken of the vertical dimension to compensate for the modest advancement; that is, the chin is elongated 9 mm. A simultaneous rhinoplasty was performed.

Postoperatively, there is a significant increase in overall skeletal volume of the lower face due to the vertical elongation. As a result of this, the labiomental fold is less deep, and there is a marked enhancement of the overall soft tissue support throughout the lower face. However, there is still a residual deficiency of projection of the lower face because of the aesthetic limitation of osseous genioplasty.

PREOPERATIVE

POSTOPERATIVE

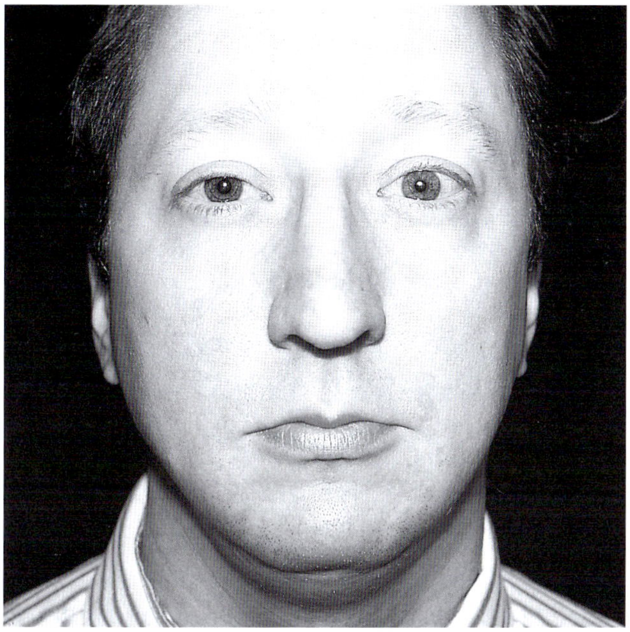

A

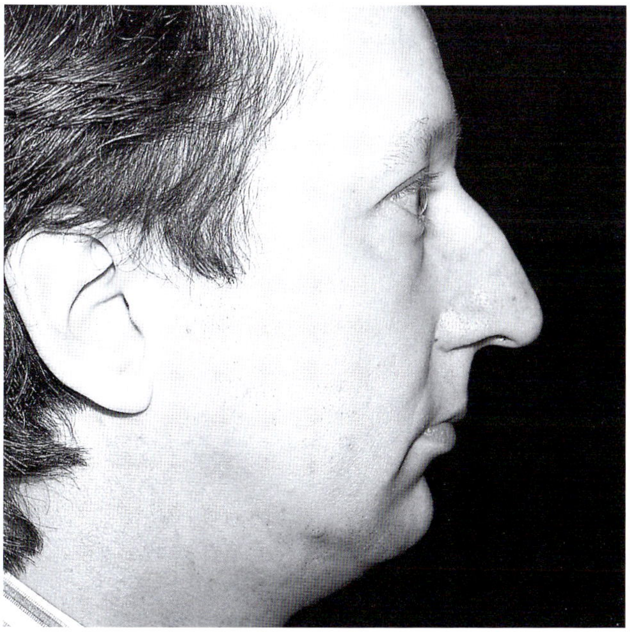

B

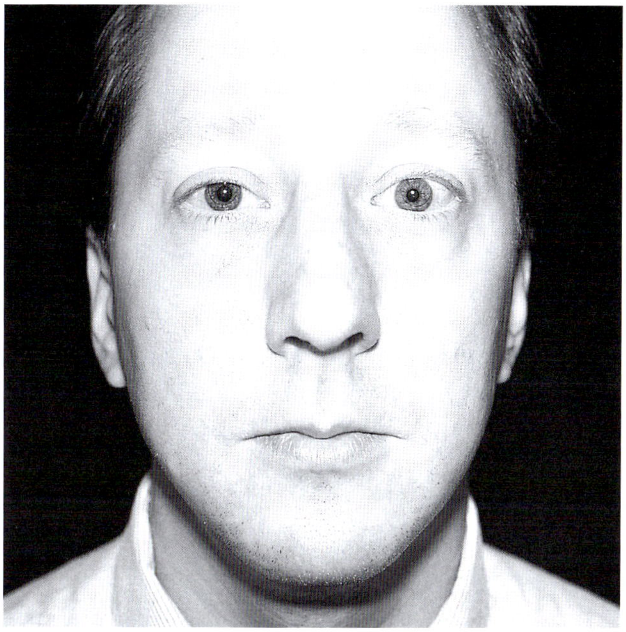

C

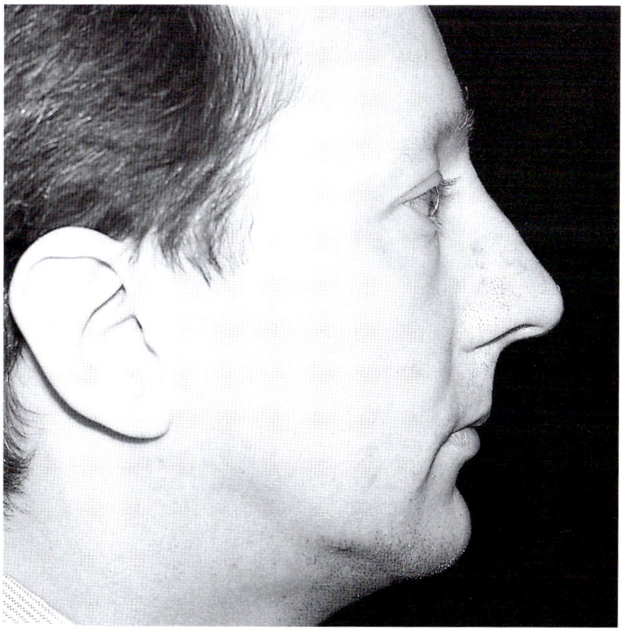

D

CASE STUDY 11-12 A & B: Preoperative appearance of Class II deep-bite patient. The chin can be advanced only 4 mm because of the retrusive lower lip. Accordingly, overcorrection in the vertical dimension of 9 mm is planned. Rhinoplasty will also be performed. C & D: The postoperative appearance. The mandible remains sagittally deficient relative to the midface. From: Rosen, H.: Aesthetic refinements in genioplasty: The role of facial disproportion. *Plast Reconstr Surg* 95:463, 1995. Reprinted with permission.

Case Study 11-13

This patient presents yet another variation of the Class II patient seeking aesthetic enlargement of the chin. She has undergone previous orthodontic correction of a Class II, Division 1 malocclusion. There is a procumbent lower lip with a deepened labiomental fold and excessive facial convexity with rather marked posterior facial divergence. This patient represents a rather difficult treatment planning dilemma because of the combination of a pre-existing deep labiomental fold and normal lower face height. Although the position of the lower lip will tolerate a 6 to 7 mm advancement, this will make the fold too deep. On the other hand, preexisting normal lower face height does not allow one the option to vertically lengthen the chin to soften the appearance of the fold. Or does it?

In this instance, one can rely on the aesthetic principle of vertical facial disproportion being potentially attractive as long as it is unaccompanied by excessive attenuation of the fold, excessive lip strain, or abnormal lip–tooth relations. Accordingly, this patient underwent a modest 4-mm advancement of the chin and 5-mm vertical elongation of the chin in an effort to compromise, that is, creating vertical disproportion and maintaining modest residual sagittal deficiency.

The postoperative photographs demonstrate marked enhancement of the appearance of the patient. This is largely due to a more angular appearance in the frontal view with enhanced soft tissue support and less procumbency of the lower lip. Although there is still sagittal weakness in the profile, there is less procumbency of the lower lip and a smoother transition between the lower lip and chin.

This is still another example of utilizing the principle of facial disproportion in both the sagittal and vertical dimensions to achieve the best possible aesthetic results. This allows us to expand the utility of chin surgery as well as orthognathic surgery. However, to do so mandates that one rely on his or her aesthetic sense of what the patient looks like as opposed to what the X-ray film looks like.

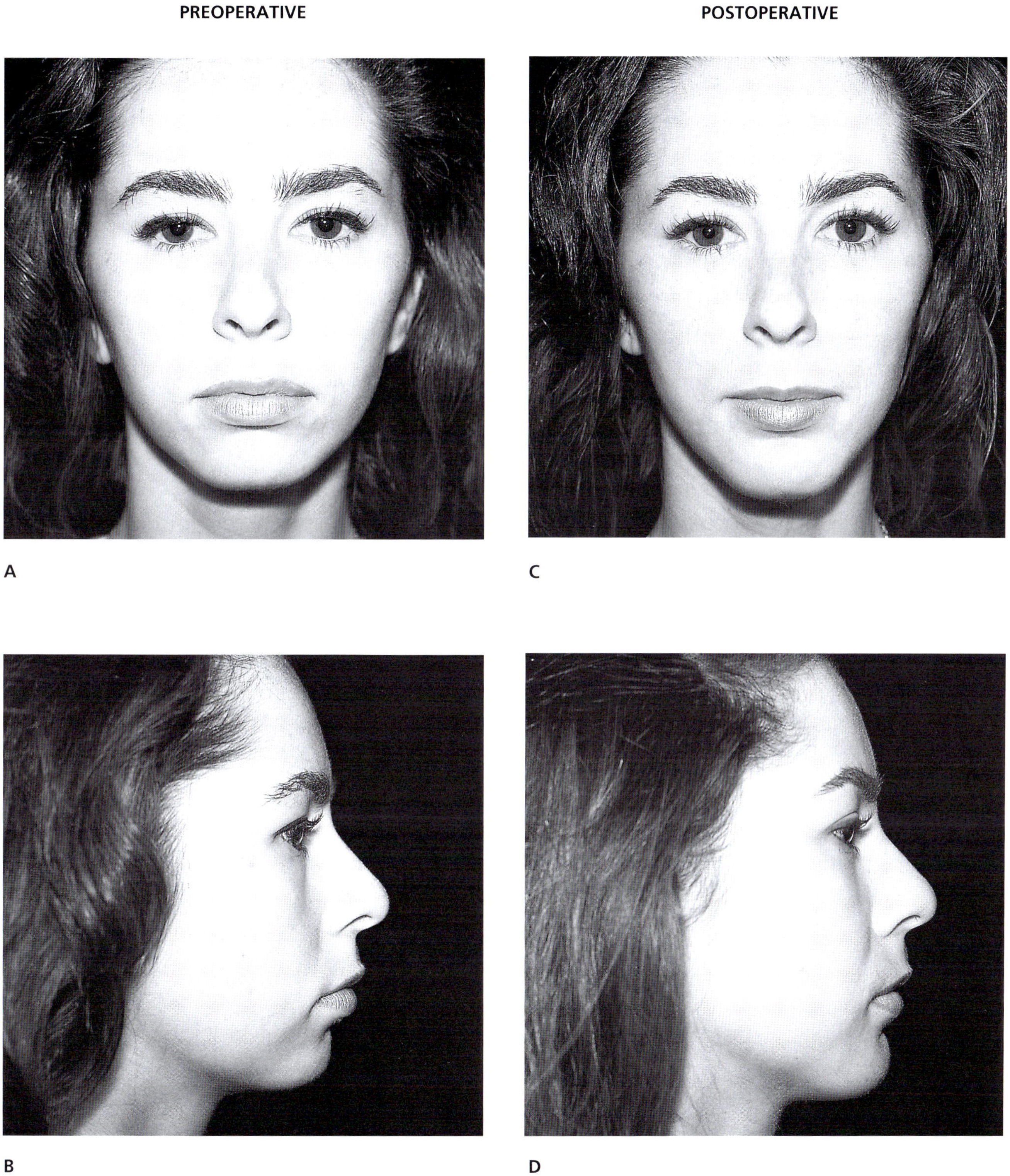

CASE STUDY 11-13 A & B: A Class II patient with normal lower face height and retrusive chin and lip and deep labiomental fold. The deep fold and retrusive lower lip limit the extent to which this chin can be advanced. Accordingly, the chin can be lengthened to camouflage the appearance of a small mandible. C & D: The postoperative appearance following 4-mm advancement and 5-mm vertical elongation of the chin. From: Rosen, H.: Aesthetic refinements in genioplasty: The role of facial disproportion. *Plast Reconstr Surg* 95:463, 1995. Reprinted with permission.

Conclusion

Osseous genioplasty is a potentially powerful technique that should be part of the surgical repertoire of anyone performing facial aesthetic surgery. However, it is best envisioned as a camouflage technique in the overwhelming majority of patients seeking enlargement of the chin. This perspective on chin surgery allows the surgeon to focus on what this technique cannot achieve as well as what it can. Although advancement genioplasty affords us direct control over the sagittal, vertical, and transverse skeletal dimensions and indirect control over the aesthetics of the labiomental fold, it has little, if any, influence over the retrusive position of the lower lip, as seen in many of these individuals. It is only with this realization that one can understand the limitations of osseous genioplasty and, therefore, use it to its full aesthetic potential.

Suggested Reading

1. Rosen, H.: Surgical correction of the vertically deficient chin. *Plast Reconstr Surg* 82:247, 1988.
2. Rosen, H.: Aesthetic refinements in genioplasty: The role of the labiomental fold. *Plast Reconstr Surg* 88:760, 1991.
3. Rosen, H.: Aesthetic guidelines in genioplasty: The role of facial disproportion. *Plast Reconstr Surg* 95:463, 1995.
4. Wolfe, S.A.: Shortening and lengthening the chin. *J Craniomaxillofac Surg* 15:223, 1987.
5. Zide, B.M. and McCarthy, J.G.: The mentalis muscle: An essential component of chin and lower lip position. *Plast Reconstr Surg* 83:413, 1989.

Adjunctive Soft Tissue Surgery

Chapter 12

Occasionally, the planned skeletal displacements of jaw surgery do not achieve the desired soft tissue outcome of support, definition, and angularity. Direct manipulation of the soft tissue may be desirable in such instances. Though a myriad of soft tissue alterations are possible, this chapter will deal only with those that might be performed simultaneously with skeletal displacements.

Liposuction

Simultaneous liposuctioning of subcutaneous fat deposits is the most commonly performed adjunctive soft tissue alteration. Areas treated in order of frequency are the submental triangle, submandibular triangle, and jowl regions. In this context, fat is never removed from the subplatysmal space.

The appropriate candidate for simultaneous liposuctioning must have adequate skin elasticity to avoid skin redundancy once the fat has been removed. It is, at least in part, for this reason that one wants to err on the side of undercorrection in removing fat. An additional reason is to prevent the skin from adhering to the platysma muscle, which is not only unsightly, but impossible to secondarily correct. A small amount of fat removed in these areas can make a dramatic difference.

Liposuctioning is performed as the initial part of the procedure before any swelling has occurred. In this way one can accurately assess the appropriate amount of fat to remove. Submental and submandibular removal can be accomplished through a 2-mm stab incision just posterior to the submental crease. Small cannulas of 2.4 mm or less in diameter should be utilized. This too will help avoid removing an excessive amount of fat in one area. Postoperatively, patients have been found to have no greater swelling or bruising as compared to those who do not undergo liposuctioning. No compressive dressings are used in any of these patients. Provided that one heeds the principle of undercorrection, the procedure should be enthusiastically undertaken because of its extraordinarily high level of patient satisfaction in the appropriate candidate.

Buccal Fat Resection

This, too, is a useful procedure for those patients in whom adequate zygomatic definition and submalar hollowing may not take place secondary to skeletal displacements alone. It is performed with equal frequency in those patients undergoing maxillary surgery, mandibular surgery, or double-jaw surgery.

When done in conjunction with maxillary surgery, no additional mucosal incisions are required. Through the maxillary circumvestibular incision, the periosteum is punctured and then spread just posterior to the root of the zygomatic buttress. The most commonly committed error in removing the buccal fat pad is failure to appreciate its position posterior to this anatomic landmark. Once the periosteum is spread, the buccal fat pad will enter the intraoral wound. It is then easily removed with a combination of suctioning and teasing. Hemostasis is rarely required. In the absence of maxillary surgery, a 1.5-cm mucosal incision is made in the maxillary buccal sulcus opposite the molar teeth. One then directs the dissection posterior to the root of the zygomatic buttress and spreads the periosteum. Following removal of the buccal fat pad, several resorbable sutures are placed.

Though the results of buccal fat pad removal are subtle, skeletal definition and highlighting can be enhanced using this simple adjunctive procedure. However, one must be prepared for a significant increase in the postoperative edema that is ordinarily associated with jaw surgery.

Nasal Surgery

The extent to which one should attempt to directly alter the appearance of the nose at the time of jaw surgery remains controversial. There is little question that full rhinoplasty can be undertaken at the time of isolated mandibular and/or chin surgery. This is predicated upon the assumption that a skilled anesthesiologist is available to convert from a nasotracheal intubation to an orotracheal intubation. It also takes for granted that rigid fixation be utilized so that postoperative intermaxillary fixation can be entirely avoided.

However, the debate remains unsettled as to the advisability of simultaneous nasal surgery and maxillary surgery. This primarily relates to the rather profound changes in nasal appearance that can take place secondary to displacement of the maxilla. Unfortunately, these changes are qualitative;

that is, the extent to which they occur are unpredictable. How then can one justify performing a rhinoplasty when one does not know the magnitude of the nasal deformity that one is attempting to correct? When this preceding factor is placed in the context of performing what many consider to be the most difficult of all aesthetic procedures in which to achieve competence, one can begin to understand the lack of legitimately compelling reasons for considering simultaneous rhinoplasty.

What have been the reasons postulated for performing simultaneous rhinoplasty and maxillary surgery? Many have stated that the aesthetic results can be of equal caliber as those achieved when performing rhinoplasty at a secondary stage. With very few exceptions, this author would argue strenuously against those making these claims. Consider those changes in nasal morphology that take place secondary to maxillary displacement. Depending on the vector of maxillary movement, they include widening of the nasal base and possible flaring of the alar rims, an increase in projection of the nasal tip, and a relative decrease in the projection of the nasal dorsum as tip projection increases. In addition, if closure of the maxillary vestibular incision includes an alar cinch suture in an effort to minimize alar base widening, further increase in the projection of the nasal tip may occur. If all of these potential changes occur, it would seem difficult to plan a rhinoplasty procedure. This is particularly relevant if one attempts to achieve with primary rhinoplasty the very same changes that occur secondary to maxillary surgery. The result is frequently an overresected nasal dorsum.

Other reasons given for performing rhinoplasty in the same sitting are economic; that is, the patient can avoid further hospital costs because it is a cosmetic procedure that would not be covered by third party payers. Lastly, some argue that the surgeon will never get a second opportunity to operate on these patients because they are, in general, quite pleased with the results of their initial surgery. Though these latter two reasons are true, one can hardly justify performing simultaneous rhinoplasty and maxillary surgery as the quality of the end result is the most important factor to be considered.

Alar Cinch Suturing

This is a worthwhile adjunctive procedure to perform at the time of maxillary surgery in an effort to minimize the alar base widening and possible flaring of the alar margins that can occur as the maxilla is repositioned anteriorly and/or superiorly. However, its significant limitations and related soft tissue changes that can occur secondary to its use must be appreciated. Simply stated, alar cinch suturing is unpredictable. In my experience, it never completely prevents nasal base widening, although it may minimize the extent to which it occurs. Contrary to what others may say, the results can hardly be quantified. Nevertheless, it is a worthwhile technique to utilize because it may help to minimize nasal changes that can otherwise detract from a good aesthetic result.

When alar base cinching is effective in narrowing the alar base, it must be accompanied by the soft tissue changes that occur as the nasal bases are brought toward one another. In this context, alar cinching should be considered somewhat similar to a Z-plasty; that is, reductions in length in one direction occur only at the cost of changes in length in another direction. As alar base width decreases, accompanying changes in the adjacent soft tissue may occur singly or in any combination. These include increased nasal tip projection, lengthening and eversion of the upper lip, and increased fullness in the area of subnasale. These changes may be beneficial in certain individuals but detrimental in others. They must be considered when employing this technique.

Secondary Nasal Surgery

Secondary nasal surgery to enhance aesthetic results should be undertaken no sooner than 3 months following maxillary surgery. The most commonly utilized procedure in this context is alar base resection or Weir excisions to narrow interalar width and reduce alar rim flaring. This is a remarkably simple and effective procedure to employ in those patients who have unattractive changes in nasal base width following maxillary surgery. This is ordinarily done under a local anesthetic in the office setting. Its use should be enthusiastically recommended to individuals who would benefit from it.

When primary total rhinoplasty is to be undertaken, it should be performed no sooner than 3 months following maxillary surgery.

Suggested Reading

1. Bell, W.H.: Le Fort I osteotomy for correction of maxillary deformities. *J Oral Surg* 33:412, 1975.
2. Bell, W.H. and Proffit, W.R.: Maxillary Excess. In W.H. Bell, W.R. Proffit, and R. White (Eds.): *Surgical Correction of Dentofacial Deformities.* Philadelphia: Saunders, 1980, p. 368.
3. Carlotti, A.E., Aschaffenburg, P.H., and Schendel, S.A.: Facial changes associated with surgical advancement of the lip and maxilla. *J Oral Maxillofac* 44:593, 1986.
4. Radney, L.R. and Jacobs, J.D.: Soft-tissue changes associated with surgical total maxillary intrusion. *Am J Orthod* 80:191, 1981.
5. Rosen, H.: Lip–nasal aesthetics following Le Fort I osteotomies. *Plast Reconstr Surg* 81:171, 1988.
6. Schendel, S.A. and Williamson, L.W.: Muscle reorientation following superior repositioning of the maxilla. *J Oral Maxillofac Surg* 41:235, 1983.
7. Schendel, S.A. and Delaire, J.: Facial Muscles: Form, Function and Reconstruction in Dentofacial Deformities. In W.H. Bell (Ed.): *Surgical Correction of Dentofacial Deformities: New Concepts.* Philadelphia: Saunders, 1985, p. 259.

Postoperative Aesthetic Considerations

CHAPTER

13

This chapter will deal with those instances in which early reoperation (within the first days following surgery) should be undertaken to avoid an unaesthetic result. A reoperation should be considered under two circumstances. The first is when the treatment plan was inadvertently not followed. The second, if appreciated in the early postoperative period, is when an error in treatment planning was committed that will have unfavorable aesthetic results.

Failure to Execute the Treatment Plan

The most commonly committed error in which one fails to execute the treatment plan involves incorrect positioning of the mandibular condyle in the glenoid fossa. This can have profound consequences on the aesthetic outcome. If it is recognized in the immediate postoperative period (it certainly should be if rigid fixation has been employed), it should be surgically corrected. This represents a significant advantage in the use of rigid fixation; failure to seat the condyle correctly can be appreciated in the immediate postoperative period. Accordingly, postoperative intermaxillary fixation should not be used in conjunction with rigid fixation because it eliminates this advantage. In addition, the combination of these modalities may be harmful in the event that the condyle has been overseated.

Prior to the widespread use of rigid fixation, failure to adequately seat the mandibular condyle was most commonly encountered in mandibular advancement surgery. The clinical context in which this occurs has dramatically changed with rigid fixation. In this author's opinion, it is now most commonly encountered in association with either surgical correction of anterior open bite or surgical correction of lateral mandibular asymmetries.

Anterior open-bite correction involving posterior maxillary impaction requires adequate removal of bone in the maxillary tuberosity and posterior lateral nasal wall regions. An inexperienced surgeon may find this to be technically demanding. If inadequate removal occurs, the mandibular condyles may be distracted as the maxilla is rigidly fixated while the mandible is temporarily wired to the maxilla in the desired occlusal relationship. Failure to recognize this will leave the patient with a residual anterior open-bite deformity, and as a result, the lower face remains excessively long with accompanying lip strain and labiomental fold attenuation. If detected in the early postoperative period, it is best to return the patient to the operating room and correct this with further impaction of the posterior maxilla. The alternative to this is unacceptable: an orthodontic attempt to extrude anterior teeth to correct the malocclusion that is not only unstable but will further impair the aesthetic result.

Correction of lateral mandibular asymmetries occasionally results in lateral displacement of the condyle on the side away from which the mandible is being shifted. As the condyle reseats itself in the immediate postoperative period, the mandible shifts back to establish the original deformity. Depending on the extent to which this occurs, immediate reoperation should be considered. Reoperation at this time will reestablish facial symmetry and avoid a malocclusion that may be difficult or impossible to correct orthodontically.

Failure to execute the assigned treatment plan may also occur when a change in the vertical position of the anterior maxilla is desired to decrease or increase the amount of incisal show beneath the relaxed upper lip. Accurate vertical positioning of the anterior maxilla may become problematic, particularly when other directional changes in the maxilla take place; that is, it is the only maxillary displacement that cannot be built into an intermediate surgical splint when performing double-jaw surgery or a final surgical splint when performing single-jaw surgery. It is the one intraoperative measurement that must be manually determined. Because of the imperfect nature of such an intraoperative determination (see Chapter 5), errors can sometimes occur. Though these are usually minimal, if a major error occurs, it can lead to a disastrous aesthetic outcome in terms of the final relationship between the maxillary incisal edge and upper lip. This discrepancy can be easily ascertained by obtaining a postoperative lateral cephalogram and comparing the position of the maxillary incisal edge to the intended position according to the prediction tracing. If a major discrepancy exists that will impair the final aesthetic outcome, it should be corrected at the earliest opportunity.

Because this author always uses an intermediate surgical splint when performing double-jaw surgery, there is little risk of deviating from the surgical plan as it relates to maxillary positioning in other vectors and attitudes other than the anterior maxillary vertical position.

Errors in Treatment Planning

Though many errors in treatment planning can be made in regard to the aesthetic outcome, bear in mind that we are now considering only those that, if recognized, should be corrected in the immediate postoperative period. This most commonly relates to two specific instances: failure to appreciate a significant discrepancy between centric relation and centric occlusion and failure to achieve good midline lower face symmetry when laterally shifting the chin.

Failure to appreciate a discrepancy between centric occlusion and centric relation will be built into any intermediate surgical splint and result in inadequate anterior movement of the maxilla or, even worse, its posterior displacement. At the time of surgery, the condyles are completely seated as the patient is no longer posturing forward. However, the intermediate surgical splint has an occlusal relationship based on the chronic, abnormal forward posturing of the patient's mandible. Accordingly, the sagittal position of the maxilla is located posterior to where it was planned to be positioned. This represents an error in treatment planning based on the failure to diagnose this discrepancy. Once again, this can easily be appreciated by comparing the position of the maxilla on a lateral postoperative cephalogram as compared to its intended desired located on the preoperative prediction tracing. If a major discrepancy exists, it should be corrected at this point. Failure to do so will result in inadequate projection of the lower face and inferior aspect of the midface, along with its subsequent adverse effects on the soft tissue support.

Finally, insufficient lateral displacement of the symphyseal segment in an osseous genioplasty can sometimes be appreciated only after surgery. If the clinical impression is confirmed with a frontal cephalometric X-ray film showing significant residual transverse asymmetry of the chin, it is a simple matter to return the patient to the operating room to correct this. Additional lateral displacement of the symphyseal segment involves a small effort that can result in a big aesthetic difference by achieving enhanced symmetry of the lower face. This is most commonly encountered when treating lower face asymmetries involving simultaneous lateral shifting of the mandible and the symphyseal segment with additional osseous genioplasty.

Suggested Reading

1. Rosen, H.: Rigid fixation in orthognathic surgery: A personal appraisal. *J Craniofacial Surg* 3:2, 1992. [Editorial]
2. Rosen, H.: Bimaxillary Surgery. In M. Habal (Ed.): *Advances in Plastic and Reconstructive Surgery*. St. Louis: Mosby, 10:93, 1993.

Index

D

E

F

H

I

L

M

T

V

W